THE ENGINEER IN AMERICA

THE ENGINEER IN AMERICA

A HISTORICAL ANTHOLOGY FROM *TECHNOLOGY AND CULTURE*

EDITED BY TERRY S. REYNOLDS

THE UNIVERSITY OF CHICAGO PRESS
Chicago and London

The essays in this volume originally appeared in various issues of *Technology and Culture*. Acknowledgment of the original publication date may be found on the first page of each essay.

The University of Chicago Press, Chicago 60637
The University of Chicago Press, Ltd., London
©1963, 1969, 1970, 1971, 1972, 1975, 1977, 1979, 1984, 1985, 1986, 1988 by the Society for the History of Technology.
©1991 by The University of Chicago.

01 00 99 98 97   6 5 4 3 2

Library of Congress Cataloging-in-Publication Data

The Engineer in America : a historical anthology from technology and
  culture / Terry S. Reynolds, editor.
      p.   cm.
    Includes bibliographical references and index.
    ISBN 0-226-71031-9 (cloth) : $35.00 (est.). — ISBN 0-226-71032-7
  (paper) : $18.00 (est.)
      1. Engineering—United States—History. 2. Engineers—United States—
  History. I. Reynolds, Terry S.
  TA23.E54   1991
  620'.00973—dc20
                                                                    91-19356
                                                                    CIP

⊗ The paper used in this publication meets the minimum requirements of American National Standard for Information Science—Permanence of Paper for Printed Library Materials, ANSI Z39.48-1984.

The Engineer in America
A Historical Anthology from
Technology and Culture

For Linda Gail Rainwater Reynolds, my wife,
and for our four sons: Trent, Dane, Brandon, and Derek

Proceeds from this book will be donated to the Society for the History of Technology for the support of graduate students.

Introduction

Engineers form one of the largest "professional" groups in the United States today, with around 2 million practitioners; it is the single largest occupation of American males. The rise of this occupational group and its place in American society furnish the focus for this anthology.

The engineer's role in the development of modern America is, in a sense, paradoxical. As the various essays in this anthology demonstrate, engineers played a major role in designing, building, and managing the infrastructure on which American material and economic prosperity has been based. In the 19th century engineers like Claudius Crozet, Andrew Humphreys, James Eads, and John Jervis designed and supervised construction of the nation's early transportation systems—roads, bridges, canals, waterways, and railroads. When cities reached the natural limits of their growth, other engineers, like Ellis Sylvester Chesbrough, designed the sewerage disposal and water supply systems that permitted continued growth. Still others, like William Sellers, designed and built the machines and machine tools that permitted mechanization of production or developed the standards that made mass production possible. In the 20th century engineers like Charles Kettering and Henry Ford revolutionized transportation by designing automobiles that were cheap, convenient, and safe to operate, while engineers associated with the Bureau of Roads adapted the nation's highways to the new means of transportation. Simultaneously, engineer-inventors like Thomas Edison were revolutionizing power transmission by developing electricity as a means of power distribution. Even before Edison left the field for other pursuits, engineer-managers were developing techniques to make large-scale electrical networks commercially profitable. In communications, engineers like Michael Pupin and George Campbell were taking Alexander Graham Bell's crude, short-range communication device and developing the means to transform it into a sophisticated, long-range communication system, while others further transformed communications and information systems by developing radio, television, and computing.

Yet, even though engineers played a major role in designing, build-

Dr. Reynolds is professor of history and head of the Department of Social Sciences at Michigan Technological University. He has authored several books, including a history of water power and a history of the American Institute of Chemical Engineers. He is currently working with Bruce Seely on a history of American engineering education.

1

ing, and managing the infrastructure on which American material and economic prosperity has been based, they have often had very limited control over how the works of their minds and their hands have been applied. This role, in America, has largely been played by entrepreneurs, financiers, politicians, and corporate executives.

Nonetheless, the contributions of engineers have been central to the way American society has evolved. To use a military analogy, while engineers have not usually been the commanding generals in American society, they have dominated general staff positions. General staffs develop the plans and options and set the parameters on which command decisions are based. Without such contributions the effectiveness of commanding generals would be sharply limited. Moreover, some engineers have left staff positions to become commanding generals themselves.

Because engineers have played an important role in the development and management of American technology, it is not surprising that the thirty-one volumes of *Technology and Culture* published since the journal's founding in 1959 contain a significant number of contributions dealing with American engineers, with American engineering practice, and with the institutions which American engineers developed to serve their needs. This reader collects, in one volume, a number of the most notable of these contributions.

The essays contained in this anthology provide an overview of the evolution of the engineering profession in America. Part I deals with the engineer and the engineering profession in 19th-century America; Part II, with 20th-century America. The turn of the century provides a logical point of division. In the 19th century, engineering was largely an individualistic profession. Independent consultants and owner-engineers dominated the profession, on-the-job training was preferred to academic training, and empiricism was regarded as far more important than theory. Around 1900 the growing scale of corporate organizations, the increasing complexity of technology, and the rising status of science began to change these patterns. Salaried engineers employed by corporations displaced independent consultants and owner-engineers as the dominant figures in the profession, and engineering training and practice became substantially more "scientific."

The two broad chronological divisions of this anthology have parallel organizations. Each is introduced by an editor's overview that reviews the basic themes in the history of the engineering profession in America and places the essays that follow into a broader historical context. After the overviews, each division contains one or more *Technology and Culture* background essays that illustrate general trends or deal broadly with the entire chronological period. These background essays are followed by case studies of engineering practice and case studies that look at the institutions that American engineers developed to reproduce the profession (educational institutions) or to diffuse knowledge to practitioners and represent the profession to outsiders (professional societies).

Although the primary criterion used in selecting essays was "fit" with the focus of this anthology—the engineer in America—a significant number of the selections have been singled out for recognition for the quality of their scholarship by the Society for the History of Technology. The Society awards its Usher Prize annually to the best essay appearing in *Technology and Culture,* selected from the past three years' issues of the journal. Among the Usher Prize winning pieces found in this anthology are James Brittain's essay on the introduction of the loading coil (1971), Stuart Leslie's essay on Kettering and the copper-cooled engine (1980), Thomas Hughes's essay on the builders of America's electrical systems (1981), and Bruce Seely's essay on the "scientific mystique" in the Bureau of Public Roads (1987). In addition, W. Bernard Carlson's essay on Dugald C. Jackson and the GE-MIT cooperative program won the 1989 IEEE Prize for the best essay on the history of electrical technology published in the preceding year. Finally, this volume contains several essays that have become standards. One piece very clearly falls in this category: Edwin Layton's essay on "Mirror-Image Twins" (1971). This essay remains one of the most influential contributions to literature dealing with the relationship between science and technology and is probably one of the most frequently cited articles published in *Technology and Culture.*

In compiling this anthology and preparing the introductory overviews, I have accumulated a number of debts. I would like to provide special thanks to Rudi Volti of Pitzer College for providing me with a copy of an unpublished book-length manuscript that he prepared in conjunction with John Rae (*The Engineer in History*). The sections of that manuscript dealing with American engineering proved helpful in refining the introductory overviews in this anthology. Robert Post, Bruce Seely, and Jeffrey Stine read and provided criticism and suggestions on the overview essays, leading me to significantly reorganize and refocus their content. What appears is much better because of their input. Finally, I owe thanks to my spouse, Linda Rainwater Reynolds, who referees everything that I do.

TERRY S. REYNOLDS
MICHIGAN TECHNOLOGICAL UNIVERSITY

Part I

The Engineer and Engineering in 19th-Century America

Overview

THE ENGINEER IN 19TH-CENTURY AMERICA

TERRY S. REYNOLDS

The American engineering profession emerged in the 19th century shaped by two traditions: a French (or continental European) tradition and a British tradition.[1] As these traditions blended in the 1800s, however, they yielded a distinctly American engineering tradition, better adapted to American economic and political conditions and American social values than either of its predecessors.

European Antecedents

The origin of the French engineering tradition, the older of the parents of American engineering, lay in the rise of a strong national monarchy in the 16th and 17th centuries. The centralized bureaucracy and tax-collection systems established by the French monarchy made possible a greatly enlarged, permanent military, which provided the first major employment base for engineers.

Engineers began to be regularly employed in the French army in small numbers in the 16th century, as the expanding use of gunpowder artillery made full-time, technical experts necessary to design fortresses to resist cannon fire, to conduct siege operations against cannon-armed fortresses, and to locate, design, and construct the roads and bridges over which armies moved. In 1676, recognizing the growing importance of this body of technically trained men, the French minister of war created a separate *Corps du génie* (corps of engineers) in the French army. This corps contained 132 military engineers by 1683.[2] Its personnel were recruited largely from the poorer elements of the French nobility, with infusions from the upper middle class.[3]

The bureaucratic agencies established by the centralized nation-state further increased employment opportunities. To reinforce central authority and provide for speedier deployment of troops, the French

---

[1]The existence of two different European traditions of engineering is touched on by Arthur L. Donovan, "Engineering in an Increasingly Complex Society," in *Engineering in Society*, ed. George S. Ansell (Washington, D.C., 1985), pp. 94–96.

[2]Frederick B. Artz, *The Development of Technical Education in France, 1500–1850* (Cambridge, Mass., 1966), p. 48.

[3]Ibid., pp. 45, 57–58.

monarchy undertook construction of a national network of roads. To provide the technical personnel to survey, lay out, supervise construction of, and maintain this network, the royal government created a second special engineering corps, the *Corps des ponts et chaussées* (corps of bridges and roads) in 1716. French engineers also worked on an assortment of other projects deemed in the national interest, including canal construction, river and harbor improvements, and water supply systems.[4]

In 1747, because of a shortage of trained personnel and because the quality and standards of construction differed immensely on different parts of the road network, the French government created the *École des ponts et chaussées* (school of roads and bridges) to systematize the training of road and bridge engineers. This school initiated the shift from apprenticeship as a means of training engineers toward formal academic education, characterized by an increased emphasis on the scientific and mathematical principles underlying practice. The new approach worked so well that the French government steadily expanded its use of formal technical schools to train engineers, founding, for example, the *École du corps royal du génie* in 1749 for military engineers and the *École des mines* in 1778 for mining engineers. The process culminated in 1794 with the creation of the *École polytechnique*, the institution that became the foundation for formal theoretical and mathematical engineering education in France.[5]

By 1800 engineering was a well-established occupation in France. It was characterized by recruitment from the ranks of the lower nobility and the upper-middle class, formal education in government-sponsored schools, employment by the state, and an emphasis on the use of mathematics and theoretical methods to guide practice.

The state (and more particularly the military) remained the primary patron of engineering not only in France but throughout Europe prior to 1750. In the mid-18th century, however, Britain's unprecedented commercial and industrial expansion had provided some individuals, partnerships, and corporations with sufficient wealth to undertake large and complex projects for private commercial purposes, projects that required full-time technical supervision and considerable technical expertise. This growth in scale of commercial operations enabled a group of self-trained, nonmilitary engineers to make careers of

---

[4]French engineering in the early modern period is discussed by W. H. G. Armytage, *A Social History of Engineering* (Cambridge, Mass., 1961), pp. 95–105; James Kip Fich, *The Story of Engineering* (Garden City, N.Y., 1960), pp. 137–83; and S. B. Hamilton, "The French Civil Engineers of the Eighteenth Century," *Transactions of the Newcomen Society* 22 (1941–42): 149–59.

[5]Artz, esp. pp. 60–181.

providing technical expertise. Before the end of the 18th century, British engineers working on civilian projects (hence the term "civil" engineer) had developed a professional identity and had begun to meet regularly to exchange technical ideas and to promote mutual interests.[6]

John Rae, in the introductory essay of this volume, "Engineers Are People," reviews the ancient roots of the engineering profession, focusing on the fragmentary data available on individual engineers in China, the ancient Mediterranean civilizations, and medieval Europe. He then studies the family background and training of Renaissance engineers, before turning to the early-modern French and British engineers discussed above. Although he does not discuss the early-modern French engineers at length, he provides statistical data on the family origins and training of British engineers of the 17th through 19th centuries. These data suggest that the British engineering tradition differed sharply from the French. Rae's data, for example, reveal that British engineers were drawn from all classes, not just the upper and middle classes, as in France. Similarly, his data show that apprenticeship training was the norm in Britain long after it had been replaced by formal education in France. And, Rae notes, in Britain civil (i.e., nonmilitary) engineers were a much more firmly established and self-conscious group than on the Continent, where state-employed, military engineers were dominant. Although Rae does not mention it, British engineers also differed from their continental counterparts in preferring a heavily practical and empirical approach to problems and in being more suspicious of mathematical and theoretical methods of engineering problem solving.[7]

Rae's wide-ranging study extends beyond the Old World to a review of information in the Smithsonian's "Biographical Archive of American Engineers." His data on early American engineering summarized in his essay suggest that it was a blend of the French and British engineering traditions. Rae's statistics indicate, for example, that early American engineers, like French but unlike British engineers, were recruited from the upper and middle levels of society; ordinary laborers were a negligible source. But, like British engineers, formal education

[6]British engineers in the early modern period are discussed in Alex Keller, "The Age of the Projectors," *History Today* 16 (July 1966): 467–74; Armytage, pp. 67–94; Samuel Smiles, *Lives of the Engineers* (London, 1904); and R. A. Buchanan, "Gentlemen Engineers: The Making of a Profession," *Victorian Studies* 26 (Summer 1983): 407–29; among others.

[7]Differences between British and French engineering practice are touched on in D. S. L. Cardwell, *Turning Points in Western Technology: A Study of Technology, Science and History* (New York, 1972), pp. 122–27. See also Armytage, pp. 95–105.

in a technical school was not part of the ordinary career path of early and mid-19th century American engineers. Among U.S. engineers born before 1830, Rae found, only a little over 30 percent were educated in an institute or college of engineering.

The Emergence of the Engineering Profession
in the United States

Rae found that only a handful of American engineers had been born before 1790. This was largely because engineering did not exist as a distinct occupation in colonial America. The appearance of engineering has traditionally been associated with the emergence of large-scale organizations (e.g., the military) or large-scale projects in which the level of capital investment and risk are high enough to justify the expense of full-time technical experts. But neither large organizations nor large projects were common in colonial America. Governments in the colonies were relatively small and weak, unable or unwilling to support large permanent military organizations or to undertake large public improvement projects. Furthermore, there were few large, private commercial enterprises. Manufacturing, when it occurred at all, was usually small in scale.

The outbreak of the American Revolution, however, created a sudden demand for engineers. Only with the "esoteric" knowledge of trained military engineers was it possible to conduct sieges, place artillery, construct fortifications to resist artillery bombardment, and build bridges and other military works quickly, efficiently, and well. Recognizing this, the Second Continental Congress on June 16, 1775, established a Corps of Engineers in the Continental army. But there were practically no experienced people to appoint to the corps. Washington was compelled to look abroad.

French military engineers by the late 18th century were recognized as the finest in the world, and French sympathy for the American cause provided a reservoir of engineering talent from which Americans could draw. Perhaps two dozen French military engineers served in the American armies during the course of the Revolution. For seven years French military engineers organized, trained, and led the American army's Corps of Engineers. When peace came in 1783, the corps was dissolved with the rest of the American army, and most French military engineers returned to Europe. During the French Revolution, however, some of the engineers who had served in the American army emigrated to America and were called on by President Washington to begin construction of a series of seacoast fortifications.[8]

---

[8]On the role of French military engineers in the American Revolution see Paul K. Walker, ed., *Engineers of Independence: A Documentary History of the Army Engineers in the American Revolution, 1775–1783* (Washington, D.C., 1981); and Sylvia G. Goldfrank,

Because of continued military threats to the new republic, Washington and his presidential successors Adams and Jefferson supported creation of both a permanent corps of engineers and a means of training new military engineers for that corps. Congress responded in 1794 by authorizing creation of the Corps of Artillerists and Engineers and assigning it to garrison West Point. In 1802 Congress separated the Corps of Engineers from the artillery, stationed it at West Point, and, simultaneously, established the U.S. Military Academy at the same site. The army appointed several senior officers to provide instruction in the elements of military engineering to cadets stationed there, although there were no formal classes, grading, or courses of instruction.[9]

As the army sought to meet its need for military engineers through West Point, the nation as a whole faced the parallel problem of securing engineers for civilian projects. Civil engineering emerged in the United States in the early years of the 19th century in response to the need of a vast and underpopulated but ambitious country for an improved transportation network. American fear of centralized government, born under English rule, produced a weak federal government. Thus, outside of military engineering, the American engineering profession could not emerge as the French profession did—the result of a strong central government's patronage. The role played by the royal government in France as patron for the engineering profession was assumed by state governments and by private enterprise in the United States, patronage more characteristic of the British tradition.

Rivalry for trade between eastern ports led state governments in the early 19th century to sponsor the earliest large-scale technological projects. For example, New York, beginning in 1816, undertook construction of the 365-mile-long Erie Canal, in large part to attract western trade to New York City and away from Pennsylvania. Pennsylvania countered in the late 1820s with the Pennsylvania Mainline canal and portage railway. Other states similarly sought engineers to work on state-financed "internal improvements" or to provide guidance for private or mixed-enterprise transportation projects.

Robert Hunter's essay, "Turnpike Construction in Antebellum Virginia," illustrates the early drive of American states to develop a transportation infrastructure to stimulate trade and the role that engineers played in that drive. The subject of Hunter's essay, Claudius Crozet, served for almost two decades as Virginia's state engineer. His

---

trans. and ed., *French Engineers and the American War of Independence* (Washington, D.C., c. 1976).

[9]*The Centennial History of the United States Military Academy at West Point, New York* (Washington, D.C., 1904), p. 263.

chief responsibility was upgrading the quality of Virginia's transportation systems, especially its public roads and turnpikes.

Hunter's essay deals with a number of central themes in early American engineering. For example, his protagonist, Crozet, provides an example of the influence of the French tradition on early American engineering practice. Born and educated in France, Crozet served as an artillery officer under Napoleon. He emigrated to the United States in 1816, serving for a time as an engineering instructor at West Point, before accepting the post of state engineer in Virginia in 1820.

In Virginia, as Hunter points out, one of Crozet's main responsibilities was applying European road-building expertise to a region where the development of internal trade was hindered by bad road location, inadequate knowledge of surfacing techniques, poor drainage, and excessive grades. Hunter's essay provides an excellent account of the technical skills required for engineering practice in early America, such as ability to locate the most economical routes, provide cost estimates, design bridges, and operate surveying and leveling instruments.

Technology, however, also operates in a social environment, and political conditions in America were far different than in Crozet's native land. In France a strong centralized government insured that engineering advice and high standards were followed in road construction. Engineers in America operated in an environment suspicious of strong government. Thus the impact of Crozet's sound technical advice, as Hunter notes, was limited by the absence of legal mechanisms giving him the power to enforce his technical judgments on the local contractors involved in Virginia's mixed-enterprise system for financing turnpikes.

Crozet was one of the few people in the United States in the early 19th century who possessed the skills to design and supervise the construction of public works like canals and high-quality roads. As late as 1816, according to one reliable estimate, there were no more than thirty engineers or quasi engineers engaged in civilian projects in the entire United States.[10]

The Development and Education of an Engineering Work Force

The engineers who ultimately constructed America's transportation infrastructure and then began work on the technological systems necessary to sustain urban growth and expand manufacturing came from a variety of sources. Poorly trained native talent had to be supple-

---

[10]Daniel Hovey Calhoun, *The American Civil Engineer: Origins and Conflict* (Cambridge, Mass., 1960), p. 22.

mented early on by foreign-born and foreign-trained engineers. Hunter's protagonist, Crozet, was only one example. The British engineer William Weston provided indispensable aid to Laommi Baldwin on the first major American canal project—the Middlesex Canal (1793–1803). Weston was not in the first rank of British engineers, but he knew so much more about canals than any American that he was involved in virtually every major American canal project between 1793 and 1800, when he returned to England.[11] Other immigrants also contributed, such as Benjamin Latrobe, working out of Philadelphia, and Hamilton Fulton in North Carolina.[12] Rae's statistics in "Engineers Are People," however, indicate that foreign-born engineers did not dominate American practice for long. Three native systems of engineering training quickly emerged to fill American engineering needs: on-the-job training, the national military academies, and civilian academic engineering programs.

Apprenticeship and On-the-Job Training

Through most of the 19th century most American engineers were practically trained, through either apprenticeship or some other system of on-the-job training, in a manner similar to the British engineering tradition.

Between 1817 and 1820, while constructing the Erie Canal, New York State developed the most widely used on-the-job training system for civil engineers. Because New York could not secure the services of foreign-trained engineers, it appointed three men—Benjamin Wright, Canvass White, and James Geddes—with limited surveying experience and marginal technical knowledge but with a good deal of adaptability and intelligence to direct the initial stages of construction. From trips to existing canals, self-education, and trial-and-error experimentation, Wright, White, and Geddes slowly developed the technical expertise necessary to guide construction. They soon recognized, however, that work on the canal was spread over too wide an area for three men to supervise effectively. Hence they requested and received authorization from the New York legislature to hire assistant engineers to supervise work on shorter sections of the canal. The problem then became, Where were the assistant engineers going to come from? Wright,

[11]Elting E. Morison, *From Know-how to Nowhere: The Development of American Technology* (New York, 1974), pp. 22–38.

[12]For Latrobe see Darwin H. Stapleton, "Benjamin Henry Latrobe and the Transfer of Technology," in *Technology in America: A History of Individuals and Ideas*, ed. Carroll W. Pursell, Jr., 2d ed. (Cambridge, Mass., 1990), pp. 34–44; on Hamilton Fulton see Neal Fitzsimons, "Wanted: One Civil Engineer," *Civil Engineering* 41 (no. 1) January 1971, p. 52.

White, and Geddes promoted their best survey crew chiefs to the positions, which created in the process, a standardized and recognized means of recruiting and training civil engineers. A beginning job as ax-man or chainman on a survey crew became the formal entry point to the profession. Nascent engineers learned the rudiments of surveying, mapping, route selection, mathematics, personnel management, and the sciences while working their way up to survey crew chief. The best then made the jump to assistant engineer. This method of recruiting and training engineers worked very effectively on the Erie Canal. Thus, when New York engineers completed the Erie in 1825 and moved to similar positions elsewhere, they carried the Erie's system of on-the-job training through survey crews with them.[13]

Civil engineers trained in this way were soon used for large engineering projects outside canal construction, such as laying out railroad lines and urban water supply systems. Louis P. Cain's essay, "Raising and Watering a City: Ellis Sylvester Chesbrough and Chicago's First Sanitation System," provides a good example of the work of one of the finest products of early American on-the-job engineering training. Chesbrough, as Cain notes, had little formal education. He entered the engineering profession around 1830 using the system developed on the Erie Canal—he began his career as a chainman on a railroad survey crew, acquiring his engineering knowledge on the job. When the pace of internal improvements slowed in the early 1840s, he secured work with the Boston Water Works, working with John Jervis, "perhaps the most competent engineer trained by the New York canal system." In 1855 Chesbrough left Boston and accepted the position of chief engineer for the Chicago Board of Sewerage Commissioners.

In the late 1840s and early 1850s Chicago's undisciplined waste-disposal methods had led to a succession of epidemics that threatened to stunt the city's growth. To solve this problem Chesbrough designed and constructed the first comprehensive sewerage system undertaken by any major city in the United States. As Cain points out, Chesbrough's novel system required literally raising the elevation of the entire city to provide the necessary gradient for an effective and economical gravity-flow sewerage network. Less than a decade later Chesbrough engineered another unique solution to the city's pressing need for an uncontaminated water supply by designing and supervising construction of a 2-mile-long subaqueous tunnel to draw uncontaminated water from far out in Lake Michigan.

Chesbrough's technical competence was paralleled by a sense of economic reality. His attention to "ability to pay," according to Cain,

[13]Calhoun, pp. 24–37.

established him in the eyes of Chicagoans as a "practical man." Chesbrough's image as a "practical" rather than a "theoretical" engineer "lent credence to his innovative ideas."[14] These ideas, put into action, eliminated the cycle of waterborne disease epidemics that threatened to halt the growth of Chicago, permitting it to develop into America's second-largest city.

As civil engineering was reaching a vigorous adolescence by directing construction of, first, the nation's transportation system and, then, its urban infrastructures, a second engineering speciality began to emerge. Near midcentury, mechanical engineering developed as a distinct speciality, prompted by the shift from the small craft shop to the more capital-intensive, mechanized factory and by the spread of steam power.

As with civil engineering, America's mechanical engineering needs were initially supplied by engineers trained on-the-job, in a manner resembling the British engineering tradition. In mechanical engineering the "field" training of the survey crew was replaced by a machine-shop apprenticeship. Most 19th-century mechanical engineers began their careers as machine-shop laborers, became apprentices, then journeyman machinists, then machinists. The small size and special-order nature of the early machine shops permitted apprenticeship to operate well as a method of training and enabled some machinists to develop wide-ranging design skills and to learn the essentials of engineering economics. Shop owners raised the best of their mechanics to superintendents or chief engineers. Some machinists left the shops in which they had been trained for appointment as superintendents or chief engineers in similar shops or to establish their own shops as owner-engineers. The leading figures in early American mechanical engineering were those machinists who established their own machine shops.

William Sellers, the mechanical engineer whose life is briefly described in Bruce Sinclair's essay "At the Turn of a Screw," exemplified the ideal mechanical engineering career in 19th-century America. Sellers came from an upper-class Philadelphia family. After private schooling, he skipped college and served an apprenticeship in his uncle's machine shop. He then superintended in another family-connected machine shop for three years, before establishing a machine tool firm of his own in partnership with another Philadelphian.

---

[14] Ibid., p. 52, has a table indicating that employers of civil engineers generally preferred "practical" engineers. Calhoun found that around three-quarters of the "chief engineers" on projects in 1837 were trained by practical means, such as the New York canal system, while only around one-quarter had been trained by academic means.

While early American civil and mechanical engineering were characterized by on-the-job training and by employment on engineering projects whose patrons were private or local, which were characteristics shared with the British engineering tradition, other elements of the emerging American engineering tradition were strongly influenced by the French model. This was especially true of the formal engineering training schools that grew up parallel with on-the-job training systems and eventually replaced them as the primary source of nascent engineers.

The Military Academies

Between 1820 and 1860 the U. S. Military Academy at West Point rivaled on-the-job systems as a source of American civil engineers. It served also as the primary conduit for French influence on the growth of the American engineering profession. As previously noted, in the years immediately following the re-creation of the Corps of Artillerists and Engineers in 1794 and its garrisoning at West Point, engineering training was carried out rather haphazardly. There was no formal course of instruction, no grading, and no degree granted at the end of a period of study. Only after the War of 1812 did the level of engineering instruction begin to improve. Claudius Crozet's appointment as an instructor in 1816 marked the first significant step. In 1817 superintendent Sylvanus Thayer introduced a standard four-year curriculum, grading, and methods of instruction (lecture, recitation) along the lines of the French *École polytechnique*, making West Point the first formal engineering school in the United States. West Point not only adopted the French engineering tradition's mathematical and theoretical approach to engineering training but also made heavy use of French textbooks and French or French-trained instructors until past midcentury.[15]

West Point's position in American society in the 19th century was paradoxical. It was an elitist, military institution in an egalitarian, antimilitary society. Yet it survived and even flourished by simultaneously serving both military and civilian ends. West Point engineers explored and mapped much of the American West, making the first climatological and mineral-resource surveys in that region. They further promoted civilian interests by charting the nation's coasts and by improving its harbors and waterways.[16] They contributed to relieving

[15]West Point's early development of an engineering curriculum is discussed in Calhoun, pp. 37–43, and James Gregory McGivern, *First Hundred Years of Engineering Education in the United States (1807–1907)* (Spokane, Wash., 1960), pp. 32–39.

[16]The best account of the early work of the Army engineers is Forest G. Hill, *Roads, Rails and Waterways: The Army Engineers and Early Transportation* (Norman, Okla., 1957).

the shortage of trained civil engineers for transportation projects through the General Survey Act. Passed by Congress in 1824, this act permitted corps engineers to assist private groups in planning internal improvements deemed to be in the "national interest." Before the act was repealed in 1838, army engineers had assisted on transportation projects all over the nation.

But these were not the only ways that West Point contributed to relieving the pressure on America's limited civil engineering supply. For several decades after 1820, West Point graduated more engineers than there were officer vacancies; hence, many West Point–trained engineers left the military for employment in the civilian sector. Still other West Point engineers served a term in the military, then resigned and joined the civilian engineering force. In the 1830s as many as 120 West Point graduates were working as civil engineers.[17] Altogether, West Point may have trained as many as 15 percent of all civil engineers in pre–Civil War America.[18] In fact, in the early 1830s, army inspectors complained that West Point's curriculum was more suited for teaching civil engineering than military engineering.[19]

Martin Reuss's essay, "Andrew A. Humphreys and the Development of Hydraulic Engineering: Politics and Technology in the Army Corps of Engineers," deals with the career of a West Point–trained engineer and illustrates several central issues in 19th-century American engineering. First, it provides a good illustration of both the French engineering tradition as transferred to America via West Point and the Corps of Engineers' contribution to American civilian engineering. Second, it demonstrates that 19th-century American engineering was not homogeneous: it was composed of several subgroups with differing and often conflicting interests and philosophies. Third, it shows how the political context in which engineering operates can affect the growth of technical knowledge.

Andrew A. Humphreys graduated from West Point in 1831, served in the artillery, and then left the military to become a civilian engineer. He rejoined the army in 1838 to work on various engineering projects. Around 1850 Humphreys and Henry Abbot made a pioneering study of the hydraulics of the Mississippi River as part of the corps' responsibility for improving navigation. They brought to this project the approach of engineers trained in the French tradition. After gathering extensive stream measurement data from the lower Mississippi, they applied the data to theoretical flow formulas developed by French and

[17]Calhoun, pp. 207–8.
[18]John Rae, "Engineers Are People," *Technology and Culture* 16 (1975): 416.
[19]Calhoun, p. 42. See also Hill, pp. 142–43.

Italian hydraulic scientists and found these formulas flawed. Reacting to this discovery, they attempted to derive a new universal theoretical formula to measure the flow of water in natural streams. This work, as Reuss points out, was very thorough and brought widespread acclaim to Humphreys and to the corps.

Humphreys's position on the relationship between engineering and the state, like his theoretical approach to engineering problems, was derived from the French tradition. He believed in a strong, well-financed Corps of Engineers serving the public good, and he strongly opposed civilian engineers trespassing on projects in the corps' realm of responsibility.

Humphreys became chief of engineers of the army in 1866 in a period in which, as Reuss notes, the corps' control over federal public works projects, especially those involving river and harbor work, was under attack from the growing body of civilian engineers nurtured in the British tradition of private enterprise. Humphreys responded to this situation by tying the accuracy of his hydraulic studies to his commitment to the corps' public service mandate. To attack one was to attack the other. This position led to major conflicts with engineers coming out of the British tradition of American engineering, notably James Eads.

Eads's training and philosophy were diametrically opposite that of Humphreys. He had little more than a grammar-school education; his development as an engineer had largely occurred on-the-job. He was distrustful of theory and operated as an entrepreneur, seeking to make a profit out of hydraulic engineering projects. Reuss describes Eads's successful confrontation with Humphreys and the corps on the effectiveness of jetties for opening the mouth of the Mississippi River, a conflict involving both the accuracy of Humphreys's theoretical work and the monopoly held by the corps over engineering work on the nation's navigable waterways.

The conflict between Humphreys and Eads in the 1870s and the earlier conflict between Humphreys and Charles Ellet, also discussed in Reuss's essay, clearly show that American engineers in the 19th century were not a homogeneous lot. Because of the different traditions that influenced their training (French and British), they had different views on the value of theory and on the role the state should play in engineering projects. And, Reuss shows, there was no unanimity of opinion even within these major subgroups. Some army engineers dissented from Humphreys's position.

Finally, Reuss' essay illustrates the interactions between technology and politics. For example, his research demonstrates how political decisions can seriously effect the growth of technical knowledge.

Humphreys, as chief of engineers, succeeded in making the conclusions of the Humphreys-Abbot report the unofficial position of the Corps of Engineers on the feasibility of hydraulic works. As a result, attacks on Humphreys's report become attacks on the Corps of Engineers. This merger froze the army corps into Humphreys's theoretical conclusions and impeded corps' research into river hydraulics for decades. As late as 1950 the corps opposed, often successfully, the use of reservoirs for flood control because that idea was contrary to the Humphreys-Abbott theory.[20]

At the very time the Corps of Engineers' monopoly on navigable waterway projects was being successfully challenged by engineers like James Eads, West Point's preeminence as an institution for formally training civil engineers also went into decline. Besides stagnation in curricula, a critical factor in this decline was the emergence of formal engineering programs at a large number of civilian colleges.

The Naval Academy at Annapolis did not play the central role for the education of mechanical engineering that the Army Corps of Engineers and West Point did for civil engineering. But it did play a role. Founded in 1845, the Naval Academy offered only a general scientific course for several decades. In the 1860s the academy began to improve its engineering instruction. In 1861 it introduced a course on steam engineering into the regular curriculum, and in 1865 the academy established a Department of Steam Engineering, one of the earliest academic departments focusing on mechanical engineering in the United States. In 1866 the Naval Academy inaugurated a two-year program focusing on steam engineering, and in 1874 expanded it to four years.[21] As a center for academic mechanical engineering training, however, it too was overshadowed by civilian academic programs by the 1870s.

Civilian Engineering Schools

The first American civilian engineering education programs emerged in separate technical institutes because American higher education was initially dominated by colleges and faculty that emphasized the classics, theology, or pure science and did not welcome training for practical occupations. The first civilian engineering school was Norwich University, founded by Alden Partridge in 1820 as the American Literary, Scientific, and Military Academy. It was initially intended to

[20]An exceptionally critical account of the Army Corps of Engineers' involvement in civil engineering is Arthur E. Morgan, *Dams and Other Disasters: A Century of the Army Corps of Engineers in Civil Works* (Boston, 1971).

[21]Monte A. Calvert, *The Mechanical Engineer in America, 1830–1910: Professional Cultures in Conflict* (Baltimore, 1967), pp. 20, 255.

train officers for state militia along lines similar to West Point but with less emphasis on theory and more on practice. Around 1825 Norwich began offering studies in engineering. At first the program required no standard course of study or amount of time. By 1834, however, Norwich had developed a formal three-year course.[22] Rensselaer Polytechnic Institute was founded in 1824 and in its first decade focused on training teachers to instruct farmers and artisans in the scientific principles behind their vocations. In 1835, however, Rensselaer began awarding civil engineering degrees to students completing a one-year course of study. By 1850 Rensselaer, too, had developed a three-year-long curriculum.[23]

Scattered attempts were made to create engineering programs in mainstream colleges, such as the University of Virginia and the University of Alabama, in the 1830s and 1840s. But these programs often faltered after a few years. In the 1850s, however, engineering programs begin to find permanent homes at some traditional institutions, such as the University of Michigan, Harvard, Yale, Union College, and Dartmouth.[24] Academic engineering programs, however, remained on the periphery of the American educational scene. Only a half-dozen programs were even modestly successful by 1860.

Ironically, the U.S. Military Academy played a major role in the emergence of the civil engineering education programs that were eventually to undermine its position as America's foremost engineering school. Engineers trained at West Point were hired as faculty in many of these programs. They brought with them not only the French tradition of formal, theoretically oriented engineering education but also academic respectability. By 1868 some 139 West Point graduates had entered the field of education, mostly higher education, and texts authored by West Point graduates were in wide use in American colleges.[25] When the curriculum at West Point stagnated after the 1850s, the Military Academy was quickly surpassed by civilian engineering schools as a source of civil engineers.

At about the same time, the Naval Academy's nascent program for

[22]McGivern, pp. 42–45; Calhoun, pp. 43–45.

[23]McGivern, pp. 48–52, 58–61; Calhoun, p. 45.

[24]On other early attempts at developing academic engineering programs see Calhoun, pp. 45–46; McGivern, pp. 63–87; and Howard L. Hartman, ed., *Proceedings of the 150th Anniversary Symposium on Technology and Society—Southern Technology: Past, Present, and Future* (Tuscaloosa, Ala., 1988). The latter work contains Robert J. Norrell, "A Promising Field: Engineering Education at the University of Alabama," pp. 120–143, and O. Allan Gianniny, Jr., "The Overlooked Southern Approach to Engineering Education: One and a Half Centuries at the University of Virginia," pp. 144–176.

[25]Hill, p. 209.

training mechanical engineers was similarly being overshadowed by civilian engineering programs. In 1871 Robert H. Thurston, a central figure in developing the academic program in mechanical engineering at Annapolis, resigned to take a position at Stevens Institute of Technology, a new institution of higher learning devoted specifically to mechanical engineering. At Stevens, Thurston created a mechanical engineering curriculum heavily influenced by the French tradition with its emphasis on mathematics and science and by the German tradition of university research. Within a few years, Stevens Institute was widely recognized as the premier academic institution for educating mechanical engineers. It was quickly imitated, however, by a host of public colleges who initiated mechanical engineering programs under the provisions of the Morrill Act.

The passage of the Morrill Act by Congress in 1862 was, in fact, one of the major factors in the decline of the federal military academies as suppliers of American engineers in the post–Civil War era. Prompted by the continued growth of American agriculture and industry and public frustration at the failure of traditional colleges to incorporate "practical" and democratic forms of higher education, the Morrill Act offered states substantial grants of federal lands to subsidize creation of college-level programs or schools in "agriculture and the mechanic arts," the latter term usually interpreted to mean engineering. State legislatures and existing colleges and universities eagerly sought the land grants, and the number of academic engineering programs in the United States shot up sharply, reaching seventy by 1872 and eighty-five by 1880.

Many of these programs were initially weak, partially because their institutions wanted the land grants but did not welcome the intrusion of "agriculture and the mechanic arts." Hence they did the very minimum necessary to satisfy the Morrill Act's requirements. For instance, they often used the act's requirement of military training to persuade the War Department to assign them West Point–trained army officers. These officers would then be used both to teach military science and to serve as the institution's entire engineering faculty.[26]

Through these officers, however, West Point continued to influence the institutions that were replacing it as a supplier of civil engineers. The Naval Academy played a parallel role in mechanical engineering. Many of the mechanical engineering programs established under the

---

[26]McGivern, pp. 88–114; and Lawrence P. Grayson, "A Brief History of Engineering Education in the United States," *Engineering Education* 67 (December 1977): 250–54 review the development of engineering education in the decades after the passage of the Morrill Act.

provisions of the Morrill Act achieved respectability only in the 1880s when Congress authorized the navy to detail surplus officers to teach the elements of steam engineering and iron and steel ship design. Between 1879 and 1896, when this practice ceased, forty-eight colleges and universities secured such appointments.[27] These naval officers widely disseminated the French tradition of formal mathematical and theoretical training in mechanical engineering, and a significant number ultimately resigned and remained at the schools to which they had been detailed.

The rapid expansion of American industry in the late 19th century insured the ultimate triumph of academic engineering education over apprenticeship and related on-the-job training methods. The latter simply could not supply trained technical personnel fast enough to meet growing demands. By the 1890s academic methods for training engineers were firmly established in the United States.

The academic programs established at both separate technical institutes and mainline colleges were more mathematically and theoretically oriented than the British-inspired on-the-job training programs but less so than the French theoretical tradition found at West Point. Much of the history of American engineering education in the 19th century, in fact, turns around the search for a compromise between the two traditions. This struggle in mechanical engineering education is one of the themes of the essay by James Brittain and Robert McMath, Jr., "Engineers and the New South Creed: The Formation and Early Development of Georgia Tech."

Brittain and McMath review the struggle in American universities between those who wanted mechanical engineering training to stress higher mathematics and theoretical science (the French tradition) and those who wanted mechanical engineering education to stress practical shop work and produce graduates immediately able to step into positions as machine-tool foremen or machine-shop superintendents (the British tradition). As Brittain and McMath point out, Georgia Tech, founded in 1888 as part of the drive to create a new, industrialized South, was part of this struggle.

Georgia Tech's program was initially patterned after that of Worcester Polytechnic Institute, a school whose compromise between the French tradition and the British tradition was much closer to the British. At Worcester, practical shop instruction took precedence over instruction in theory and mathematics; as part of their engineering education students were expected to work long hours in the school shop and produce items for commercial sale. When Georgia Tech opened in

27Calvert, p. 50.

1888, it adopted similar policies. By 1896, however, Georgia Tech had begun to shift in the direction of the French educational tradition, abandoning the construction shop and shifting emphasis to courses on science and mathematics. Nonetheless, the curriculum that was ultimately adopted represented a compromise between the two traditions. Georgia Tech's mechanical engineering program was not wholly mathematical, theoretical, and research oriented, even after 1896. Shop work continued to be required, supplemented by engineering laboratory work. Brittain and McMath skillfully analyze the factors that led Georgia Tech to shift from an emphasis on the British tradition in education to more of an emphasis on the French tradition.

What had emerged in American engineering education in general by 1900 was a system that much more closely paralleled the French tradition than the British tradition but that retained elements of the latter's emphasis on hands-on instruction. These elements were most visible in the form of continuing shop work requirements in American engineering curricula and, to a lesser extent, in the development of laboratory instruction in American engineering courses, a development that distinguished American from French engineering education.[28]

Creating Professional Institutions

The development of formal means of professional training is one important element in the development of a profession. Another is the creation of professional institutions to promote and diffuse new knowledge, to represent occupational groups on issues of common concern, and to standardize and improve practice.

In Britain, John Smeaton organized informal meetings of civil engineers in the 1770s, and these led to the creation of the first formal engineering professional society, the Institution of Civil Engineers, in 1818. In America professional institutions for engineers on a national level emerged much later because of numerous problems. Before 1860 engineers were few in number and spread too thinly; transportation facilities were too primitive and service was too infrequent to permit convenient travel to a central meeting point. Sectionalism, urban rivalries, and philosophical differences on organization further complicated matters. This combination led to the failure in the late 1830s of the first attempt by American engineers to organize a national en-

---

[28]The use of laboratory work for students in engineering courses was an American innovation first developed and widely used by Robert H. Thurston in the 1870s. It can be considered a blend of the German tradition of academic research and the British tradition of hands-on training for engineers-to-be. Grayson, p. 256, cites it as one of the major American pedagogical accomplishments of the late 19th century.

gineering society and insured that only local technical societies survived before the 1860s.

The most successful of these local societies was the Franklin Institute of Philadelphia. Its early development is briefly described in Bruce Sinclair's essay, "At the Turn of a Screw: William Sellers, the Franklin Institute, and a Standard American Thread." Founded in 1824, the Franklin Institute was originally part of the mechanics' institutes movement—an attempt through free libraries, evening schools, public lectures, contests, exhibits, and museums to educate craftsmen in the scientific principles behind their work. The self-help philosophy of the mechanics' institutes had distinctly British roots. By the 1840s, however, the movement was in sharp decline because artisans had little time or energy for night schools and because a knowledge of scientific principles did not necessarily produce superior artisans.

In part because its leaders early recognized the shortcomings of traditional mechanics institutes, the Franklin Institute had deviated by 1830 from the organization's original purpose and had begun an ambitious program of engineering-oriented research. At the same time, they sought to make the Institute into a national center for civil engineering and an official adviser to the state and federal government, activities more in the French than British engineering tradition. Although ultimately unsuccessful in these endeavors, by the 1850s and 1860s the Franklin Institute had evolved into the closest thing the American engineering profession had to a national center and, as Sinclair's essay demonstrates, it played an important role in the diffusion and adoption of what became the standard American screw thread.[29]

The burning off of sectionalism by the War for Southern Independence and the expansion of America's transportation and communications systems allowed American engineers to create truly national engineering societies after 1865. The American Society of Civil Engineers (ASCE), after reorganization in 1867, became the first truly national American professional engineering society. American mining engineers split off from the ASCE in 1871 to form a second national engineering organization, the American Institute of Mining Engineers. In 1880 mechanical engineers organized as the American Society of Mechanical Engineers, and in 1884 electrical engineers formed the American Institute of Electrical Engineers. All four embarked on programs involving periodic national meetings and regular publication of new knowledge, and several began to experiment with other programs of professional importance, such as standards set-

[29]A detailed history of the Franklin Institute is provided by Bruce Sinclair in *Philadelphia's Philosopher Mechanics: A History of the Franklin Institute, 1824–1860* (Baltimore, 1982).

ting.[30] They soon displaced the Franklin Institute, whose local ties were always more important than its national ties.

Development of an American "Style" of Engineering

While American engineers were struggling to establish professional institutions, they were simultaneously developing a "style" of engineering that differed from both the French and British traditions from which they drew.[31] The American "style" of engineering that emerged in the 19th century placed more emphasis on reducing labor costs and on economy of construction than the parent traditions had and placed less emphasis on strength, permanency, aesthetic appeal, and safety.[32]

Bruce Sinclair's essay, "At the Turn of a Screw," provides an excellent case study of the American "style" of engineering. William Sellers, a shop-trained American mechanical engineer, developed in the 1860s a new form of screw thread that, using the Franklin Institute as a forum, he urged should be adopted as the American standard. Sinclair points out that, by comparison to the British standard screw thread, the Sellers standard was better adapted to American conditions. The Whitworth, or British, standard thread required considerable labor and craft skill for accurate production. But in America labor costs were high and the craft tradition weak. Sinclair points out that the Sellers standard, while technically no better than the Whitworth standard in terms of strength, durability, and range of pitches, was better adapted to American conditions because it could be produced more easily by ordinary mechanics and because it was less complicated to manufacture and easier to maintain. These characteristics sharply reduced labor costs and the use of skilled labor. Sellers's success in developing a standard screw thread for American conditions was one of the foundations on which American mass-production technology was based.

Conclusion

By 1900 engineering had sunk deep roots into American soil and had become a well-established profession with nearly 40,000 practi-

[30]The early histories of the four "founder" societies are discussed in Edwin T. Layton, *The Revolt of the Engineers: Social Responsibility and the American Engineering Profession* (Baltimore, 1986), pp. 25–46.

[31]The notion of technological "style" has been used by several scholars; see, e.g., Thomas Hughes, *Networks of Power* (Baltimore, 1983), pp. 363–460.

[32]I review some of the elements of the American engineering "style" noted by British engineers in Terry S. Reynolds, "American Engineering and British Technical Observers," *Transactions of the Wisconsin Academy of Sciences, Arts, and Letters* 64 (1976): 83–108. The *American Railroad Journal*, one of the earliest engineering journals, contains a number of articles dealing with differences between American and British practices in locomotive design and roadbed construction.

tioners.[33] Institutionally, the engineering profession had erected solid foundations. It had four widely recognized and active national engineering professional societies, a number of smaller societies, and around 100 university and college programs for training its practitioners. While elements of the two traditions from which it had been born were still visible in its institutions and educational practices, American engineering had developed a "style" of its own, well adapted to the political, geographical, economic, and social context of the United States.

But toward the end of the 19th century portents of significant changes to come were clearly visible, most notably the growing number and prominence of engineers employed by large enterprises. For most of the 19th century, engineering had been a very individualistic enterprise in America. The most prominent American civil engineers, figures like Roebling, Eads, and Ellet, were not deeply enmeshed in organizational hierarchies. They were retained to plan and direct construction of a project. Once a project was complete, they sought out new clients, or new clients sought them out. A similar situation prevailed in mechanical engineering. Through most of the 19th century, independent owner engineers of machine shops, like William Sellers, dominated mechanical engineering just as independent consultants dominated civil engineering. The dominant image in the profession, and the professional ideal, was that of the independent consultant or the owner engineer, who was not tied down in a large organizational bureaucracy.

The intimate association of engineering with large-scale projects and organizations, however, had early set in motion the incorporation of engineers within corporate or government hierarchies. Beginning with the state public works organizations established in the 1820s and 1830s and continuing with the early railroad companies and rapidly expanding manufacturing enterprises of the late 19th century, engineers worked increasingly as employees of large organizations rather than as independent operatives. By the 1890s it had become clear that the future of the profession no longer lay with the independent consultants and the owner-engineers who had dominated it in the 19th century, but with the engineers employed by large organizations. Adapting to this new situation was to become one of the critical problems faced by the American engineering profession in the early 20th century.

[33]U.S. Department of Commerce, *Historical Statistics of the United States: Colonial Times to 1970* (Washington, D.C., 1975), p. 140.

Background

ENGINEERS ARE PEOPLE

JOHN B. RAE

The proper way to introduce this topic is to announce that the engineer is a neglected figure in history and that I propose to remedy this defect. The claim is only partially accurate, but accurate enough for me to feel that my own objective can be something more than a rehash of work already done. Many eminent engineers of modern times have had biographies written. We know something of the engineers of the ancient world through L. Sprague de Camp and of the Renaissance through Bertrand Gille, and for the United States we are fortunate to have four first-class scholarly studies of American engineers as a professional group, by Daniel H. Calhoun, Monte Calvert, Edwin T. Layton, and Raymond H. Merritt.[1] Yet it remains a general truism that we know a good deal more about engineering works than we do about engineers and much more about who promoted engineering works than about who actually designed and built them. To give a conspicuous example, most Americans would unhesitatingly name De Witt Clinton as the builder of the Erie Canal, if they could name anyone at all. The number who have ever heard of James Geddes, Benjamin Wright, or Canvass White is probably contained in this room.

But at least we know something about who these three were. Through most of history engineers appear only occasionally as

JOHN B. RAE delivered this paper as his presidential address at the annual meeting of the Society for the History of Technology in Chicago on December 28, 1974. He was for many years professor of the history of technology at Harvey Mudd College, and he authored a number of books on the automobile and aircraft industries. Dr. Rae died on October 24, 1988.

[1]The books referred to are: L. Sprague de Camp, *The Ancient Engineers* (Cambridge, Mass., 1960); Bertrand Gille, *Engineers of the Renaissance* (Cambridge, Mass., 1966); Daniel H. Calhoun, *The American Civil Engineer* (Cambridge, Mass., 1960); Monte A. Calvert, *The Mechanical Engineer in America, 1830–1910* (Baltimore, 1967); Edwin T. Layton, Jr., *The Revolt of the Engineers* (Cleveland, 1971); and Raymond H. Merritt, *Engineering in American Society* (Lexington, Ky., 1969).

This essay originally appeared in *Technology and Culture*, vol. 16, no. 3, June 1975.

names, with a tantalizing absence of information about their personalities, social and economic background, training and education, and status in their societies. I am using the term "engineer" for those who did engineering as we understand the word now, regardless of the title used at the time, such as *architekton* in ancient Greece or *architectus* in Rome. This dearth of information seems to run through the standard histories of the early civilizations. This is not a field in which I can claim authority, but, as an example, I found in Joseph Needham's volume on Chinese technology some twenty individuals clearly identified as engineers and perhaps as many more who might be, but with little biographical information. One was a woman who built a mountain road in Fukien in A.D. 1314;[2] another, a general, was credited with the invention of the wheelbarrow about A.D. 232.[3]

Somewhat more detail is given in Needham's description of Chhihkan A-Li, appointed chief engineer by the Emperor Pho-Pho in A.D. 412 and authorized to "mobilize" 100,000 workers for the building of a new capital:

> A-Li was extremely skilled and clever, but also cruel and violent. He caused the workers to bake bricks to make the city wall. He used to test the bricks and if a hammer blow would make a depression as much as an inch deep, he would have the worker responsible killed and buried inside the wall. Pho-Pho thought that this showed much loyalty on the part of A-Li, and gave him the responsibility of all buildings and repairs.[4]

This must be one of the earliest recorded systems of quality control.

Egyptian civilization offers some fragmentary insights into the personalities of the men who built its massive structures because some of them left obelisks behind that have survived. A conspicuous example is Ineni, who served several Pharaohs about 1500–1450 B.C. The inscription on his obelisk tells us that he was a noble and a first-class engineer, lists his achievements and titles, and concludes:

> I became great beyond words: I will tell you about it, ye people. Listen, and do the good that I did, just like me. I continued powerful in peace and met with no misfortune; my years were spent in gladness. I was neither traitor nor sneak, and I did no wrong whatever. I was foreman of the foremen and did not fail. I never hesitated but always obeyed superior orders, and I never blasphemed sacred things.[5]

[2]Joseph Needham, *Science and Civilisation in China* (Cambridge, 1971), 4, pt. 3:31.

[3]Joseph Needham, *Clerks and Craftsmen in China and the West* (Cambridge, 1970), p. 32. Needham adds the claim that the wheelbarrow actually dates back to the 1st century B.C.

[4]Needham, *Science and Civilisation*, 4, pt. 3:42.

[5]Reginald Engelbach, *The Problem of the Obelisks* (New York, 1923), pp. 95–96.

The British Egyptologist who quotes this epitaph has commented that, if Ineni did indeed handle Egyptian labor for some forty years without blaspheming, it was certainly not the least of his accomplishments.

We still know very little about Ineni, except that he was not burdened by modesty, but even this much light on an ancient engineer is exceptional. Most of them remain frustratingly impersonal. It would be interesting to know something about Eupalinos of Megara, who in the 6th century B.C. built a water-supply system on the island of Samos that included tunneling through a 900-foot mountain of rock. The tunnel, which was described by Herodotus and found during an excavation about 100 years ago, was 3,300 feet long and about 5½ feet square. It was bored from both ends and had to have a jog in the middle to compensate for an error of twenty feet horizontally and three feet vertically.[6] Since it was just a conduit for water, the jog was unimportant, and considering the instruments and tools available to him, the surprising feature is not that Eupalinos made this error but that he was able to do the job at all. But all Herodotus tells us about Eupalinos himself is that his father was a Megarian. Was he (the father) also an architect? What professional status and recognition existed this early in Greek civilization to account for a Megarian being selected to execute a difficult project on Samos? Unfortunately, the answers to questions of this kind have probably vanished forever. Or who was Harpalos, the Greek engineer who built two bridges of boats to pass the army of Xerxes across the Hellespont? We know from Herodotus that Harpalos did the job after unidentified Egyptian and Phoenician engineers had tried, had seen their handiwork destroyed by a storm, and had been penalized by beheading at the orders of an angry emperor, who had the Hellespont whipped as well.[7] But we know nothing of Harpalos himself, although we can be reasonably sure that he did his stress analysis on his bridges very carefully.

Roman engineering has attracted abundant attention from historians—more so, as usual, than have Roman engineers. Yet in Rome we do have clear indications that the person we would call an engineer enjoyed greater prestige than in other early civilizations. While individual engineers do not stand out in the records, with Rome we are justified in assuming that Appius Claudius, for instance, really did plan and supervise the construction of the Appian Way, because every Roman official, certainly during the republic and to a

[6]Herodotus, *History* (Great Books of the Western World), 3. 60. 102. I note with interest that Eupalinos does not get into the index. See also de Camp, p. 100.
[7]Bernard Brodie and Fawn Brodie, *From Crossbow to H-Bomb* (New York, 1962), p. 14; Herodotus, *History* 3. 35–36.

lesser extent in the empire, was likely to have had some practical training in military engineering. The Roman attitude is well illustrated in the fact that the most honorific title for the chief priest was Pontifex Maximus—the boss bridge builder.

The esteem which the architect-engineer enjoyed in the Roman world appears in two items. First, the specifications laid down for the ideal *architectus* by Marcus Vitruvius Pollio in the 1st century B.C.:

> The mere practical architect is not able to assign sufficient reasons for the forms he adopts; and the theoretic architect also fails, grasping the shadow instead of the substance. He who is theoretic as well as practical, is therefore doubly armed; able not only to prove the propriety of his design, but equally so to carry it into execution.
>
> An architect should be ingenious, and apt in the acquisition of knowledge. Deficient in either of these qualities he cannot be a perfect master. He should be a good writer, a skilful draftsman, well versed in geometry and optics, expert at figures, acquainted with history, informed on the principles of natural and moral philosophy, somewhat of a musician, not ignorant of the sciences of both law and medicine, nor of the motions, laws, and relations to each other of the heavenly bodies.[8]

This excerpt has a particular attraction for me because it foreshadows the curriculum of Harvey Mudd College.

Second, some three centuries later the emperor Constantine issued this instruction: "We need as many engineers as possible. As there is lack of them, invite to this study persons of about 18 years, who have already studied the necessary sciences. Relieve the parents of taxes and grant the scholars sufficient means."[9]

Note the clear implication in this order that the education of the *architectus,* if not up to Vitruvius's standards, was definitely more than training at the bench. The recruits were to have studied "the necessary sciences," and they are referred to as "scholars."

Much the same pattern continues through the Middle Ages, which we know to have been a technologically significant period, summed up in Lynn White's observation that the chief glory of the Middle Ages was not its cathedrals or its scholastic philosophy but rather "the building for the first time in history of a complex civilization which rested not on the backs of sweating slaves or coolies but primarily on non-human power."[10] But the makers of this power revolution were

[8]*The Architecture of Marcus Vitruvius Pollio,* trans. Joseph Gwilt (London, 1826), pp. 3–4.

[9]W. T. Sedgwick and H. W. Tyler, *A Short History of Science* (New York, 1925), p. 142.

[10]Lynn White, jr., *Machina ex Deo* (Cambridge, Mass., 1968), p. 71.

anonymous, or at any rate unrecorded craftsmen. Occasionally a name emerges, like Villard de Honnecourt, but we know considerably more about de Honnecourt's writings than we do of the man himself. And it would be nice to know more about Ailnolth, a 12th-century Englishman who worked on the Tower of London and Westminster Bridge and was referred to as an *ingeniator*.[11] The Middle Ages did recognize the engineer as a military figure, and the distinction between military and nonmilitary engineering has always contained an element of artificiality. After all, if a road was built or a harbor improved for military reasons, it could hardly be picked up and taken away just because the theater of operations had moved somewhere else.

It may be that intensive research would uncover the kind of information I am seeking about the engineers of these early societies, although it is probably more likely that the information either is irretrievably lost or never existed. Even the enlightened Greeks of classical civilization were regrettably negligent about keeping vital statistics. With the period of the Renaissance we reach more fruitful terrain. Engineering was acquiring a recognizable professional status, and the engineers themselves begin to become better known. There is, of course, Leonardo da Vinci, but Leonardo is so much a special case that I will simply quote the statement made by Ladislao Reti to Lynn White upon the recent discovery of the Leonardo manuscripts in Madrid: "Now at last people will begin to believe me when I tell them that Leonardo was an engineer who occasionally painted a picture when he was broke."[12]

There is enough information about other Renaissance engineers to whet the appetite and to suggest that further research would be rewarding. Bertrand Gille in *Engineers of the Renaissance* describes fifty-eight individuals besides Leonardo. Table 1 attempts to collate the information on family background and education. The statistical base is of course far too small to permit drawing conclusions, especially in view of the preponderance of "unknowns," but the figures offer some interesting speculations. Among the fourteen whose fathers' occupations are known there is a clear preponderance of middle- and lower-class origins, with seven from business and professional families (this category includes law, medicine, architecture, and teaching), and four from skilled craftsmen. For professional training,

[11]W. G. H. Armytage, *Social History of Engineering* (Cambridge, Mass., 1961), p. 45.
[12]Lynn White, jr., "The Flavor of Early Renaissance Technology," in *Developments in the Early Renaissance*, ed. B. A. Levy (Albany, N.Y., 1971), p. 38.

TABLE 1

FAMILY AND EDUCATIONAL BACKGROUND OF
RENAISSANCE ENGINEERS

Background	No. Engineers
Occupation of father:	
Farmer	1
Landowner	1
Laborer	1*
Craftsman	4
Business	2
Professional...............	4
Military	0
Engineer	1
Unknown..................	44
Total....................	58
Form of training:	
Self-taught.................	5
Apprenticeship.............	10
Formal education	3
Private	1
Trained in family	6
Unknown..................	33
Total....................	58

SOURCE.—Bertrand Gille, *Engineers of the Renaissance*
(Cambridge, Mass., 1966).
*Nicolo Tartaglia; *Encyclopaedia Britannica* says, "childhood
passed in poverty"; no family identification.

sixteen out of twenty-five passed through some form of
apprenticeship—"trained in family" falls into this category.

As we go forward in time we find indications of more adequate
information. Table 2 contains some data on a miscellaneous assortment of British engineers for the period from the 17th to the 19th
century, ranging from Sir Cornelius Vermuyden, born in Holland
about 1595, to I. K. Brunel, who died in 1859. The sample is too
limited and random to be really meaningful, but it suggests that information on the social history of British engineers is available and
affords a potentially profitable field for research. The individuals on
my list cover a broad spectrum, from one of the landed gentry like Sir
Hugh Myddleton to a Scottish crofter's orphan, Thomas Telford. My
categories in this and the other tables are quite broad, and they do not
show actual economic status. For example, Telford's Scottish contem-

TABLE 2

FAMILY AND EDUCATIONAL BACKGROUND OF
BRITISH ENGINEERS, 17TH–19TH
CENTURIES

Background	No. Engineers
Occupation of father:	
Farmer	8
Landowner	2
Laborer	0
Craftsman	5
Business	2
Professional	3
Military	0
Engineer	10
Unknown	4
Total	34
Form of training:	
Self-taught	5
Apprenticeship	13
Formal education	9
Private	0
Trained in family	9
Unknown	2
Total	38*

*Four of the 34 individuals had both formal technical education and apprenticeship.

porary John Rennie was also a farmer's son, but Rennie senior was able to send his son to Edinburgh University.

To digress a little, John Rennie evidently attempted to develop other than professional skills. He ventured into poetry, in the following lines (1797):

Barren are Caledonia's hills,
Unfertile are her plains.
Bare-leggéd are her Brawny Nymphs,
Bare-arséd are her Swains.[13]

Inspiration apparently ran out at this point, because Rennie's biographer, Cyril Boucher, testifies that this appears to be the only piece

[13]Cyril T. G. Boucher, *John Rennie, 1761–1821: The Life and Work of a Great Engineer* (Manchester, 1963), p. 21.

of verse that John Rennie ever wrote. For the sake of both poetry and engineering, it is probably just as well.

Rennie's sons George and John, the latter having the distinction of being the second British engineer to be knighted (in 1830; the first was Sir Hugh Myddleton 200 years earlier), are included among the ten engineers' sons in the table, as are Robert Stephenson and I. K. Brunel. The high proportion of these in the total—almost exactly a third—is partly a reflection of the length of time represented, over 250 years. More important, it indicates the evolution over at least this period of a recognizable nonmilitary engineering profession. John Smeaton may have been the first to call himself a civil engineer, but a study of the group that called itself the Smeatonians comes to this conclusion:

> There was in Great Britain during the 18th century a much more firmly established and self-conscious profession of civil engineering than has been supposed. The men who were practising their profession did not call themselves Civil Engineers until near the end of the century, but they did call themselves engineers. Also, they were indigenous and they grew up entirely apart from military engineering practice. To the Thames, the Dee, the Clyde, and the sluggish rivers of Fenland they owe their being.[14]

Civil and mechanical engineering were not yet distinct, but we know that millwrights contributed heavily to the mechanical developments of the Industrial Revolution. Of these, A. E. Musson and Eric Robinson observe that most had a good knowledge of arithmetic, geometry, and theoretical as well as practical mechanics, and they add: "It appears, in fact, that these millwright-*engineers* [my italics] were not—as is often suggested—rough, empirical, illiterate workmen, but had usually acquired somehow a fairly good education or training."[15] It would be well worth looking into the "somehow."

Since France at this time was not only producing brilliant engineers but was setting the pattern of formal engineering education through the founding of the Ecole des ponts et chaussées (1747) and the Ecole polytechnique (1794), French engineers of this same period should offer a fruitful field of study. Some are famous: the Duc de Sully, Henri IV's great minister, and Lazare and Sadi Carnot. Others, like the 18th-century highway engineer P. M. J. Trésauget, have received

[14]Esther C. Wright, "The Early Smeatonians," Newcomen Society, *Transactions* 18 (1937–38): 107.

[15]A. E. Musson and Eric Robinson, *Science and Technology in the Industrial Revolution* (Toronto, 1969), p. 429.

less credit, at least in English-language publications, than they appear to deserve. However, this is an area where my knowledge is too limited to warrant discussing it further.

The one group I have been able to study with any intensity—but neither exhaustively nor definitively—is American engineers of the 19th century. The information I am presenting here comes principally from the Biographical Archive of American Engineers in the Smithsonian Institution's National Museum of History and Technology, and I wish to take this opportunity to thank Robert Vogel and Robert Post of the museum staff for the willing cooperation and assistance that they have given me. This archive was assembled by the museum's Division of Mechanical and Civil Engineering, with assistance from the American Society of Civil Engineers and the American Society of Mechanical Engineers. It now contains about 2,500 names of individuals who might be classified as engineers and who were born before 1861. The compilation began with civil engineers but now includes mining, mechanical, and electrical engineers.

The amount of information on each of these people varies markedly. Some folders have detailed data; others contain only a name. In my own tabulation I have reduced the number to be considered to 1,672. Most of the exclusions are names about which there was too little information to justify including them. The others represent individuals who do not seem properly classifiable as engineers. I used two criteria: (1) There had to be evidence of some systematic professional study of engineering, whether academic, through a form of apprenticeship like shop training, or by self-education. This criterion excludes principally the cut-and-try inventor. (2) The individual in question had to have engaged in the practice of engineering for a significant part of his career, thereby excluding men who may have had some engineering training but whose careers were predominantly in management.

Table 3 requires little comment. It indicates that most 19th-century American engineers were home grown (the pre-1790 category contains very few names of men who were engineers in the 17th or 18th century—not surprising, since civil engineering was barely beginning to achieve professional recognition even in Europe). There is very little information on European engineers who may have worked for a while in the United States and then returned to Europe. The archive and the ASCE's *Biographical Dictionary of American Civil Engineers* (New York, 1972) include one or two of the more eminent of those, such as Marc I. Brunel, and military engineers, like Thaddeus Kosciusko and Louis du Portail, who served in the Revolutionary War, but there is no way at present to tell how many lesser-known engineers might fall in this category.

TABLE 3

BIRTHPLACES OF AMERICAN ENGINEERS BORN BEFORE 1861

| | BIRTHPLACE | | | | | | |
| | United States* | | Europe | | Other** | | |
PERIOD BORN	N	%	N	%	N	%	TOTAL N
Before 1790 ...	62	83	13	17	...	...	75
1790–1830	277	86.5	40	12.5	3	0.9	320
1831–60	1,078	84.4	166	13	33	2.6	1,277
Total	1,417	85	219	13	36	2	1,672

SOURCE.—Biographical Archive of American Engineers, National Museum of History and Technology, Smithsonian Institution.
*Includes individuals born before the American Revolution in the area that became the United States.
**Predominantly Canada, some from Latin America.

Table 4 represents an effort to identify social origins for 19th-century American engineers through the father's occupation. The principal drawback is, as the table shows, the large category "unknown," amounting to half the total number for the whole period studied. An analysis of places of birth might provide some clarification. Small towns and villages predominate, as would be expected for this period of American history, so that a good many of the unknowns were probably farmers, but I have preferred not to guess.

Some of the categories have indistinct borderlines: farmer and planter or estate owner, for example. I have followed the terminology of the sources of information; where this was uncertain, as where the father had more than one occupation. I have used the one that seemed most important.

With due allowance for inadequacies in the evidence, some features stand out. It appears quite clear that the largest single proportion of American engineers has been recruited from the upper and middle levels of society: the categories listed as planter or estate owner, business or finance, professional (law, medicine, and clergy, specifically), military, and engineer. If these groups are combined, their proportion of the whole appears thus:

	Before 1790	1790–1830	1831–60	Total
% of total	33.3	44.3	36.5	37.7
% of group where father's occupation is known	48.9	70.7	79.3	76.4

TABLE 4

FAMILY BACKGROUND OF AMERICAN ENGINEERS BORN BEFORE 1861

	PERIOD BORN											
	Before 1790			1790–1830			1831–60			COMBINED PERIODS		
OCCUPATION OF FATHER	N	% of Total	% of Known	N	% of Total	% of Known	N	% of Total	% of Known	N	% of Total	% of Known
Farmer............	16	21.3	31.3	35	11	17.4	80	6.2	13.7	131	8	15.7
Laborer...........	1	1.3	1.9	2	0.6	1	7	0.5	1.2	10	0.6	1.2
Craftsman........	9	12	18.2	22	7	11	33	2.6	6	64	3.8	7.7
Planter or estate owner...........	1	1.3	1.9	6	1.9	3	20	1.7	3.4	27	1.6	3.2
Business or finance........	10	13.3	19.6	66	20.6	33	181	14.2	31	257	15.4	30.8
Professional......	5	6.7	9.8	43	13.4	21.4	177	14	30.4	225	13.5	27
Military	7	9.3	13.7	14	4.4	7	32	2.5	5.5	53	3.2	6.3
Engineer.........	2	2.7	3.9	13	4	6.3	53	4.1	9	68	4	8.1
Unknown.........	24	32	...	119	37	...	694	54.3	...	837	50	...
Total N	75	...	...	320	...	...	1,277	...	...	1,672	...	...

SOURCE.—Biographical Archive of American Engineers, National Museum of History and Technology, Smithsonian Institution.

In summary, at the very minimum, if all the unknowns are excluded from these classes, they still produced between one-third and three-eighths of the engineers. At the maximum, if the unknowns are assumed to divide in the same proportion as the knowns, the middle and upper classes provided about three-fourths of the engineers. It seems a safe assumption to give them at least half.

At the other end of the scale, it appears evident that ordinary laborers were a negligible source of engineers. Identification of the unknowns might make a difference, but, as stated above, for this period of American history a far more likely result would be enlargement of the farmer category.

Some explanations are needed for table 5.

1. The totals are larger than for the others because of conscious duplication. Some engineers completed an apprenticeship as well as an academic technical training; this was, in fact, standard for European-educated engineers and to a considerable extent still is. In these cases I have counted both types of training.

2. The "secondary" category includes individuals who completed a secondary education and went no farther.

3. "Institute or college of engineering or science" includes schools of engineering or science in universities, as well as independent polytechnics or institutes of technology. In general, I have included here all the recipients of engineering degrees.

4. "Military academy" includes all service institutions.

Some general observations can be made about this table. It could be expected that the proportion of academically educated engineers would increase, but the leap between the 1790–1830 and the 1831–60 periods is spectacular. It reflects the proliferation of land-grant and privately supported technical colleges after 1860, but the growth rate is surprisingly high. Part of the explanation may be that records of college-trained engineers are easier to locate.

It also appears that the United States Military Academy made a relatively small *quantitative* contribution to the supply of engineers even during the first half of the 19th century. The eminence of the West Pointers in the period says much for the quality of the education.

Finally, the proportion of engineers educated abroad has some interesting points. For the pre-1790 period it is 32 percent of all those receiving some form of higher education, but this is eight persons out of twenty-five. The comparable percentages for the later periods are 7.8 for 1790–1830, 12.4 for 1831–60, and 12.3 overall. Most of these were persons born abroad who completed their education before coming to the United States. There is little indication of Americans

TABLE 5

EDUCATIONAL BACKGROUND OF AMERICAN ENGINEERS BORN BEFORE 1861

EDUCATION OR TRAINING	PERIOD BORN						COMBINED PERIODS	
	Before 1790		1790–1830		1830–60			
	N	%	N	%	N	%	N	%
Self-taught	16	17.4	40	12	42	3	98	5.4
Apprenticeship	12	22.8	60	18	203	14.7	284	15.7
Secondary school	1	1.1	49	14.6	120	8.7	170	9.4
College or university, nontechnical:								
U.S.	8	8.7	35	10.4	124	9	167	9.3
Other	2	2.2	...	...	6	0.4	8	0.04
Institute or college of engineering or science:								
U.S.	3	3.3	43	12.8	606	44	652	36.1
Other	1	1.1	11	3.3	94	6.8	106	6
Military academy:								
U.S.	6	6.5	53	15.6	79	5.7	138	8
Other	5	5.4	...	...	15	1.1	20	1.1
Privately educated	13	14.1	16	4.8	37	2.7	66	3.6
Unknown	16	17.4	28	8.4	52	3.8	96	5.3
Total N	92	...	335	...	1,378	...	1,805	...

SOURCE.—Biographical Archive of American Engineers, National Museum of History and Technology, Smithsonian Institution.

TABLE 6

Method of Entry into Engineering of American Engineers Born before 1861

Method of Entry into Engineering	Period Born								Combined Periods	
	Before 1790		1790–1830		1831–60					
	N	%	N	%	N	%			N	%
Trained on job	35	46.7	115	36	296	23.2			446	26.6
Apprenticeship	15	20	58	18.1	176	13.8			249	14.9
Technical education (academic)	15	20	128	40	752	59			895	53.5
Family association	4	5.3	15	4.7	38	3			57	3.4
Unknown	6	8	4	1.25	15	1.2			25	1.5
Total N	75	...	320	...	1,277	...			1,672	...

SOURCE.—Biographical Archive of American Engineers, National Museum of History and Technology, Smithsonian Institution.

seeking formal engineering education in Europe; they might and did go to Europe to study engineering works but not in any great number to attend engineering schools.

Table 6 probably involves the greatest degree of subjective judgment on my part. I tried to determine how each of these 1,672 individuals got into the actual practice of engineering. The apprenticeship and technical education categories are the most readily defined; it is usually easy to see the individual go from the completion of his education or training directly into engineering work. "Trained on job" applies to those men who with no identifiable previous technical training began working on an engineering project and in time became recognized as professional engineers. "Family association" is used for those who started their careers working in a family business connected with engineering, or for relatives who were engineers. The results suggest some validity for the classifications, in that the proportion entering from on-the-job training and apprenticeship shows a steady decline while the proportion of academically trained engineers correspondingly increases.

These tables are, of course, only part of the story of the 19th-century American engineer—and I should add that a good many of the men born in the latter part of the period covered by the archive had careers running well into the 20th century. I had to decide what was the most important information to look for in a limited time and with limited resources. There are undoubtedly some errors in recording and counting, but on the whole I believe these tables to be reasonably accurate for what they purport to present.

In effect, I have done a fairly rapid strip-mining operation on the archive, although I hasten to maintain that I have restored the terrain to approximately its original contours. Most of the ore is still there, and my own partial sampling indicates a rich vein.

To conclude, trying to pursue the engineer as an individual through history is a large, frequently frustrating, but thoroughly fascinating topic. While I have not explored it more than superficially so far, I believe the topic to be important as well as interesting. Knowing who the engineers were may help us to understand how and why they functioned as they did. To put it another way, I am suggesting that, if discernible patterns can be found for such features as family background, education, and social status for the engineers in a given society, then such patterns might help to explain the nature and level of the technology attained by that society. I can't guarantee results, but I offer this as a promising field for research.

Practice

TURNPIKE CONSTRUCTION IN ANTEBELLUM VIRGINIA

ROBERT F. HUNTER

STUDENTS OF TRANSPORTATION in the antebellum United States have given their attention almost exclusively to railroads and canals and have neglected turnpikes.[1] It is true that turnpikes were relatively unimportant as main lines of transport, yet most American farmers, and many townsmen, were not fortunate enough to live on a waterway or a railroad line. To these people, the quality and cost of common road transportation was a matter of everyday concern.

In the nineteenth century, as in the twentieth, the turnpike was essentially an expedient method of highway financing—and an even less satisfactory method then than now because of the relative ease with which users could avoid paying tolls. Yet it did succeed in improving roads in many states and lowering the cost of road transportation from between $0.20 and $0.25 per ton-mile on common roads to about $0.15 per ton-mile on the best macadamized turnpikes. Some of the best ones, such as the Lancaster Turnpike in Pennsylvania and the Valley Turnpike in Virginia, achieved renown for their high quality.

The construction of roads lost its historical status as a highly developed craft during the first half of the nineteenth century and was advanced to the fast-growing list of exact sciences, for the best engineering standards in road construction, developed first in France and England, were exacting enough to demand considerable training. Although the United States could claim the first professional school of civil engineering in the English-speaking world (Rensselaer Polytechnic Institute, established in 1823), and many other American universities entered the field early, Virginia turnpikes were not built often by trained engineers. Most of them were built by amateurs who applied their own notions to the actual work.

COL. ROBERT F. HUNTER is professor of history emeritus at the Virginia Military Institute. He has published, with coauthor Edwin L. Dooley, Jr., *Claudius Crozet: French Engineer in America, 1790–1864* (University Press of Virginia, 1989).
[1] An exceptionally well-balanced study which does give due attention to turnpikes is Edward C. Kirkland, *Men, Cities, and Transportation: A Study in New England History, 1820-1900,* 2 vols. (Cambridge, Mass., 1948.) See also Wheaton J. Lane, *From Indian Trail to Iron Horse: Travel and Transportation in New Jersey, 1620-1860* (Princeton, 1939.)

This essay originally appeared in *Technology and Culture,* vol. 4, no. 2, April 1963.

Early in 1816, when New York and Pennsylvania were showing signs
of renewed interest in the western trade, the Virginia General Assembly passed an act (5 February 1816) which created a Fund for Internal
Improvement and a Board of Public Works to administer the fund.
Soon afterward, the Assembly passed the General Turnpike Law (7
February 1817) designed to guide the Board of Public Works in its
dealings with private turnpike companies. Those companies approved
by the Board were permitted to sell two-fifths of their authorized
capital stock to the State.

What sort of highway system was produced by this system of mixed
enterprise? How good was the quality of the roads built by private
enterprise with State aid?

The Virginia Board of Public Works concerned itself with the task
of transmitting to those untrained contractors who built turnpikes as
much as possible of the latest scientific knowledge that they might
put to practical use, and for that purpose used an official known as the
Principal Engineer. The most competent man to fill that office, Colonel
Claudius Crozet, expended an immense amount of time, energy, and
patience in attempting to educate the amateur road builder, too often
a thankless and frustrating job.

Crozet was an interesting person. Born in France in 1789, he was an
artillery officer under Napoleon. Emigrating to the United States in
1816, he joined the academic staff at West Point, and in 1823 became
Principal Engineer of the State of Virginia. In that capacity he acquired a thorough knowledge of the state and its internal improvements program and probably did more than any other person to write
the history of the turnpike movement in Virginia, both with pen and
theodolite. His letters to the Board of Public Works, printed in their
annual reports, are indispensable as source material.[2]

As Crozet instructed his charges, the construction of a turnpike
involved three preliminary steps. First was the location of the best
line between the two terminals, no simple task. Second was the esti-
mation of the cost, and third was the negotiation of a contract or
contracts with the builders.

Location was a task that called for skill, especially in rough terrain.
If a road were to be built between two towns located on a level plain,
the problem could be as simple as finding the straight line between
those points. Most of the 180-odd turnpike companies in Virginia were
found west of the Blue Ridge, however, in terrain that ranged from
moderately hilly to mountainous. Crozet remarked upon inspecting the

[2] For a biographical sketch of Crozet, see Colonel William Couper, *Claudius
Crozet* (Charlottesville, 1936; *Southern Sketches*, Number 8, First Series.)

Lexington and Covington Turnpike that it was a shame such skill in road-making had not been used on a better location.[3] He and other engineers made the same comment so often that it can be assumed that faulty location was a chronic defect of Virginia turnpikes.

In theory, the line of location should stay as close to the direct line between the terminals as possible, and the line that did, other things being equal, would naturally be preferred. But other things were not often equal, and deviations from the direct line might stray some distance to secure an advantage of some kind, such as an easier grade, a lower crossing of a ridge, avoidance of bridging a stream, avoidance of soft or marshy ground, locating on a preferred type of soil, avoidance of excavation or embankment, locating closer to sources of surfacing materials, or securing a better exposure of the road surface to sun and wind.

Contemporary engineers did not seem interested in prescribing an optimum percentage of increase of distance over the direct line. Apparently the factors determining the best location were too variable. However, study of available evidence indicates that if there were as little as 15 per cent increase, in moderately rough terrain, the engineers were pleased, but if there were as much as 25 or 30 per cent, explanation was in order. A road from Danville to Wytheville, for example, was devious enough to add a 36 per cent increase to the distance. It crossed the direct line a dozen times between Danville and the mountains, and in meandering its way across the Blue Ridge, strayed as far away from the direct line as 6¾ miles, the reasons for which Crozet went to some length to explain to the Board of Public Works.[4]

But if excessive deviation called for explanation, insufficient deviation called for sharp criticism. Crozet harped on this. The grade of Ashby's Gap Turnpike in places, he observed, was too steep, the stones too large, and the carriageway too narrow. "The two last defects may be easily remedied; the first would require a change of location; or, in other words, a new road." This was a result, he said, of "the general error of aiming at making straight roads."[5] In similar comment on the road between Franklin and Beverly (in present-day West Virginia), Crozet remarked, "It is always with regret that I thus see an expense for so useful a purpose fail of its object by an error of location; and labour, which might otherwise have been durable, almost thrown

---

[3] Crozet to the Virginia Board of Public Works, 12 March 1838, *Annual Reports of the Board of Public Works*, Vol. XXII, p. 97. (Cited hereafter as *ARBPW*.)

[4] *ARBPW*, XII, pp. 319-20.

[5] *ARBPW*, XIII, p. 515.

away." [6] The existence of a newly-constructed, badly-located road would retard for a long time the making of a better one.

In mountainous terrain, the locating of the line of a road could sometimes be deceptive. The direct line between points on opposite sides of a mountain might well be longer, when the necessary deflections had been made to furnish tolerable grades, than a road around the mountain. When Engineer Peter Scales surveyed in 1834 for a road between Huntersville and Warm Springs, he noted some of the local citizens contended " that it would be more eligible to pass over the Big Back Creek Mountain, than to subject the travel [sic] to what they deemed an inadmissible increase of distance, incurred by running round its southern end." Scales proceeded to make a survey across the mountain along the old county road then in use, discovered grades as steep as 15 degrees, and concluded that to relocate the road across the mountain with reasonable grades (maximum: 3½ degrees) " would have called for several zigzags on the face of the mountain upon a north exposure . . . [increasing] the distance to an extent nearly equal, if not greater than the line now thrown round the end, upon easy grades, favorable ground and a good exposure." [7]

Where the mountain ridge remained unbroken for miles, as is characteristic of the Blue Ridge, there was no choice between going over or around the mountain. Other things being equal, the gap nearest the direct line would be preferred, but other factors usually influenced the choice. For example, the General Assembly passed an act (1 March 1826) directing the Principal Engineer to survey the " shortest and best practicable route from Covington to Richmond." Five years later, having sandwiched this survey by bits and pieces in between a multitude of other duties, Crozet submitted his report. " The main difficulty in this business," he wrote, " is the passage of the Blue Ridge." East of Lexington there were four gaps in the range, two of which, White's Gap and Irish Gap, were more favorable than the other two, Indian Gap and Robertson's Gap. White's Gap was nearest the direct line, but having crossed this (going east), " it would require the crossing of the valley of Pedlar River, and another high mountain beyond it." Irish Gap would require 390 feet more ascent than White's Gap, but it could be reached with a smaller grade, the streams on the eastern descent would all be headed (i. e., the road would avoid bridges because it would pass above the heads of all streams), and there would be no additional mountain to cross. On the other hand, the descent on the slope east of Irish Gap involved an uninterrupted slope five miles

[6] *Ibid.*, p. 508.
[7] *ARBPW*, XVIII, pp. 297-99.

long at a grade exceeding 4½ degrees, "to which should always be preferred alternate ascents and descents," to rest the horses, which the descent east of White's Gap allowed. All things considered, it amounted to a choice of the lesser evils, but so evil were the choices, concluded Crozet, "it appears that no advantage would be obtained by forcing a turnpike from Lexington through either Irish or White's Gap, and that it would be better, in every respect, to intersect the Staunton and James River Turnpike at Waynesborough." [8] The result of this survey was probably the chief reason why no Covington-Richmond turnpike was built. Crozet did conclude in the same report, however, that a turnpike between Covington and Lexington was feasible, and soon after his survey the Lexington and Covington Turnpike Company was formed.

Also to be considered in locating the line for a road was the existence of other roads or turnpikes in the vicinity and the potential effects of the new construction on their traffic. This consideration was especially important to the turnpike system, in which traffic meant tolls, and tolls meant the very existence of the road. Crozet pointed to this in reporting on the Middleburg and Strasburg road in 1829. Middleburg, east of the Blue Ridge, and Strasburg, in the Valley, were about 40 miles apart. A few miles to the north, the Ashby's Gap Turnpike traced a parallel line. Crozet agreed that the proposed location of the new road was a good one, "in the abstract . . . certainly the best." But the distance saved would not be worth the effort, because "instead of a desirable concentration of interest, an injurious competition would be created." Crozet admitted that there was no guiding principle to be used as a rule of thumb in such cases and that judgment and foresight were required sometimes to an impossible degree. "These competitions between roads leading in the same direction," he wrote, "advocate strongly the importance not to undertake short turnpikes, without having previously enquired into the most favorable direction that a general road would ultimately assume; or else, a more extensive improvement will frequently be the destruction of those that have preceded." [9]

Once the line of the road had been surveyed by Crozet or another state-appointed engineer, there seemed to be no way to compel the turnpike company officials to use the line recommended. Crozet and other engineers complained about this frequently to the Board of Public Works. "I cannot too often bring such instances to your notice," he wrote:

---

[8] *ARBPW*, XV, pp. 157-60.
[9] *ARBPW*, XIII, pp. 505-06.

... persons not aware of the carefull [sic] attention with which the different considerations which constitute a good road are balanced by the engineer who locates it, will not always appreciate his motives; and thus modifications are frequently made with the best intentions, which result, however, to the disadvantage of the road. It is only when steep grades or muddy places appear, that the error is, though too late, discovered.[10]

Even the elaborate Valley Turnpike was not immune to the discovery of an error in location by the ubiquitous and perceptive Crozet. He was aware that the construction had been in competent hands (those of Joseph Reid Anderson) and was inclined to believe that the changes of location had been made for the purpose of saving the expense of paying damages to farmers along the line. "Farmers prefer straight fences very naturally to others, and frequently, also, offer to relinquish damages on particular locations of their own choice." In complying with such proposals, the company might save damages, but in the end the road and the public would suffer, for a departure from the best principles of location "seldom fails to lead the company itself into greater expense than the original saving." [11]

Curiously, this principle was recognized clearly by a lesser man than either Anderson or Crozet. In the course of building the Brandonville, Kingwood, and Evansville Turnpike, E. M. Hagans wrote the following note to his brother:

... we have got I think the right of way cleared ... excepting Col. Fairfax. have not made an examination to know whether we can change to suit him. he says if we can change the location between his house and gate he will give way and 150$. but if this change is not made he will give nothing and claim damages. now we will make the change if we can without injuring the road, but if not we must go through on the location.[12]

An indeterminate number of smaller turnpikes and at least one long one (the Cumberland Gap and Price's Turnpike Road) were not merely handicapped but ruined by mistakes made in the location. The General Assembly made the initial mistake in the case of the Cumberland Gap and Price's Turnpike Road by authorizing a grade of six degrees.[13] Most, if not all, turnpikes were handicapped to some extent by location errors, the most difficult type of error to correct.

[10] *ARBPW*, XXII, pp. 76-77.
[11] *ARBPW*, XXV, p. 500.
[12] E. M. Hagans to Harrison Hagans, 27 July 1839, Hagans Papers, University of West Virginia Library.
[13] *Acts of Assembly*, 1831-1832, p. 62; 1833-1834, pp. 106-09.

After the location had been determined, the next step was to prepare an estimate of the cost. Information on the accuracy of estimates made in the building of Virginia turnpikes would be interesting if available, but unfortunately it is not. In only four instances are figures available indicating both the estimated cost before construction and the actual cost when built. Although these scanty figures are insufficient for generalization, they do suggest that with all of Crozet's abilities, money matters were not his forte.[14]

Making an accurate estimate was beset with all sorts of difficulties, especially in cases where an estimate was made by one person, the letting of contracts by someone else, and the construction done by still others. But even if everything were done by the same group, contingencies arose right and left.

First, the land itself had to be bought, and there were abundant complaints that turnpike companies were forced to pay high prices for the exercise of eminent domain. Excavations and embankments were expensive operations, especially in rocky areas. Mechanical structures, such as bridges and culverts, were expensive items. Engineering expenses, including surveys, superintendence of contractors, and keeping of records, were usually estimated at ten per cent of the amount of the other items. Surfacing the finished road was a more or less expensive item, depending on the resources of the company after the other work was finished and the availability of materials. More often than not, this item was dispensed with in Virginia. Finally, at least ten per cent of the total amount must be added for contingencies. "Even then," wrote W. M. Gillespie, "the actual expense will generally exceed the estimate." First, the price of labor, as the work proceeds, may rise higher than what it was at the time of the estimate. Second, in deep excavations, rock may be encountered where earth was expected, which increases the cost tenfold. Third, many improvements in detail are suggested and adopted as the work proceeds, almost always with an increase in cost. Fourth, minor incidental expenses are nearly always overlooked in the estimate, "trifling in themselves, but considerable in their aggregate."[15]

Before an advertisement was placed for bids, a specification had to be prepared: a complete, exact, detailed description of the desired road, based on the plans and field notes of the locating engineer. The pros-

---

[14] He estimated the Charleston and Point Pleasant Turnpike would cost $12,500; it actually cost $55,235. He estimated the Giles, Fayette, and Kanawha Turnpike would cost $68,400; it actually cost $25,429.

[15] William M. Gillespie, *A Manual of the Principles and Practice of Road-Making* (New York, 1852), pp. 147-48.

pective contractor would go over the plans and specifications carefully, as well as the ground itself, before submitting his bid. With the larger enterprises, the usual practice was to advertise for bids in local and Richmond newspapers. Road contractors seldom came from out of the state.

Advertisements were fairly specific. In February and March 1826, for example, the directors of the Staunton and James River Turnpike Company advertised in the Richmond *Enquirer* for bids on the eastern 30 miles of their road:

> 1. There is to be a carriage way of earth—22 feet wide at least—the centre of which is to be raised 15 inches above the base line—to be well dressed up to a smooth and even surface.
>
> 2. On each side of the carriage way, there is to be a regular and smooth ditch cut, sufficient to carry off the water that may drain from the road and adjacent lands.
>
> 3. Culverts constructed of stone, and built in a substantial and durable manner, are to be placed wherever the President and Directors, or the superintendent they may appoint, may consider necessary.
>
> 4. On each side of the ditches, the timber is to be grubbed, cut and cleared away, 14 feet at least.
>
> 5. No part of the carriage way is to exceed five degrees of elevation . . .[16]

Other things being equal, the contract would go to the lowest bidder, but other things had to be considered. To have made a sound estimate of the cost before advertising for bids was important to the company at this point. Armed with an accurate knowledge of the lowest cost at which the work could be done, the company was to some degree protected against the trickery of underhanded contractors. Cases were not unknown in which a contractor would obtain a large supply of tools and provisions on credit and suddenly depart with this as loot.

Crozet, Gillespie, and other engineers emphasized the importance of making the contract thoroughly specific. It should contain " copious and stringent " stipulations as to when and how the work should be done, in what sequence of parts, at what rate of progress, and when it should be completed. It should be made clear which expenses were to be borne by the company and which by the contractor, when payments should be made for work done, what penalties were to be imposed for neglect, and so on, " always remembering that everything

[16] Richmond *Enquirer*, 24 March 1826, p. 4.

must be expressed, and nothing left to be inferred." [17] In a report on the Cumberland Gap and Price's Turnpike Road, Crozet observed a lack of necessary ditches and drains. It was the turnpike's loss. " As neither specifications, notes, or contracts designated the necessary number of drains, they could not be expected from the contractors." [18]

The services of the Principal Engineer to private turnpike companies did not end with the survey and location. If the company wished, he would draw up specifications on every detail of construction, which the company officials could expect their contractors to follow. Then during and after construction, the Principal Engineer would inspect for faithfulness in following his specifications. He was often disappointed. For example, upon inspecting the Red and Blue Sulphur Springs Turnpike two years after he had drawn up specifications, he commented:

> . . . In the first place, the road is seldom 22 feet wide.
> Secondly, the shape of the road along hill-sides, is not as required by the specifications, flat and inclining to the hill, but the surface is rounded and raised in the centre . . .
> Thirdly. The drains have been much neglected, and particularly along graded hills. The contractor told me, indeed, that it had been concluded to grade the road at first, regardless of drains, and to make them afterwards, a plan highly objectionable . . .[19]

However, if turnpike officials and the Principal Engineer had their difficulties with contractors, the contractors had their difficulties from the time bids were requested. The prospective contractor would have to keep his price low enough to be competitive, making allowances for constructing the road according to specifications in the face of *known* difficulties, plus a moderate profit, which could be more than absorbed by contingencies.

The most persistent problem facing the contractor, once employed, was securing and holding a labor force. References to this problem abound in the records. Contractors recruited most of their road workers from the poorer whites of the towns and villages, set up work camps as soon as the road was out of commuting distance, and provided food and a small wage. Unfortunately, the records do not abound with information on the wages that were paid; such information that does exist is too fragmentary to be useful. Apparently 75¢ per day was considered a high wage in western Virginia in the early 1840's.

[17] Gillespie, *loc. cit.*
[18] *ARBPW* Supplement, XXVI, pp. 10-13.
[19] *ARBPW*, XXV, pp. 465-71.

M. Geary, State Commissioner for the Charleston and Weston Road, reported that amount as the wage rate for those who were being paid in stock subscriptions and boarding themselves on the job.[20]

In summer, the best season for road construction, road workers would often be tempted to abandon their jobs for a more adventurous life, at least temporarily. Elisha Hagans complained to his brother, "Our folks here at least some of them are troubled with the hunting and fishing fever—every summer, it comes on generally about the first of July and then it cannot be cured but by spending whole days & nights in the woods. this [sic] has operated against our road." [21]

Contractors, turnpike officials, and Principal Engineer alike shared the common burden of legislative intrusion and incompetence. The General Assembly of Virginia had concerned itself since very early times with roads and transportation. It was a matter of long-standing habit developed during the colonial period for road legislation to specify certain details of construction, particularly the width. Even after the techniques of road construction had become so elaborate as to be beyond the competence of most legislators, the General Assembly at times wrote road legislation that continued to specify details of construction; in so doing it issued some impracticable instructions. In most cases, the road builders ignored the Assembly's instructions and followed their own judgment.

The General Turnpike Law of 1817 provided, on the subject of width, that the company officials " shall make the road in every part thereof, sixty feet wide at least, eighteen feet of which shall be well covered with gravel or stone, where necessary." [22] Thus the law required an 18-foot road in a 60-foot roadway. The Assembly often allowed exceptions to this in granting charters of incorporation, so it was not a uniform rule.

In Crozet's opinion, the width of a road might vary, depending upon where it was located, the potential traffic upon it, and the financial circumstances of the agency building it. In the western portions of Virginia, where " new country " was being opened to settlement and internal improvements for the first time, Crozet thought that a width of ten or twelve feet was adequate at first. It could always be widened later. " The main point," said Crozet, " is to locate them well in the first instance . . . the disposition and means for widening them afterwards, will naturally result from their benefits." It happened too frequently, he added, that " great exertions and sacrifices were made to

[20] *ARBPW*, XXVII, pp. 415-16.

[21] E. M. Hagans to Harrison Hagans, 27 July 1839, Hagans Papers.

[22] *Revised Code of the Laws of Virginia, 1819*, Vol. 2, Ch. 234, p. 216.

open roads that later proved so faulty that they had to be abandoned and the same expense incurred for a new road, "the country in the mean time remaining in its unimproved state for a much greater period."[23]

On one occasion, Crozet made a statement suggesting that he had reduced the relationship between the width and the cost of constructing a road to a formula, which he may have used as a rule of thumb. In a report on the survey between Staunton and Parkersburg, in 1823, he wrote: "The expense of a road being nearly proportional to the square of its width, I think that this road ought not to be made wider than 12 feet: with this width, 8 or 9 miles of it would not cost more than one mile of a road 30 feet wide."[24]

It was possible for the width of the roadway to be out of proportion to the width of the road. In examining the Salem and Pepper's Ferry Turnpike in 1841, Crozet found the road generally 24 feet wide, except for a difficult stretch of about 150 yards where it had been reduced to 18 feet. This was nothing to criticize, but the entire roadway was authorized by law to be only 30 feet wide. Crozet noted that "for a road 24 feet wide, this is quite too narrow; it leaves no room for the ditches, and deprives the road of ground to repair it. The width of 60 feet required by the general law is everywhere proper."[25]

At the beginning of the turnpike movement in Virginia, the construction of "summer roads" was a dying but not quite dead practice. It involved the extending and smoothing of the shoulders of the roadway. At the end of the spring season, the road surface had been subjected to bad weather and heavy traffic, so that the road by May or June would be rutty and unpleasant to travel. With the coming of drier summer weather, the shoulders would be smoother than the road itself and would be travelled in preference to the road by horsemen and by the light, high-wheeled carriages that could clear the occasional stumps. By the late eighteenth century, this practice had become so general that laws were passed requiring the stumps to be cleared out. A "summer road" was thus provided for even the heavier vehicles.

With the advent of the practice of paving the center surface, the continuation of the summer road was thought desirable by some persons, particularly legislators, to save the expensive surfacing "from the wear and tear occasioned by the lighter kind of vehicles during the summer," although how this could be enforced was unclear. D. H. Mahan was opposed to the use of summer roads on the ground that "it has the inconvenience of forming during the winter a large quan-

[23] *ARBPW*, XI, p. 84.
[24] *ARBPW*, VIII, p. 140.   [25] *ARBPW* Supplement, XXVI, pp. 15-16.

tity of mud which is very injurious to the road-covering." [26] Turnpike officials objected to them on the ground that they were an unnecessary expense, the removal of stumps being a tedious and labor-consuming business. Not until the middle 1830's, however, did it become customary for the General Assembly to waive the requirement.

Nineteenth-century engineers, not only in Virginia but elsewhere, were much concerned with the problem of grades. The amateurs who did most of the construction were not, and Crozet was constantly complaining to the Board of Public Works about their ignoring his recommended locations and sacrificing easier grades for straighter lines. The General Assembly in its turnpike company charters usually specified the maximum allowable slope, most often three degrees, but sometimes hindered the Principal Engineer by allowing four, five, or even six degrees.

The General Turnpike Law did not mention grades, probably because the General Assembly's long-standing habit was to prescribe the maximum allowable grade in the case of each road it authorized. Nineteenth-century engineers usually used one of three methods of measuring grades: (1) by the angle the road made with the horizontal; (2) by the proportion between the vertical ascent and the horizontal distance; or (3) by the ascent in feet per mile. It is also possible to express grade in terms of percentage (number of feet of ascent per 100 feet of horizontal travel), which is the same as the second method except for the form of expression. Most professional engineers expressed grade by the second method, e. g., one foot of ascent to 57 feet of horizontal travel, or 1/57, probably because it was more precise than the others. In Virginia, however, the General Assembly and the Board of Public Works expressed grade in terms of the angle the road made with the horizontal and continued to use this method during the turnpike era.

The criterion used by professional engineers in determining the optimum grade was what they called the " angle of repose." This was the angle at which a vehicle would just remain at rest on the slope, or " if placed in motion, would descend by the action of gravity," any increase in the angle *causing* the vehicle to roll downhill. This angle would vary with the condition of the road (and in a practical demonstration, the condition of the vehicle, but the hypothesis assumes a well-lubricated vehicle in good condition). The better the road, the smaller the angle of repose. Engineers considered this the optimum angle for

[26] Dennis H. Mahan, *An Elementary Course of Civil Engineering, for the Use of Cadets of the United States Military Academy* (6th ed., New York, 1869), pp. 299-300.

the maximum allowable grade, excessive steepness making the uphill pull too hard and reducing the speed at which the slope could be travelled with control going downhill. One stretch of excessively steep slope on a road would make it necessary for wagoners to employ on the whole line the number of horses necessary to negotiate it.

Gillespie estimated the angle of repose on the best roads (those with a broken-stone surface) to be 1/35, or slightly more than 1½ degrees. Mahan said between 1/35 and 1/49. However, most American roads were not the best. Gillespie estimated that on the average American road the angle of repose was nearer 1/20 than 1/35. This could allow a steeper grade, but in anticipation of improved road surfaces, slopes not exceeding 1/30 (about two degrees) might be a "just medium." [27]

A maximum of two degrees was a far cry from what was actually attained by the turnpike movement in Virginia. Crozet was as grade-conscious as anyone, so it was not for lack of exhortation. In discussing the Staunton and James River Turnpike in 1828, for example, Crozet wrote to the Board of Public Works:

> . . . Considering what slight reasons frequently induce the increase of the grade of a road, sometimes as much as one degree or over, it would seem that the influence of one degree on the draught of horses, and on the preservation of roads, is not generally appreciated by those who locate them; and yet, even on a good turnpike, the load of a waggon may be made at least 100 lbs more per horse up an ascent of 4 degrees than along one of 5 degrees.[28]

Testimony from turnpike company officials supporting engineers' views on grades is not abundant, for like most people, they preferred not to admit their mistakes if they could avoid doing so. One did go so far as to admit that relatively low grades on his turnpike would have saved repair expenses and suggested that, " If the legislature, in granting charters for turnpike roads, had confined them to 3½ degrees they would be much better, and the expense for keeping them in repair greatly reduced." [29]

In summary, professional engineers advised the use of two or 2½ degrees as a maximum, while Crozet would permit three. Careful study of the reports indicates that 3½ degrees were a rarity in actual practice, four were considered good by turnpike officials, five were considered normal, and grades often exceeded five degrees, sometimes greatly.

[27] Gillespie, *op. cit.*, p. 41.
[28] *ARBPW*, XII, pp. 335-36.
[29] Report of Henry Alexander, President of the White and Salt Sulphur Springs Turnpike Company, to the Board of Public Works, 24 December 1837, *ARBPW*, XXII, pp. 340-41.

Virginia had no monopoly on excessively steep grades. In New Jersey, turnpikes "which led through generally level territory were permitted a deviation of three degrees from the plane of the horizon, while others which encountered a rougher relief were allowed six." [30] E. C. Kirkland informs us that in New England, road engineers were grade-conscious, and "with great elaboration . . . calculated the savings in horsepower and gains in loads secured on roads with proper grades." But the gap between theory and practice was probably as great there as in Virginia. A Connecticut man wrote to Secretary Gallatin that in his state an opinion had prevailed that no ascent greater than five degrees should be allowed. "Nothing, however, is more certain than no such principle has been adhered to." [31]

If road engineers were quite grade-conscious, as evidence indicates they were, the same evidence does not indicate that they were more than slightly curve-conscious. Their attention to problems involving the radius or other attributes of a curve is rarely encountered and even then is characterized by little or no elaboration. Why this was true may only be surmised. Perhaps wagons and carriages did not attain sufficient speed to confine the radius of curves to a certain minimum length.

A simple curve is an arc which, if extended fully, would describe a circle. Therefore, the sharpness of a curve is indicated by the radius of the hypothetical circle of which it forms an arc. Railroad-building necessitated the development of more complex multi-centered curves, which are used also in highway construction today. But nothing more complex than the simple curve was used on ninetenth-century turnpikes.

Even the term radius as applied to curves was used in the nineteenth century almost exclusively in railroad-building, very seldom in the construction of turnpikes. Nowhere does Crozet or any other Virginia engineer mention a standard length for the minimum allowable radius. Even so, there were limits. Thomas H. DeWitt relocated a road across the Alleghany Mountains in Pendleton and Pocahontas counties in 1834 and noted that a departure had been made from a previous location made by Crozet, "more particularly, on the eastern face of Shaw's Ridge, where a turn has been made so sharp, that a wagon and four horses cannot turn without unhitching." [32] Clearly, this was carrying the turnpike's facility for employing curves of short radius too far.

[30] Lane, *From Indian Trail to Iron Horse*, p. 154.

[31] *American State Papers, Miscellaneous*, I, pp. 869-70, quoted in E. C. Kirkland, *Men, Cities, and Transportation*, I, pp. 39-40.

[32] *ARBPW*, XIX, p. 390.

The problems raised by excavations and embankments were not referred to frequently in the records of the turnpike movement in Virginia, not even on such a turnpike as the Northwestern, for which there exists a nearly complete record of construction. Upon at least one occasion Crozet made comments that suggested a general lack of experience in this operation on the part of the road contractors. It appears also that Crozet himself was averse to using any more excavation and embankment than necessary, preferring to deviate from the direct line.

All road engineers agreed that it was vitally important to keep a road as dry as possible. In order to convey rain water off the road and to keep intermittent water courses away from it, a system of thorough drainage was necessary, composed of such devices as gutters, ditches, and culverts. Such drainage systems, said Mahan, should " cut off from the soil beneath the roadway, to a depth of at least three feet below the bottom of the road-covering," the water that filters through the adjacent ground, as well as convey off " that which falls upon the surface . . . before it can filter through the road-covering." [33]

Gutters were hollowed-out depressions in the surface of the road itself, for carrying off the rain as it fell on the surface. These were not necessary on level ground, where the road surface was rounded for that purpose, but they were necessary on slopes, where the road was properly built with a flat surface. The lowly " thank-you-M'am " (so-called because of the jolt it gave the vehicle) could become the more elaborate cross-mitre drain, a covered drain built of wood or masonry, flush with the road surface, in a herringbone pattern upslope, the V's spaced about 60 yards apart and pointing in the direction of ascent.

The common ditch, to drain the soil beneath the roadway and on either side of it, as well as to catch the run-off from the gutters, was dug to a depth of three feet below the road surface, parallel with the road, and was properly termed an open side drain. The ditch or drain conveyed the water downhill to the bottom of the depression, at which point one of two things could happen to it. If the road at that point were higher than the adjacent land on both sides, the water would simply run off from the two parallel ditches in opposite directions into the nearest natural watercourses. If the road were not higher than the adjacent land on one side, however, the water in the ditch on the high side would have to be conducted under the road to the low side by means of a culvert, a structure made either of wood, brick, or stone, usually arched at the top for added strength.

[33] Mahan, *op. cit.*, pp. 286-87.

The construction of drainage systems on Virginia roads and turn-pikes was of poor quality as a rule. A few turnpikes built elaborate drainage systems, such as the Manchester and Petersburg, which spent $12,000 on stone and brick culverts out of a total investment of $76,000. But this was highly exceptional. Some of the common defects may be observed in sample comments of Crozet. Of the White and Salt Sulphur Springs Turnpike in 1840, he remarked, " Its ditches are mostly too shallow; they choke up, and the water spreading over the surface, tears it into deep ruts and holes." [34] Of the Lynchburg and Pittsylvania Turnpike in 1842, he wrote, " It is not well drained, a number of culverts in flat places and gutters on breaks on slopes are wanted." [35] Of the Giles, Fayette, and Kanawha Turnpike he said, " What had been done was generally pretty good except the drains, which are as usual imperfect and too few in number, the gutters being too narrow and occasionally even not at the lowest places." [36] Of the Wellsburg and Washington Turnpike, " For want of good original drains, it is now intercepted by breaks which convey the water to the outside; but form very objectionable obstacles to the rapid motion of carriages, and are an impediment in the way of heavy wagons." [37] And of the Natural Bridge Turnpike in the same year, " Its ditches have all disappeared and its drains are few and bad." [38]

One of Crozet's unfailing sources of irritation was the road builder's rounding the surface on hillsides as well as on level ground. He seemed never to tire of explaining to the Board of Public Works his reasons for considering this a poor practice. On the Staunton and James River Turnpike, he observed, for example,

> Wherever a road slopes forward, a rounding is perfectly useless: for the water will run along the ruts whatever the elevation may be in the middle; and therefore, it is only on occasional paved gutters that one should rely to turn off the water from the road into the ditches. A rounded road rather causes ruts to be cut deeper, by the weight of the waggons bearing mostly on the wheels towards the edge of the road.[39]

Paving roads with crushed or broken stone would, of course, com-pensate for defective or inefficient drainage by protecting the road surface from such rut-cutting. However, the antebellum turnpike movement in Virginia (or any other state) should not be thought of as the means by which the macadamized road became ubiquitous.

[34] *ARBPW*, XXV, p. 503.
[35] *ARBPW*, XXVII, p. 602.
[36] *Ibid.*, pp. 586-89.
[37] *Ibid.*, p. 591.
[38] *Ibid.*, pp. 594-95.
[39] *ARBPW*, XI, pp. 87-88.

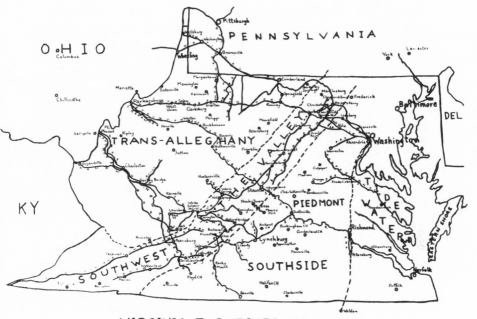

VIRGINIA TURNPIKES IN 1848

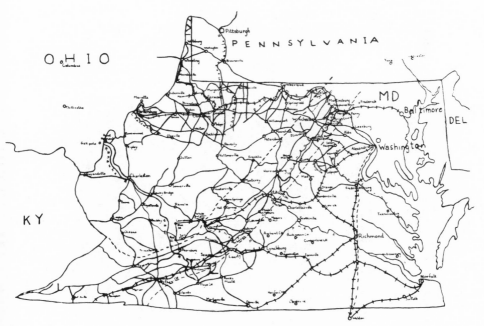

VIRGINIA TURNPIKES IN 1860

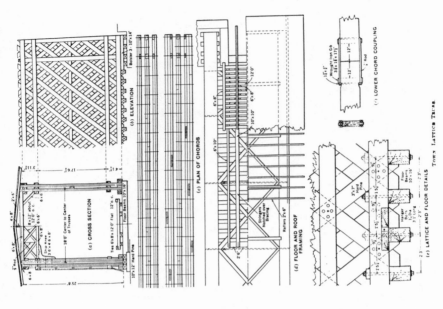

TOWN LATTICE TRUSS

Figure 2. Details of Town Lattice-Truss bridge construction. (Reproduced by permission of the American Society of Civil Engineers, from Robert Fletcher and J. P. Snow, " A History of the Development of Wooden Bridges," *Trans-actions of the American Society of Civil Engineers*, Vol. 99

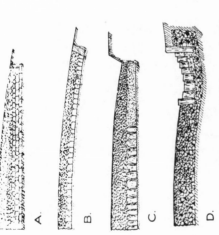

FIGURE 1. *Cross-sections of various roads.* (A) *French, previous to 1775.* (B) *Trésaguet (1764). The foundation layer, which was parallel to the road surface, was formed of large stones placed on edge and hammered in. The next layer was of smaller stones, also hammered in, and the surface, which was curved, was of small hard stones.* (C) *Telford (1824). A layer of large stones was placed edgeways on a level bed. Smaller stones were then added, to a depth of 7 in at the centre and sloping off to 3 in deep 15 ft away. The centre 4 ft were covered with small stones, well trodden down by horses. Finally the whole road was covered with 1 to 1½ in of good clean gravel.* (D) *McAdam (c 1820). No large stones were used, the only foundation being the well drained natural soil. A layer of hand-broken stones, about 1½ inches in diameter and none weighing more than 6 oz, was spread over and worked in. No binding material was used.*

Figure 1. Construction of road surfaces. (Reproduced by permission of the Editors, from Charles Singer et al., eds., *A History of Technology*, Vol. IV

The stone-surfaced turnpike (or portion of a turnpike) remained an oasis in a desert of mud roads. Two factors were generally recognized as determining whether or not an earth surface would be adequate: the character of the soil and, more important, the amount of traffic. Once the decision in favor of hard surfacing was made, questions concerning types and methods of surfacing arose.

Not even the experts agreed exactly how the ideal road surface was constructed. Among road engineers of the early nineteenth century, broken-stone roads (as distinguished from mere gravel roads) were the subject of sometimes heated partisanship, one school of thought following the lead of John McAdam, the other devoted to principles expounded by Thomas Telford.[40] Differences between engineers did not end with this simple divergence, but many French, English, and American engineers had their own notions that differed from both McAdam and Telford. Some believed there were superior merits in the older Trésaguet road. However, the basic alignment with McAdam or Telford remained the essential pattern, despite minor variations. The most important difference between them concerned the necessity of a paved foundation beneath the coating of broken stones (see Figure 1). Telford insisted it was necessary, while McAdam insisted it was not. As for the two leading American engineers, Gillespie gives the impression of having favored McAdam, while Mahan seems to have favored Telford.

Engineers in Virginia, Crozet in particular, gave less thought to problems of surfacing than to other aspects of road construction. In all of Crozet's voluminous correspondence with the Board of Public Works he did not once expound the relative merits of the McAdam, Telford, Trésaguet, or any other road-surfacing method. In fact, certain of his remarks suggest that he regarded road surfacing as incidental. More often than not, his only criticism would be that the stones were not broken fine enough. Probably neither McAdam nor Telford would have found fault with such criticism, but neither of them would have

---

[40] John Loudoun McAdam (1756-1836) was not a professional engineer, but a layman who developed a keen interest in roads upon being appointed a road commissioner at Bristol. In 1819, he published *A Practical Essay on the Scientific Repair and Preservation of Roads* and, in 1820, *Remarks on the Present System of Road Making*, which had gone through a ninth edition by 1827. In that year, the House of Commons appointed him general surveyor of roads and paid him an indemnity for his personal expenses. He was also offered knighthood, which he declined. Thomas Telford (1757-1834), a professional engineer, was interested primarily in bridge construction but made important contributions to road construction. He wrote his own autobiography (1838), but for Telford as a road maker, see also Sir Henry Parnell, *Treatise on Roads* (2d edition, London, 1838).

approved of such a suggestion as he made when he wrote, "For a considerable distance, it has been paved with large broken stones which make it very rough: they ought to be covered with smaller materials or earth." [41] Crozet was far more attentive to faults of location, width, grades, and drainage. A Virginia engineer named James Herron, however, made an extended comment on McAdams *versus* Telford roads which was perhaps characteristic of engineering opinion in Virginia in the middle thirties. Herron had been ordered to survey a route for a turnpike between James River at Lynchburg and the Tennessee line. In discussing the possibilities for surfacing, he wrote to the Board of Public Works:

> . . . Mr. McAdam's plan of road making, called for by the law, has been very generally exploded in England; or at least the indiscriminate application of a thin capping of finely broken stones to every variety of soil and circumstance. On a dry firm soil it is quite sufficient, but on clays and wet soil, it has generally failed. The famous new roads, recently constructed by the British government, on the plans of the experienced Telford, are formed with a rough pavement foundation, which is said to cost less in the first instance, to last one-third longer, and from experiments made with a dynamometer, to give much less resistance to carriages passing over them than roads constructed entirely of broken stone.[42]

The first Virginia turnpike to use the McAdam plan for surfacing was the Fauquier and Alexandria, in 1824. The charter required the company to pave or gravel the road to a width of 20 feet, but since McAdam's plan was expensive, the company petitioned the General Assembly to allow a width of 16½ feet. As often happened, for cause, the General Assembly was presented with a *fait accompli* when President J. H. Hooe informed the Assembly the change in width had been made and the work completed before the Assembly had made its decision. Crozet, however, was not exactly pleased with it, and wrote to the Board:

> . . . I was informed by the President of the company, that this section had been contracted for at $28,000; in consideration whereof, it was to be overspread with a bed of broken stones, 12 inches thick, and 18 feet wide . . . the stones to be broken to 6 ounces weight. The macadamized portion of the road lay between Warrenton (Fauquier C. H.) and Buckland, a distance of 8½ miles.

---

[41] Report on the Lewisburg and Blue Sulphur Springs Turnpike, 12 March 1838, *ARBPW*, XXII, p. 95.
[42] *ARBPW*, XIX, pp. 375-76.

They exceed, however, much those dimensions: Their present size will certainly prevent their crushing sufficiently to become soon cemented: So that for a long time, they will only form a bed of rolling stones extremely fatiguing for draught horses: They should be broken smaller, or else the largest should be raked out of the road.[43]

At the end of 1827, the company treasurer reported to the Board that " the operations of this road during the last year, have been principally confined . . . to the taking up, re-laying and repairing upon M'Adam's plan, the old part of the road [Buckland to Fairfax C. H.] . . . the object being to make it correspond with, and equal to, the new road from Warrenton to Buckland, which is acknowledged to be the best road in Virginia." Five miles of the old road, he said, had been macadamized " at an expense not exceeding $2000 per mile." [44] A year later, Crozet had the opportunity to criticize the company's self-evaluation.

This road is in good order from Warrenton to about four miles beyond Buckland, a distance of about 12 miles. Some of the stones, however, are too large; and, in places, earth has been mixed with them, to the disadvantage of the carriage-way: this plan, which, in the first instance, obviates the roughness of newly broken stones, proves in the end an obstacle to their getting firmly cemented and smooth.[45]

The Fauquier and Alexandria Turnpike officials may have intended to follow faithfully the McAdam plan of surfacing, but their performance fell short.

Another turnpike company that took an early interest in macadamizing was the Lynchburg and Salem. William Radford, president, reported in 1826 that most of the paving had been done upon a plan admittedly " injudicious." The company had required in all contracts that " a bed of large rock or stone should be laid at the bottom, upon which should be a bed of pounded rock or gravel, of about five inches." The large stones continually worked up from the bottom, preventing the pavement in the middle from acquiring " the form of a solid mass." The company then made a change on a part of the fifth section, by which the McAdam plan was " in some measure adopted, requiring that the whole pavement should be of rock or stone pounded fine, and seven inches deep, instead of five inches of large rock at bottom and five inches of pounded stone or gravel at top." [46]

[43] *ARBPW*, XI, pp. 95-97.

[44] *ARBPW*, XII, p. 196.

[45] *ARBPW*, XIII, pp. 513-14.

[46] *ARBPW*, XI, pp. 33-34.

A year later (21 December 1827), Radford reported that the Lynchburg and Salem Turnpike had been "generally made in a permanent manner, with a pavement of quartz or flint rock, from seven to ten inches deep." Expenses had impoverished the company, however, and Radford appealed to the Board not to require immediately the paving of the road west of Liberty (Bedford). Crozet, in his report on the Lynchburg and Salem a few months later (22 February 1828) made no comment upon the fact that quartz and flint had been used (McAdam recommended these be avoided), but merely remarked that the stones used in capping had not been broken fine enough.

The experiences of the Fauquier and Alexandria Turnpike and the Lynchburg and Salem Turnpike were characteristic of those few turnpike companies that did any stone surfacing at all. They usually exhibited a singular lack of comprehension of the true principles of McAdam's method, which, as Herron had mentioned, required in the first place the careful preparation of a well-drained roadbed, and in the second place, a careful laying and rolling of four successive layers of the right type of stone, containing the right amount of moisture. This "water-bound" macadamized surface, properly made, was literally a four-layer sheet of stone, impervious to water and resistant to great pressure from vehicles. A very few Virginia turnpikes, such as the Little River Turnpike and the Valley Turnpike, had top-quality macadamized surfaces.

A popular cheap substitute for the broken-stone surface during a period of several decades after about 1830 was the plank road, which was more widespread in the Northern than in the Southern states. The plank road movement in Virginia was of very limited extent. The exact number of miles actually planked is difficult to determine, for while there were some companies chartered specifically as "plank road companies," some of the turnpike companies experimented with planking, as did some of the public road commissioners.

Only ten plank road companies chartered by the General Assembly became operating companies. The earliest petition for a charter came from Amelia County in 1833 (in Southside Virginia, between Lynchburg and Richmond), where some citizens wanted state aid in building one to Petersburg (about 25 miles). This road did not materialize, but it at least indicated the section of Virginia in which the plank road movement was to take place. Of the ten companies, every one was in the Piedmont section, and most of them south of the James, an area in which there were almost no turnpikes. Three of the ten were built as railroad-feeders; the others were mainly farm-to-market roads. Also notable are the brevity of the movement (slightly over three years

from the first charter to the last between 1850 and 1853) and the fact
that five of the ten companies were chartered during the final month.
On a miniature scale, then, the plank road movement in Virginia was
analogous to the turnpike movement in that it started slowly, gained
little momentum, then burned out swiftly in a flare-up of activity.

The typical bridge found on most Virginia roads and turnpikes was
a wooden structure set upon abutments and piers made of stone.
Stonework was expensive, but most turnpike officials felt it was worth
the expense; bridges that fell down were more expensive in the long
run. Superstructures were usually of truss-and-arch construction, such
as the Burr truss and, later, the Town lattice truss [47] (Figure 2).
Superstructures were covered with sides and a roof if the expense
could be met, which protected the flooring and framing from the
weather and lengthened the useful life of the bridge. The best bridges,
however, dispensed with the wooden superstructure and were built
entirely of stone arches. There were not many of these on Virginia
turnpikes, and the company that owned one was duly proud of it.

Even some wooden bridges were impressive structures, as for example one across the Staunton River owned by the Lynchburg and
Pittsylvania Turnpike Company. Over 500 feet long, it was supported
by four piers, and cost $8,000. Unfortunately, it had one span of 126
feet which was "somewhat depressed in the middle." Furthermore,
the masonry was laid without mortar, and the carriageway was a mere
14 feet wide. Last but not least of its defects, it was built on the so-
called "lattice plan," introduced from New England and of short-lived
popularity in Virginia in the 1840's. The record does not describe the
misfortune that destroyed this bridge, but the Valley Turnpike's experience with the lattice plan leaves little doubt that the Lynchburg
and Pittsylvania Company suffered a serious headache with it.

In December, 1840, when the Valley Turnpike was being completed,
Crozet was uncritical of the new lattice plan for bridges. One bridge,
over Cedar Creek, was a single span of 130 feet, praised by Crozet as
a sample of "excellent workmanship." In 1841 and 1842, three more
bridges were built by the Valley Turnpike on the same plan. The
Shenandoah River bridge was completed in December, 1841, but before
the contractor had been paid in full, it was dicovered that the bridge
would fall unless "timely supported from below." Despite the construction of three supporting trestles, the bridge still sank. In May,

[47] For a brief discussion of bridge construction, see Richard S. Kirby, et al.,
Engineering in History (New York, 1956), pp. 220-44. For a more extended
technical treatment, see Robert Fletcher and J. P. Snow, "A History of the
Development of Wooden Bridges," *Transactions of the American Society of
Civil Engineers*, XCIX, pp. 314-408.

1842, high water carried away first the trestles, then the bridge, depositing it upon an island about a mile downstream. By December, Crozet had examined the damage to this and other bridges on the Valley Turnpike, and reported his analysis of the difficulty. It was the lattice plan,

> . . . the many defects of which have been well demonstrated by numerous failures, that it is a matter of wonder that this plan is yet so often chosen. The chief defect . . . is the impossibility of restoring its level and direction, whenever it thas once cringed, without a labour equal to the original building . . . Probably this plan has been favored on account of its being within the comprehension and capacity of any common carpenter, which I would myself rather consider an objection, as it not unfrequently [sic] becomes a guarantee of inferior workmanship . . . While theoretically, the lateral trusses may be possessed of sufficient vertical strength, they are ill calculated to resist lateral strains . . . the joist bearers or cross ties, whose dimensions are only 12 by 9 or 10 inches, are too weak for a length of bearing of 19 feet, to sustain, without great deflection, the whole weight of one of the heavy wagons used on this road, some of which, with their load, cannot weigh much less than 5½ tons.[48]

Building bridges according to defective plans was expensive, but probably no more expensive than building bridges that were not needed. For example, Crozet, upon examining the Buffalo Springs Turnpike in 1842, noted the details of two bridges the company had built, then concluded, "The cost of both might have been avoided by making the road all along on the same side of the river." [49]

Unduly expensive also was the hiring of bridge builders of doubtful integrity. The Northwestern Turnpike Board (not a "company" because it was a purely state project) thought they had scored a high mark when they employed Lewis Wernwag to build a bridge over the Potomac in the 1830's. He was the builder of the famous "Colossus" over the Schuylkill River at Fairmount, Pennsylvania, with a span of 340 feet, in 1812. Apparently Wernwag had gone downhill and was coasting on his earlier reputation, for he defrauded the Northwestern Turnpike by filling his stone-veneered piers with rubble and building his superstructure with green timber, all of which was revealed when the first high water toppled it into the river.[50]

[48] *ARBPW*, XXVII, pp. 596-97.
[49] *Ibid.*, p. 603.
[50] Northwestern Turnpike Road, Minute Book "A" (29 March 1831-30 April 1835), Virginia State Library.

Even the choice of sites for bridges could trap the unwary and inexperienced into costly mistakes. "It is a very common mistake," wrote Crozet, "to choose narrow sites for bridges." Other things being equal, the narrowest site might be the most desirable, but narrow places in rivers were usually deep places as well. If the river could be crossed at such a place with a single span, the site would be favorable, but if it were necessary to build a pier, two or more piers across a wide, shallow place would be less troublesome than one pier in deep water. This was especially true in flood times, when the depth of water at the narrow places was much greater.

In summary, can we, from the foregoing discussion, attempt an answer to the question posed at the outset? How good was the quality of Virginia's antebellum turnpikes? Did the mixed enterprise system and the toll road succeed or fail in producing a superior road, with a consequent reduction in the cost of transportation? Only tentative and partial answers to such questions are possible, but even these may be of some value.

By comparison with other states, Virginia's turnpikes were probably average in quality. There were a few outstandingly good ones, such as the Southwestern, the Little River, the Manchester and Petersburg, and the famous Valley Turnpike. But in the main, the tolls that were paid on the 180-odd Virginia turnpikes were for the use of roads of middling quality, and sometimes of very poor quality.

Many mistakes were made in the construction of Virginia turnpikes, probably in the following order of seriousness. First, errors of location were the worst, for a bad location could be corrected only by rebuilding in a new one. A good example was the Lexington and Covington Turnpike, one of several ruined by this error. Crozet recommended that the line should go well to the north of the direct line to take advantage of a lower crossing of the mountain on an easier grade. The directors of the company ignored his advice and built a shorter road, but one that twisted its way upward over a steep escarpment near Collierstown, which served only as a bottleneck to long-distance, heavy freight. The result was that this decision damaged not only the Lexington and Covington Turnpike, but also the Kanawha Turnpike, going westward from Covington.

Errors in drainage systems were probably second in seriousness, for water is a road's worst enemy. Some turnpikes, such as the Manchester and Petersburg Turnpike, had excellent drainage systems, but more often, little or no attention was given this problem by amateur road builders. Grades were third in importance, for the steepness of a single small section of a road could determine the amount of freight carried by a wagon between its terminals. With the relative absence of excavation and embankment, however, it is difficult to distinguish between an excessive grade and an error in location.

Surfacing has been overestimated in importance, both by contemporaries and by historians, although not by Crozet. The McAdam surface was particularly expensive in the United States, where the cost of manual labor was higher than in England. The Telford surface was even more expensive, since it called for even more hand labor in setting the base, stone by stone. Most of the Virginia turnpikes that were stone-surfaced were rough by today's standards, but some of them were remarkably durable. The Valley Turnpike, for example, accommodated horse-drawn, steel-tired wagons (tires were usually four inches in width), sometimes weighing between five and six tons. Even modern concrete or asphalt roads would do well to stand up under such stress. However, this was probably due more to the high quality and superabundance of Valley limestone than to skill in construction.

Did Virginia's taxpayers, tollpayers, and turnpike stock investors receive their money's worth for their outlay? An analytical answer to this question is not feasible from the evidence considered in this essay, nor probably from any historical sources available. Yet a strong impression emerges that with but a slight improvement in the State government's management of these affairs, Virginia could have enjoyed a much better road network for the same expenditure. Virginia was fortunate beyond many, if not all, her sister states in having Claudius Crozet as Principal Engineer. The General Assembly employed this highly-paid ($3,000 per year) expert, let him spend his entire time advising turnpike, canal, and railroad officials how to construct their works, encouraged him to inspect for faithfulness in following his instructions, and failed utterly to pass any law requiring turnpike officials to pay the slightest heed to his advice. Crozet urged the legislature repeatedly to pass such a law, but with never a response. The shortcomings of Virginia's turnpikes could not then be blamed on poor administration at the top. Rather, the General Assembly must bear the burden of responsibility for failing to take fuller advantage of its good fortune.

RAISING AND WATERING A CITY:
ELLIS SYLVESTER CHESBROUGH AND
CHICAGO'S FIRST SANITATION SYSTEM

LOUIS P. CAIN

The engineers responsible for invention and mechanization in agriculture, manufacturing, and transportation are prominent historical figures, but few people are aware of the men who pioneered the sanitation systems so crucial to urbanization. As cities grew, their initial approaches to waste disposal and water supply proved unacceptable. As early as 1798 Benjamin Latrobe noted in his journal that the fresh groundwater which located the site of Philadelphia was befouled by the city's increasing population concentration. In Latrobe's opinion, Philadelphia's existing water-supply strategy was a major source of disease. Even before he assumed the responsibility for the city's new waterworks, Latrobe was convinced of the project's utility: " The great scheme of bringing the water of the Schuylkill to Philadelphia to supply the city is now become an object of immense importance, . . . though it is at present neglected from a failure of funds. The evil, however, which it is intended collaterally to correct is so serious and of such magnitude as to call loudly upon all who are inhabitants of Philadelphia for their utmost exertions to complete it."[1]

The emerging concentrations of population and manufacturing in the 19th century necessitated a reexamination of sanitation strategies. With urbanization, the haphazard approaches of the past could not guarantee pure water supplies and adequate waste disposal. Urban growth inevitably required the implementation of sanitation systems, and these systems, in turn, permitted further growth.

Students of Chicago's formative decades inevitably encounter the name of Ellis Sylvester Chesbrough; by studying Chesbrough, a student can focus on the truly unique character and contribution of Chicago's sanitation system. Chesbrough's works were the innovations most responsible for Chicago's unrestricted urban growth; they freed the city from the limitations imposed by an unfavorable

LOUIS P. CAIN is professor of economics at Loyola University of Chicago and adjunct professor of economics at Northwestern University.

[1]Benjamin Henry Latrobe, *The Journal of Latrobe* (New York, 1905), p. 98.

This essay originally appeared in *Technology and Culture*, vol. 13, no. 3, July 1972.

69

natural topography. A flat, nonporous terrain, slightly elevated from Lake Michigan and the Chicago River, made drainage and absorption nearly impossible. In rainy weather, the topsoil became swamplike. Urban growth required a drainage system which could remove both surface water and household wastes. The natural depository for such a drainage system was Lake Michigan; however, the lake was simultaneously the city's natural water-supply source. Lake water had to be conserved if it was to be potable, and this meant it had to be protected from urban wastes. Fortunately, beginning in the 1850s, Chicago's city fathers recognized pollution as a serious threat to the city's health and took immediate action. This paper investigates how Chesbrough responded to Chicago's anomalous water-supply and waste-disposal needs in the 1850s and 1860s, and inquires into his engineering education to discover the antecedents of his innovative ideas.

I

Ellis Sylvester Chesbrough was born of Puritan ancestry in Baltimore County, Maryland, in July 1813. An unsuccessful business venture exhausted the family's means and suspended young Sylvester's education, and so, at nine years of age, he went to work. Between his ninth and fifteenth birthdays Chesbrough spent only a year in a classroom, but he did find time outside his countinghouse duties to pursue his studies. Chesbrough acquired most of his basic education without the benefit of formal training or a regular teacher, and the same was true of his engineering education.

In 1828 Chesbrough's father took a job with a railroad engineering company employed by the Baltimore and Ohio Railroad Company. Through the father's influence, the son gained employment as a chainman with a similar company engaged in preliminary surveying work in and about Baltimore.[2] Chesbrough's company was under Lt. Joshua Barney, U.S. Army, and most of the engineers were army officers, many of them graduates of the U.S. Military Academy's practical, as opposed to theoretical, engineering course.[3] Chesbrough was fortunate in being affiliated with several of the

[2]The engineering education of E. S. Chesbrough began in this company, and he quickly proved an apt student. See *Biographical Sketches of the Leading Men of Chicago*, written by the Best Talent of the Northwest (Chicago, 1868), p. 192; see also *Journal of the American Society of Civil Engineers* 15 (November 1889): 161.

[3]See Daniel H. Calhoun, *American Civil Engineer: Origins and Conflict* (Cambridge, Mass., 1960), p. 38; and Forest Hill, *Roads, Rails, and Waterways* (Norman, Okla., 1957), pp. 12 ff.

army's most prominent engineers. In 1830–31 he worked as an assistant engineer to Col. Stephen H. Long.[4] Near the end of 1831, Chesbrough joined the engineering corps of Capt. William Gibbs McNeill, where he served immediately under Lt. George W. Whistler.[5]

The Panic of 1837 and the resulting depression dealt a hard blow to the country's internal improvement's bubble, and Chesbrough, like many other engineers, found himself out of work as the flow of funds dried up in the early 1840s. He went to his father's residence in Providence, Rhode Island, where, during the winter of 1842, he spent his leisure time in the workshop of a nearby railroad learning the practical use of tools. The following year he purchased a farm adjacent to one owned by his father in Niagara County, New York. His venture into farming was mercifully brief; after an unsuccessful year, Chesbrough gladly returned to engineering.

In 1846 Chesbrough was offered the position of chief engineer on the Boston Water Works' West Division. This position completed his engineering education. Up to this time, all his experience was related to railroad engineering, and he had mastered many civil engineering essentials, such as grading, tunneling, and surveying. Chesbrough was reluctant to accept the Boston position because he considered himself unacquainted with hydraulic engineering. His friends and Boston's water commissioners implored him to accept

[4]Of particular interest to Chesbrough's later career is the fact that Long had carried out extensive exploratory surveys in the West. In 1816 Long was asked to report to the federal government on the physiographic features in the region of a proposed canal between Lake Michigan and the Illinois River. Although it is only speculation, one wonders how much knowledge of Chicago's topographical peculiarities Long passed on to Chesbrough. It is known that Long prepared detailed reports of his visit to Chicago. See Richard George Wood, *Stephen Harriman Long* (Glendale, Calif., 1966).

[5]The major supply of engineers developed from what Calhoun called "the persisting pattern of on-the-job training." The supply provided by the leading scholastic source, the U.S. Military Academy, and the leading civilian source, the New York State canal system, was insufficient. The engineers of that day were active builders; thus, some form of on-the-job training had to be inaugurated to increase the supply and meet the demand. What developed was a hierarchical engineering corps. Lacking any formal education, Chesbrough learned every phase of his job by working his way up the civil engineering hierarchy.

In addition to the books by Hill and Calhoun (see n. 3 above), other recent books which discuss the oral transmission of engineering knowledge are: Stephen Salsbury, *The State, the Investor, and the Railroad: The Boston and Albany, 1825–1867* (Cambridge, Mass., 1967); Harry N. Scheiber, *Ohio Canal Era* (Athens, Ohio, 1969); and Ronald E. Shaw, *Erie Water West* (Lexington, Ky., 1966).

the position, and, after being assured John Jervis's counsel, Chesbrough assented.

There was good reason for Chesbrough to consider an association with Jervis valuable. Jervis had been active in every phase of engineering, particularly those dealing with hydraulics. Jervis was a product of the New York canal system and had learned hydraulic engineering on the job by working on the Erie Canal. In 1846 Jervis was appointed consulting engineer on the Boston Water Works, with Chesbrough the chief engineer. Jervis had the responsibility for designing both the Cochituate Aqueduct and the Brookline Reservoir; Chesbrough, the responsibility for supervising the execution of Jervis's plans.[6] In 1850 Chesbrough became sole commissioner of Boston's waterworks, and a year later, he became Boston's first city engineer.

The United States' early experience with internal improvements and the education of engineers coalesced in Chesbrough's career. He learned civil engineering from some of the army's most competent engineers. He learned hydraulic engineering from Jervis, perhaps the most competent engineer trained by the New York canal system. The education and experience which Chesbrough utilized in freeing Chicago from its topographical liabilities and in implementing an effective sanitation system grew out of his first-hand experience with many of the country's internal improvements.

II

In the early 1850s Chicago's random waste disposal methods led to a succession of cholera and dysentery epidemics. The Illinois legislature created the Chicago Board of Sewerage Commissioners on February 14, 1855, to combat what was generally conceded to be an intolerable situation.[7] The commissioners sought "the most competent engineer of the time who was available for the position of

6Chesbrough's role in the Cochituate works is mentioned in a study of the waterworks of Boston, New York, Philadelphia, and Baltimore by Nelson M. Blake (*Water for the Cities* [Syracuse, 1956]).

7 The board was empowered to (1) supervise the drainage and sewage disposal of Chicago's three natural divisions; (2) plan a coordinated system for the future; and (3) issue bonds, purchase lots, and erect buildings implementing their plan. The board's actions were made subject to the Chicago City Council's approval. The act is summarized in several works including G. P. Brown, *Drainage Channel and Waterway* (Chicago, 1894), p. 50.

chief engineer."[8] Their selection, E. Sylvester Chesbrough, resigned his position as Boston's city engineer and came to Chicago.[9] Immediately after accepting the position, Chesbrough submitted a report in which he outlined his plan for a sewerage system designed to solve Chicago's drainage and waste-disposal problem. His plan represents the first comprehensive sewerage system undertaken by any major city in the United States. He had learned about sewer construction, grading, and "building-raising" from different sources. Now he merged them and "pulled Chicago out of the mud."

Prior to Chesbrough's arrival, Chicago's sewerage commissioners solicited the public for plans and suggestions. Thirty-nine proposals were received, and, although the board claimed Chesbrough utilized many of these suggestions, he did not use any of the proposals in its entirety.[10] Chesbrough's task was to construct a sewerage system whose main objective was to "improve and preserve" the city's health. In his opinion, the existing privy vaults and drainage sluices were "abominations that should be swept away as speedily as possible," and that "to construct the vaults as they should be, and maintain them even in a comparatively inoffensive condition, would be more expensive than to construct an entire system of sewerage for no other purpose, if the past experience of London and other large cities was any guide for the future of Chicago."[11]

Chesbrough's 1855 report to the Board of Sewerage Commissioners made several references to the sewers of New York, Boston, and Philadelphia. Additionally, the report showed that Chesbrough was

[8] A.T. Andreas, *History of Chicago from the Earliest Period to the Present Time* (Chicago, 1884), 1:191; Bessie Louise Pierce, *A History of Chicago* (New York, 1940), 2:330; Soper, Watson, and Martin, *A Report to the Chicago Real Estate Board on the Disposal of the Sewage and Protection of the Water Supply of Chicago, Illinois* (Chicago, 1915), p. 69, hereafter referred to as the *CREB Report.*

[9] It is quite possible that Jervis played a significant role in Chicago's choice, for during the early 1850s Jervis was professionally engaged in the Chicago area. Chicago's city fathers would have been aware of Jervis's engineering reputation, and it is probable that he was consulted regarding chief engineer candidates. Because he had worked with Chesbrough just prior to this, it is likely that Jervis gave Chesbrough an excellent recommendation.

In 1881 Chesbrough, serving as consulting engineer of the New Croton Aqueduct, employed Jervis, who discussed the work with Chesbrough daily. This indicates the esteem in which Chesbrough held Jervis, for Jervis was then eighty-six years old. Chesbrough, at sixty-eight years of age, belonged to another generation.

[10] Although the commissioners' report mentions the public's proposals, it does not indicate what they were, or even which parts of Chesbrough's plan were adapted from these suggestions.

[11] Brown, p. 53.

familiar, through his reading, with the sewers of London, Paris, and other European cities. It is important to remember, however, that not one U.S. city at that time had a comprehensive sewerage system, even though most had sewers. Consequently, Chesbrough had to rely on his training and intuition in assessing sewerage system alternatives.

Chesbrough's 1855 report considered four possibilities: (1) drainage directly into the Chicago River and then into Lake Michigan; (2) drainage directly into Lake Michigan; (3) drainage into artifical reservoirs to be pumped and used as fertilizer (sewage farming); and (4) drainage directly into the Chicago River, and then by a proposed steamboat canal into the Des Plaines River. Although this fourth possibility was the method which Chicago eventually adopted (the Chicago Sanitary District's Sanitary and Ship Canal), the city's 80,000 inhabitants in 1855 did not warrant the expense which this alternative involved.

Chesbrough recommended the first plan.[12] This is not to say he failed to realize that his preferred method was a potential health hazard, particularly during the warmer months, and might obstruct river navigation by making the waterways shallower.[13] Chesbrough discussed the objections to his recommended alternative:

> It is proposed to remove the first [health hazard] by pouring into the river from the lake a sufficient body of pure water into the North and South Branches to prevent offensive or injurious exhalations . . . The latter objection [obstruct navigation] is believed to be groundless, because the substances to be conveyed through the sewers to the river could in no case be heavier than the soil of this vicinity, but would generally be much lighter. While these substances might, to some extent, be deposited there when there is little or no current, they would, during the

---

[12] *Report and Plan of Sewerage for the City of Chicago, Illinois,* adopted by the Board of Sewerage Commissioners, December 31, 1855, hereafter referred to as the *1855 Report.* Also quoted in *CREB Report,* p. 71. Chesbrough had a systematic approach to costs, but a very general approach to benefits. This evidently was consistent with the approach adopted on other U.S. internal improvement projects. See Lawrence G. Hines, "The Early Nineteenth Century Internal Improvement Reports and the Philosophy of Public Investment," *Journal of Economic Issues* 2 (December 1968): 384-92.

[13] Chesbrough planned to pump sufficient lake water into the north and south branches of the Chicago River to flush offensive solid pollutants. He also proposed flushing the sewers as well. See reprinted article, Langdon Pearse, "Chicago's Quest for Potable Water," *Water and Sewage Works* (May 1955): 3.

seasons of rain and flood, be swept on by the same force that has hitherto preserved the depth of the river.[14]

Apparently, Chesbrough did not realize that spring freshets and floods might force the sewers' accumulations into the lake in such a way as to pollute the city's water supply. This is somewhat surprising, as the basic sanitation principle of the day was to locate the eventual sewage outlet as far from the water-supply source as possible.

Chesbrough had three objections to the second possibility, drainage directly into Lake Michigan. First, it would require a greater sewer length and, consequently, would incur greater cost. Second, he supposed that this plan would seriously effect the water supply, if any sewer outlets were located near the pumping station. At this time, Chicago's water-supply intake was located a short distance offshore at the Chicago Avenue lakefront, approximately ½ mile north of the Chicago River's mouth. Chesbrough did not elaborate on this objection. Third, he felt drainage into the lake would create difficulties in preventing sewer outlet injury during stormy weather, or snow and ice obstruction during winter.[15]

Sewage farming was rejected in part because of the uncertainty whether future fertilizer demand would be sufficient to cover distribution costs. Further, Chesbrough was uncertain as to both the needed reservoir capacity and the expense of building the necessary reservoirs. Finally, Chesbrough thought there would be a great health hazard created by foul odors emanating from sewage spread over a wide surface.

Chesbrough termed the use of a steamboat canal not yet constructed to flush the sewage into the Des Plaines River, the fourth possibility, "too remote." Although he was aware of the "evils" which would result when raw sewage passed into Lake Michigan, Chesbrough felt it impossible to create an outlet to the southwest. Brown claims, however, that "he appears to have believed that this would be the ultimate solution of the sewerage problem," as, in fact, it was.[16]

With regard to the fourth plan . . . which would divert a large

[14] *1855 Report.* Also quoted in Andreas, 1:191.

[15] *CREB Report*, p. 72; Pearse, p. 3.

[16] Brown, p. 53. To be precise, the Sanitary District's Sanitary and Ship Canal was the last step in Chicago's adoption of the dilution method. Ultimately, Chicago's growth was sufficient to require sewage treatment in addition to dilution.

and constantly flowing stream from Lake Michigan into the Illinois River, it is too remote a contingency to be relied upon for present purposes; besides the cost of it, or any other similar channel in that direction, sufficient to drain off the sewage of the city, would be not only far more than the present sewerage law provides for, but more than would be necessary to construct the sewers for five times the present population. Should the proposed steam-boat canal ever be made for commercial purposes the plan now recommended would be about as well adapted to such a state of things, as it is to the present.[17]

Certainly his plan was readily adaptable to such a scheme. The Sanitary District of Chicago was created in 1889 for the express purpose of implementing this fourth possibility. The Sanitary District then constructed the "proposed steamboat canal," which unquestionably was beyond the means of Chicagoans in 1855.

In December 1855 Chesbrough submitted his plan for Chicago's sewage disposal and drainage. Under this plan, all of the sewage of Chicago's west division, all the sewage of the north division except for the lakefront area, and about one-half the sewage of the south division was deposited in the Chicago River. This sewage passed from the river into Lake Michigan. The dividing line in the south division was State Street; the area east of State Street drained directly into the lake. As the area east of State Street was primarily residential, Chicago's business district was sewered into the river. This district, west of State Street, included the majority of Chicago's packinghouses, distilleries, and hotels. Thus, the river received large quantities of pollutants daily.[18]

The sewers themselves were outstanding phenomena. Brick sewers, 3–6 feet in diameter, were laid above the ground down the center of the street. Chicago's topography, being unusually flat, was unfavorable to sewer construction. The Chicago River banks were only 2 feet above the water level. Near the river's north and south branches, the ground level reached a maximum of 10–12 feet above the lake. In reality, the task of constructing underground sewers required raising the city.[19] From the beginning, Chesbrough insisted

[17] *1855 Report.* Also quoted in Brown, p. 55.

[18] R. Isham Randolph, "A History of Sanitation in Chicago," *Journal of the Western Society of Engineers* 44 (October 1939): 229; Richard S. Kirby and Philip G. Laurson, *The Early Years of Modern Civil Engineering* (New Haven, 1932), p. 234; George W. Rafter and M. N. Baker, *Sewage Disposal in the United States* (New York, 1894), pp. 169–70.

[19] *CREB Report,* p. 69.

that a high grade was necessary for proper drainage and dry streets. Chicago lacked this high grade, and, thus, the decision to raise the city's level, concomitant with sewer installation, was one which solved the waste disposal and drainage problem in the context of Chicago's existing topography and future necessities.[20]

The Chesbrough plan called for an intercepting sewer system which emptied into the Chicago River. The sewers were to be constructed on the combined system; that is, they would collect sewage from both buildings and streets. This was consistent with the best contemporary thinking and practice. As sewer construction progressed away from the river, the streets had to be raised beneath the sewers. After the sewers were laid, earth was filled in around them, entirely covering them. The packed-down fill provided roadbeds for new, higher streets. These streets were rounded in the center, with gutter apertures leading to the sewer. Such streets would stay dry and could be paved, as contrasted to the mud which had plagued the city previously.

A second facet of Chesbrough's sewerage plan involved dredging the Chicago River. The river had been dredged previously, but it was still too small to handle the anticipated sewage load. Chesbrough planned to widen and deepen the river, as well as to straighten its meandering course. Contracts for this work had been let to the partnership of John P. Chapin and Harry Fox. It was Fox who suggested using the dredgings from the river as fill around the sewers.[21]

It is interesting to digress on the consequences of Chesbrough's plan to raise the city. Where vacant lots existed, they were filled to the new level. A few old frame buildings were torn down, and the lots filled. It proved relatively easy to raise frame buildings to the new level, if the owners could afford it. The city's newer buildings were brick and stone, however, and they were constructed on the old level. These newer buildings would not be torn down, and many of

[20] The grade which the city council adopted was lower than Chesbrough advocated, but it was sufficiently high to permit the construction of 7-8-foot cellars. The council's decision was to raise the grade to 10 feet on streets adjacent to the river; Chesbrough's higher grade was rejected because the city fathers felt there would be difficulties in locating sufficient fill. See *CREB Report*, p. 70.

[21] *Biographical Sketches*, p. 482. Fox's company was responsible for almost every topographical improvement in the Chicago area. The company deepened the Chicago River, developed the Chicago Harbor, installed road and railroad bridges, dredged the Illinois and Michigan Canal, and then performed similar services throughout the Midwest.

Chicago's homes and offices were to be left "in the hole." When new buildings and sidewalks were constructed on the higher level, Chicago increasingly became a city built on two levels.[22] Legal attempts to maintain the lower level were uniformly settled in favor of the city and its new level.[23]

The raising of brick buildings proved to be a difficult proposition. George Pullman, who later became famous for his "Palace cars," devised and instituted a method to raise brick buildings.[24] Pullman first used his method in connection with the Erie Canal enlargement of the 1850s, so Chesbrough would have known that the problems concomitant with raising the city's grade were surmountable. One of Pullman's biographers described his activities during those years:

> He made contracts with the State of New York for raising buildings on the line of the enlargement of the Erie Canal, which occupied about four years in their completion. At the end of that time, in 1859, he removed to Chicago, and almost immediately entered upon the work, then just begun, of bringing our city up to grade by the raising of many of our most prominent brick and marble structures, including the Matteson and Tremont Houses, together with many of our heaviest South Water street blocks. He was one of the contractors for raising by one operation, the massive buildings of the entire Lake street front of the block between Clark and LaSalle streets, including the Marine Bank and several of our largest stores, the business of all these continuing almost unimpeded during the process—a feat, in its class, probably without a parallel in the world.[25]

The Tremont Hotel was the first brick building which Pullman raised in Chicago. Soon his method was utilized to raise all Chicago's brick buildings from their former muddy level. The work required years. No one knows the cost, but it has been estimated at $10,000,000.[26]

[22] Randolph, p. 229. For many years, some sewers lay wholly above the ground, at the same level or higher than adjoining buildings.

[23] "Up from the Mud: An account of how Chicago's Streets and Buildings were Raised," compiled by Workers of the Writer's Program, WPA in Illinois for Board of Education, 1941. The raising of cities was relatively common. It was pointed out to me that all of downtown Atlanta was "raised" by the construction of roadways.

[24] Ibid.

[25] *Biographical Sketches*, p. 472. See also Seymour Currey, *Chicago: Its History and Its Builders* (Chicago, 1962), vol. 3; and Stanley Buder, *Pullman: An Experiment in Industrial Order and Community Planning, 1880–1930* (New York, 1967).

[26] Lloyd Wendt and Herman Kogan, *Give the Lady What She Wants* (Chicago, 1952), p. 57. Wendt and Kogan do not say how they arrived at this number, and give no reference. Pullman reportedly received $45,000 for raising the Tremont Hotel. At

In December 1856 the sewerage commissioners sent Chesbrough to visit several European cities in order to discover if their sewage disposal techniques were relevant to Chicago's needs.[27] Chicago was taking an open-minded approach to this question, and, evidently, the city was prepared to adopt an unconventional approach if it proved to be the best solution. The report of this trip, which Chesbrough submitted in 1858, represents one of the first sanitary engineering treatises.[28] Chesbrough visited and reported on the sewerage of Liverpool, Manchester, Rugby, London, Amsterdam, Hamburg, Paris, Worthing, Croydon, Leicester, Edinburgh, Glasgow, and Carlisle. He concluded that none of these cities furnished an exact criterion to judge the effects of disposing sewage directly into the Chicago River, but he felt their collective experience suggested that it probably would be necessary to keep the river free of sewage accumulations.

Chesbrough ended his report by relating the European experience to Chicago's sewerage needs. Two points which Chesbrough made in this concluding section are worthy of special mention.[29] The first is the experience of Worthing, "a small watering

---

$45,000 per brick building, $10,000,000 will raise over 200 buildings. This is probably an overestimate of the number of buildings raised, but the large number of other expenditures, including Chicago River dredging and legal expenditures, suggest that the $10,000,000 figure is an underestimate.

[27] *Report of the Results of Examinations Made in Relation to Sewerage in Several European Cities, in the Winter of 1856-57*, published in Chicago by the Board of Sewerage Commissioners (1858), p. 3, hereafter referred to as the *1858 Report*. See also Randolph, p. 229; Brown, p. 57.

[28] *1858 Report*, p. 92. Chesbrough's memorialist in the *Journal of the American Society of Civil Engineers* (15 [November 1889]: 162), unhesitatingly assessed the significance of Chesbrough's European trip report: "The importance of this report and the influence it exerted . . . can hardly be estimated. At the time the report was written, there was not a town or city in the United States that had been sewered in any manner worthy of being called a system. This being, perhaps, the first really exhaustive study which the subject had received on this side of the water, and Chicago being the first city to adopt a systematic sewerage system, the Chicago system soon became famous and Mr. Chesbrough, for twenty-five years, was the recognized head of sanitary engineering in this country." Modern usage would limit the term "sanitary engineer" to those men involved with water and sewage treatment. Apparently, the American Society of Civil Engineers at that time considered a sanitary engineer to be a man involved with sanitation works. Thus, while Chesbrough was not concerned with sanitary engineering as that discipline is currently defined, he must be considered a precursor of the modern sanitary engineer and, in fact, was called one by his peers and contemporaries.

[29] On this trip Chesbrough visited Zaardam, near Amsterdam, to investigate the possibility of using windmills to pump flushing water for Chicago's sewers; he decided in favor of steam pumps (*1858 Report*, p. 29).

town on the southern coast of England." At one time this town of 5,000 had drained directly into the sea, "but owing to offensive smells caused by this practice, and the consequent injury to the reputation of town as a watering place, upon which its prosperity very much depends," Worthing decided to find an alternative sewerage scheme.[30] Chesbrough concluded that Worthing's experience "shows that the mere discharge of filth into the sea gives no security against its being cast back in a more offensive state than ever, especially when the prevailing winds are toward shore," and that this suggests "the possibility of creating on the lake shore as great a nuisance as would be taken from the river."[31]

Second, Chesbrough included a prophetic paragraph which could serve as a summary to Chicago's sanitary history for a half century thereafter:

> Under these circumstances it seems advisable to do nothing with regard to relieving the river at present, nor towards carrying out that portion of the plan which provides for forcing water from the lake into it, during the summer months. Should the Canal Company [the Illinois and Michigan Canal] not be obliged to pump enough during warm weather to keep the river from being offensive, it is understood that they would pump as much as they could for a reasonable compensation. This would furnish some criterion by which to judge of the probable effect of a still greater quantity driven in from the lake, according to the plan. The thorough [*sic*] cut for a steamboat canal, to the Illinois River, which the demands of commerce are calling more and more loudly for, if ever constructed, would give as perfect relief to Chicago as is proposed for London by the latest intercepting scheme.[32]

The Chicago River's south branch became quite polluted shortly after sewage was admitted into it. The Illinois and Michigan Canal's pumps, however, utilized south branch water to provide the canal's summit level, and, consequently, the pumps relieved a portion of the river's pollution load. The real significance of Chesbrough's statement lies in the fact that, as early as 1858, Chicagoans recognized the Illinois and Michigan Canal's sewage disposal potential.[33] In

[30] *1858 Report*, p. 39.

[31] Ibid., p. 93.

[32] Ibid., p. 94.

[33] Nevertheless, in 1863, the Board of Public Works issued a report on purifying the Chicago River. This is discussed in Brown, chap. 6. The report recommended the construction of flushing canals along the lines of Fullerton Avenue and Sixteenth Street. Therefore, although the Illinois and Michigan Canal's potential was realized, city officials evidently were not ready to pursue it.

following years, the canal's pumps were used regularly to relieve the pollution load. Further, the canal itself was deepened and additional pumps were installed to increase the canal's capacity for handling sewage. Finally, the Chicago Sanitary District was formed in order to construct a new and enlarged canal to service Chicago's waste disposal needs, as Chesbrough had prophesied.

In 1861 the Board of Public Works was formed by incorporating the duties of the Board of Sewerage Commissioners, the Board of Water Commissioners, and other miscellaneous departments. Chesbrough was named chief engineer of this new board and, consequently, inherited the water-supply problem in addition to the waste-disposal problem. His inheritance was the "vicious circle" created by Lake Michigan's dual role as water supplier and eventual waste disposer.

III

Chicago's continued population growth through the decade of the 1850s, the new sewerage works, and the expansion of packinghouses and distilleries had increased the number of pollutants drained into the Chicago River. Lake Michigan soon became fouled by the river's influx, and Chicagoans began to complain of the public water supply's offensiveness and pollution. The existing water intake was a wooden pipe which extended a few hundred feet out into Lake Michigan, $1/2$ mile north of the Chicago River's mouth. In 1859, one of Chicago's water commissioners "proposed to sink a wrought iron pipe . . . one mile out into the lake, to obtain the supply from a point which could not be affected by the river."[34] Chesbrough was asked to study and report on the commissioner's plan, and to do the same on "erecting additional pumping works, in such locality as shall secure a supply of pure water."

Chesbrough's report discussed several methods without making a specific recommendation. Even at this early date, however, he considered a tunnel under the lake to be the most desirable alternative. Chesbrough was not afraid to combine grading, tunneling, and hydraulic principles to create a new water-supply system. When he later offered plans for a lake tunnel, his innovative proposal drew considerable opposition at the start and unmitigated acclaim when it proved successful.

Shortly after its formation in 1861, the Board of Public Works adopted as its goal the acquistion of an unpolluted water supply. Consequently, the board requested Chesbrough to make a canvass of

[34] Brown, p. 32.

the various water-supply possibilities and to investigate several filtration methods. Chesbrough dismissed the existing filtration methods as inadequate; his studied opinion was that the tunnel method was the most desirable:

> The engineer of the Board [E. S. Chesbrough], after much doubt and careful examination of the whole subject, became more inclined to the tunnel plan than any other, as combining great directness to the nearest inexhaustible supply of pure water, with permanency of structure and ease of maintenance. The possibility, and, in the estimation of many, the probability of meeting insuperable difficulties in the nature of soil, or storms, or ice on the lake, were fully considered. One by one the objections appeared to be overcome, either by providing against them, or discovering that they had no real foundation.[35]

Chesbrough continued to explore the tunnel plan's potential. When he had worked out the details, a proposal was submitted to several engineers, all of whom considered the tunnel plan to be feasible. Nevertheless, the 1861 board was against adopting the project. After a new board was elected and additional soil examinations had been made, Chesbrough's water-supply tunnel plan was adopted. The new board reported:

> What is most to be desired by the city is, that the supply should be drawn from the deep water of the lake, two miles out from the present Water Works. . . . The careful investigation of the subject has satisfied us sufficiently to say, that with our present knowledge, we consider it practicable to extend a tunnel of five feet diameter the required distance under the bed of the lake, the mouth or inlet to such a conduit being the outmost shaft, protected by a pier [crib], which will be used in the construction of the tunnel.[36]

In their 1863 report, the Board of Public Works noted that three projects had been considered, any one of which would have afforded Chicago a healthier and better protected water supply. These were (1) a 2-mile lake tunnel, (2) a filtering or settling basin, and (3) a 1-mile lake tunnel located 5 miles to the north.[37] The board had two principal objections to the second plan. First, they commented:

[35] Reported in Brown, p. 33.

[36] *Second Annual Report of the Board of Public Works to the Common Council of the City of Chicago* (April 1, 1863), p. 5, hereafter referred to as *1863 Report.*

[37] Cost estimates for each of the projects were as follows: 2-mile lake tunnel exclusive of light house, $307,552; a filtering or settling basin, $300,575; a 1-mile lake tunnel 5 miles to the north, $380,000 (*1863 Report,* p. 9).

For settling and filtering the water from sediment, we are of the opinion that the basin would be found effective, and would continue to be so, but that for filtration it is not safe to rely upon it. There have been filtering basins of the character in other places. Some of them appear to have continued to work well during long use, and others have failed and become useless.[38]

Second, the board objected to the basin scheme because the water supply intakes would still be in the shallow water close to shore, and would not be located in a deeper point where the water was considered to be better.

Chesbrough's 1863 report acknowledged that the board had considered the three most promising possibilities and had rejected one; he was to assess the remaining two. Almost immediately he dismissed, on the grounds of greater cost, any project which required moving the existing water works, such as the board's third proposal:

> Other projects, such as erecting a new pumping works at Winnetka, or going to Crystal Lake and bringing a supply thence by simple gravitation, as is done for cities of New York, Boston, Baltimore, and Albany, have been considered, but their great cost, as compared with that of obtaining an abundant supply of good and wholesome water at points much nearer the city, is deemed a sufficient apology for not discussing their details here.[39]

Chesbrough concerned himself only with those plans which would bring water from a point 2 miles east of the existing Chicago Avenue Water Works, and there were two of these:

> Of the plans proposed for obtaining water from the lake, where it will be free from not only the wash of the shore, but from the effects of the river, two classes only have been considered; one, an *iron pipe with flexible joints*; and the other, *a tunnel under the bottom of the lake*.[40]

Although the cost of the iron pipe project was slightly less than the tunnel project, Chesbrough chose between them on other than an initial cost basis:[41]

> In consequence of the possibility of such a pipe being injured by anchors, by the sinking of a heavily loaded vessel over it, or by

[38] Ibid., p. 8.
[39] Ibid., p. 39.
[40] Ibid.
[41] Chesbrough roughly estimated the iron pipe scheme to cost $250,000. The choice seems to have been made on the basis of expected cost. Ibid., pp. 40–41.

the effect of an unusual current in the lake moving it from its place, it has been thought preferable to attempt the construction of a tunnel under the bottom of the lake.[42]

His research had convinced him that the tunnel's construction would be less difficult than was generally supposed. Lill and Diversey's brewery, adjacent to the waterworks, was the site of artesian borings which showed that, between 25 and 100 feet deep, the ground at the lake shore was a clay which was also found on the lake bottom where the water was 25 feet deep. A tunnel could easily be constructed in this type of clay, if it were continuous. Chesbrough was confident that the clay was continuous, but he admitted he was uncertain whether beds of sand might not be interspersed with the clay.[43]

The lake shaft was to be formed by sinking iron cylinders to the desired depth. Chesbrough noted that this was not a difficult problem in that the pneumatic process had been successfully employed on "the Theiss bridge in Hungary, and the railroad bridge across the Savannah River, . . . and recently the Harlem bridge in New York."[44]

In giving cost estimates for the tunnel project's component parts, Chesbrough clearly showed the sources of his research. The principal source was the Thames tunnel, and Chesbrough noted that the first thoughts of most people were the great construction difficulties and "enormous" costs which had been encountered on the Thames project. He was quick to refute these thoughts and countered that "as we have every reason to believe, the clay formation here would shield us from such inroads of water as were met within the Thames tunnel operation."[45] In estimating excavation costs, Chesbrough made the same point: "There is good reason to believe that nothing in the soil here would be more difficult than that through which the sewers of London are sometimes tunneled."[46]

Chesbrough also used the Thames experience, plus that of the Boston Water Works tunnel, to estimate masonry costs. Cribs had been used principally in pier and breakwater construction, and Chesbrough based his crib cost estimates on figures which had been made for a proposed breakwater in Michigan City, Indiana, at the bottom of Lake Michigan.

After reaching his cost estimate for masonry and excavation,

[42] Ibid., p. 41.
[43] Ibid.
[44] Ibid. Originally, Chesbrough planned on four shafts.
[45] Ibid., p. 43.
[46] Ibid., p. 45.

Chesbrough compared it with figures which had been reached for other major tunnel projects.[47] In particular, Chesbrough referred to reports from (1) the commissioner of the Troy and Greenfield Railroad, and (2) the Hoosac Tunnel. Included in the commissioner's report was the report of Charles Storrow, who had been sent to investigate European tunnels. Because the tunnels which Storrow had studied were for railroads, they were all much larger than the one which Chesbrough was planning. Therefore, Chesbrough estimated the cost of each tunnel had it been constructed with a 5-foot width. From these estimates, he concluded that his cost estimate for the proposed water tunnel was reasonable.

The engineering achievement involved in constructing the water-supply system was no less significant than that represented by Chesbrough's sewer system. As conceived, the task was to dig a shaft near the lake shore to a depth significantly below the lake bottom and then burrow 2 miles beneath the lake. A similar shaft was to be dug at the lake end and was to be protected by a crib. The engineering problem was to connect the shore and lake points by a straight line 69 feet below the surface of Lake Michigan. Contemporary compasses could not be used since, below ground level, local attraction rendered them inaccurate. To a worker in the tunnel, the only place where the direction of the line drawn between the two shafts could be observed was at the top of either shaft. Consequently, when the engineers attempted to run the tunnel's axis parallel to this imaginary line on the lake's surface, they ran into difficulties affecting the turn from shaft to tunnel.[48]

When the lake shaft was completed, workers were lowered to begin burrowing westward to meet with the other workers burrowing eastward. The tunnel was sloped 2 feet per mile from the lake end to the shore so that it could be emptied should repairs prove necessary; the water would be shut off at the lake end. Although the methods were primitive — the tunnel was dug entirely by manual labor — it was claimed that the workers caused the two tunnel sections to meet within 1 inch of achieving a perfectly smooth wall.[49]

Chesbrough's engineering competence was coupled with a sense of economic reality, and these traits combined to insure the reputation he earned in Chicago. His 1863 report contained a section on "plans for improving the Chicago river." Chesbrough knew that

[47] Chesbrough estimated the cost to be $13.54 per linear foot. Ibid., p. 48.
[48] J. M. Wing, *The Tunnels and Water System of Chicago* (Chicago, 1874), p. 33.
[49] Ibid., p. 76.

moving the water-supply intake farther into the lake would not improve the river's offensive condition. In the 1855 sewerage report, he had argued that flushing canals would be necessary in both the north and south branches to purify the river, and he restated this position in several reports thereafter. By raising the issue once again, Chesbrough not only demonstrated the completeness of his approach, but also what one memorialist called "the characteristic firmness of conviction and modest persistence of Mr. Chesbrough."[50]

As before, Chesbrough's methodology was to enumerate and evaluate the possibilities for improving the river: (1) north and south branch flushing canals, (2) Des Plaines River diversion into the south branch, and (3) drainage southwest into the Illinois River Valley. The first was preferred because Chesbrough felt it was "undoubtedly feasible, would be completely under the control of the city, and there is every reason to believe [it] would be effectual."[51] He considered the second plan "defective" in that the Des Plaines River's flow was least when the Chicago River's pollution was greatest. Although Chesbrough correctly assumed that the third project would be the ultimate solution, he rejected it as "requiring much larger means than the Board can at present control."[52] Chesbrough's attention to Chicagoans' ability to pay established him as a practical man and lent credence to his innovative ideas. His consideration of a sanitary canal connecting the Chicago and Illinois Rivers indicates Chesbrough had learned that water-supply and waste-disposal problems are interdependent and must be solved simultaneously.

IV

Chicago is an urban center which had, and still has, serious water pollution problems. Lake Michigan's present pollution problem is primarily the result of industrial discharge in the Calumet and Indiana Harbor areas and the discharge of inadequately treated sewage by the North Shore Sanitary District (Lake County, Illinois) and several Wisconsin cities. Under normal circumstances, the Metropolitan Sanitary District of Chicago diverts the sewage and the

[50] *Journal of the American Society of Civil Engineers* 15 (November 1889): 162.

[51] *1863 Report*, p. 57. Chesbrough was concerned with a definite planning period which seems to reflect a longer time than the Marshallian short run, and a shorter time than the Marshallian long run.

[52] Ibid., p. 57.

treated effluent from Lake Michigan. Presently, Chicago is meeting its responsibility with respect to Lake Michigan pollution. On the other hand, both the Chicago River and the Illinois River valley are polluted because some industries in the Chicago area still discharge their wastes into the water and the Sanitary District falls short of 100 percent treatment. Approximately 10 percent of the sewage goes untreated at this time, but it is the district's stated objective to achieve 100 percent treatment in the 1970s. While these few sentences oversimplify a very complex situation, the outline is apparent. Chicago must seek outside help to reduce Lake Michigan pollution and the consequent threat to the city's water supply. Chicago and its Cook County suburbs, by themselves, could significantly reduce pollution in the Chicago, Des Plaines, and Illinois rivers.

When faced with Lake Michigan and Chicago River pollution in the 1850s and 1860s, Chicagoans had sought the best solutions available. Cost considerations had entered the argument only in deciding among equally effective methods; Chicagoans were not reluctant to pay the price necessary to secure sanitary conditions. They indebted the city through bond issues and themselves through tax assessments in order to finance these public works. Muddy streets and impure water were manifest physical representations of the city's problems, and solutions to these benefitted the city's residents, individually and collectively. The public's acceptance of an increased tax burden to finance these works must be viewed as public recognition of the problems' dimensions. If the city's water supply had not been conserved, and if the city's natural topography had not been improved, Chicago's urban growth would have been severely limited.

When the pollution problem is explored in a historical context, students will find that the objectives which Chesbrough sought — minimize pollution and obtain a pure water supply — are the same as today's objectives. Nineteenth-century engineers, however, were not faced with the imminent "death" of large bodies of water; they were faced only with protecting urban populations from polluted water supplies.

In studying Chesbrough's works in Chicago, one gets the impression that today's pollution problem is not the result of ignorance as to pollution's effects, but ignorance as to how deadly the pollution load has become. In many cases, techniques first utilized in the 1850s and 1860s are still used today. Although these techniques no longer solve the problems for which they were intended, their inadequacies did not become apparent until recently. Perhaps this is because the

88   *Louis P. Cain*

demands on these techniques were much less heavy during the earlier period than they now are. Perhaps it is because the engineers of Chesbrough's generation made such dramatic innovations that the declining effectiveness of these techniques and improvements just recently became evident to sanitary engineers and laymen. Or perhaps it is because the 20th-century sanitary engineers who recognize the problem are unable to communicate the necessity for action. While the technology and technicians have been available, an uninformed and apathetic public has not invested sufficient capital in pollution control. Whatever the case, through inaction, the cost of proper treatment has reached a price which may be greater than the public is willing to pay. Unfortunately, the 20th century has been unable to find a sanitary engineer with the same farsightedness in his method, and resoluteness in seeing his proposals adopted, as that characteristic of Ellis Sylvester Chesbrough.

ANDREW A. HUMPHREYS AND THE DEVELOPMENT OF HYDRAULIC ENGINEERING: POLITICS AND TECHNOLOGY IN THE ARMY CORPS OF ENGINEERS, 1850–1950

MARTIN REUSS

In 1861, Captain Andrew A. Humphreys of the United States Army Corps of Topographical Engineers and his young assistant, Lieutenant Henry L. Abbot, completed their *Report upon the Physics and Hydraulics of the Mississippi River*.[1] The report was based on a thorough investigation of the lower Mississippi basin. Congress had authorized the survey in 1850, following two disastrous years of flooding and an appeal from the Louisiana legislature. It was far from being the first river survey, but it was the most extensive ever undertaken at that time. In it the authors challenged earlier hydraulic theories and introduced entirely new formulations to explain river flow.[2]

Although much of the theory introduced in the report was later disproved, the conclusions decidedly influenced the development of

MARTIN REUSS is a senior historian in the Office of History, Headquarters, U. S. Army Corps of Engineers. He specializes in the history of water resources development. He wishes to acknowledge the assistance of John R. Ferrell, Robert Kelly, Raymond H. Merritt, Frank N. Schubert, and Todd Shallat. This article is a revision of a paper presented at the annual meeting of the Society for the History of Technology in Philadelphia on October 31, 1982.

[1]A. A. Humphreys and H. L. Abbot, *Report upon the Physics and Hydraulics of the Mississippi River; upon the Protection of the Alluvial Region against Overflow; and upon the Deepening of the Mouths: Based upon Surveys and Investigations Made under the Acts of Congress Directing the Topographical and Hydrographical Survey of the Delta of the Mississippi River, with Such Investigations as Might Lead to Determine the Most Practicable Plan for Securing It from Inundation, and the Best Mode of Deepening the Channels at the Mouths of the River*, Professional Papers of the Corps of Topographical Engineers, United States Army, no. 4 (reprint, Washington, D.C., 1876).

[2]Hunter Rouse and Simon Ince, *History of Hydraulics* (Iowa City, 1957), p. 177; Hunter Rouse, *Hydraulics in the United States, 1776-1976* (Iowa City, 1976), pp. 43–46; L. W. Mosby, "First Step in Big River Hydraulics," *Military Engineer* 53, no. 354 (July–August 1961): 262–63; T. A. Lane and E. J. Williams, Jr., "River Hydraulics in 1861," *Journal of the Waterways and Harbors Division, Proceedings of the American Society of Civil Engineers* 88, no. WW3 (August 1962): 1–12.

This essay originally appeared in *Technology and Culture*, vol. 26, no. 1, January 1985.

river engineering and the evolution of the Army Corps of Engineers. The authors believed that "levees only" could control flooding along the lower Mississippi. Neither costly reservoirs nor cutoffs were needed. The Corps of Engineers accepted these conclusions for nearly sixty years, not just for the lower Mississippi but for other large rivers as well. The "levees only" policy profoundly affected the manner in which the United States developed its water resources. Indeed, the influence of the Humphreys-Abbot report extended past World War II, despite the fact that by then Congress had authorized hundreds of reservoir projects.

One reason why army engineers accepted the conclusions of the Humphreys-Abbot report for so long is the thoroughness with which the survey was conducted. From just south of the junction of the Mississippi and Ohio rivers to where the Mississippi empties into the Gulf of Mexico, Humphreys and Abbot obtained data on river flow, channel cross sections, and general topographical and geological features. Survey teams also took similar measurements on the major tributaries of the lower Mississippi. The two officers then examined all available literature on channel resistance and water flow. Altogether, they checked fifteen formulae, including those of the French engineers Antoine Chézy, Richy de Prony, and Pierre Du Buat, and the German engineers Johann Eytelwein and Julius Weisbach. They found every calculation lacking in some respect. While some formulae described the characteristics of small rivers well enough, none fitted the data that Humphreys and Abbot had conscientiously gathered on the Mississippi. Therefore, they decided to develop their own formula.[3]

Humphreys, who made all major decisions about the report, was ambitious. He wished to develop a universal formula that would allow engineers to measure the flow of water in natural rivers anywhere, not just in the Mississippi's channel. Unfortunately, Humphreys and Abbot's formula fell short of his goal, for it was a cumbersome and imperfect affair. Most significantly, it failed to take into account the degree of roughness of the slopes of a river channel. Still, this formula stimulated other hydraulic engineers, and further research led to important theoretical discoveries.[4]

In Switzerland, Emile Ganguillet, chief engineer of the Department of Public Works at Berne, and a member of his staff, Wilhelm Kutter, tested the Humphreys-Abbot formula on Swiss mountain streams and found it lacking. They published their results in 1869. At approxi-

[3]Humphreys and Abbot (n. 1 above), pp. 213–28, 302–49; Mosby (n. 2 above), p. 263.
[4]Mosby, p. 263; Rouse and Ince (n. 2 above), p. 177.

mately the same time, Philippe Gauckler, an engineer of Le Corps des Ponts et Chaussées in France, experimented with different formulae, including that proposed by the two Americans. He concluded that a single all-purpose formula was impossible to derive. Nevertheless, attempts continued. Finally, the Irish engineer Robert Manning published an equation in 1889 that is both simpler and more accurate than Humphreys and Abbot's. It is a formula still in use today.[5]

Although the Humphreys-Abbot "universal formula" proved flawed, their report obtained the respect of engineers around the world. No one could fault the authors' ambition, intelligence, and diligence. In this, Humphreys and Abbot clearly surpassed their fellow army engineers. Since 1824, army engineers had been involved in various kinds of navigation improvements. Most of these engineers had been educated at the Military Academy at West Point, and their reports made many references to the French and Italian engineering theory they had learned at the academy. But Humphreys and Abbot, also West Point graduates, had gone one step further. Through systematic observation and collection of data, they had *tested* and revised the European theories they had learned. Moreover, they did so in an unprecedentedly comprehensive, lucid, and organized fashion. A hundred years later, two hydraulic engineers concluded that "Humphreys and Abbot developed a method for measuring and computing discharge that, but for its cumbersome form, could with possible modifications, be used today."[6] They are probably the only American army engineers mentioned in monographs on the evolution of hydraulic engineering.

Humphreys received numerous international honors for his work. He was made an honorary member of the Imperial Royal Geological Institute of Vienna in 1862 and a fellow of the American Academy of Arts and Sciences in 1863. The following year, the Royal Institute of Science and Arts of Lombardy appointed him an honorary member, and in 1868 Harvard College conferred on him the degree of doctor of laws.[7] Naturally enough, army engineers vicariously shared in Humphreys's honors. To them, the Humphreys-Abbot report seemed to

---

[5]Rouse and Ince, pp. 177–80; Mosby, p. 263.

[6]Lane and Williams (n. 2 above), p. 12.

[7]For more on Humphreys's career, see Henry H. Humphreys, *Andrew Atkinson Humphreys: A Biography* (Philadelphia, 1924), pp. 56–140; Todd A. Shallat, "Andrew Atkinson Humphreys," *APWA Reporter* 49, no. 1 (January 1982): 8–9; Harold F. Round, "A. A. Humphreys," *Civil War Times Illustrated* 4 (February 1966): 22–25; John Watts De Peyster, "Andrew Atkinson Humphreys," *Magazine of American History* 16 (October 1886): 347–69; *Dictionary of American Biography*, s.v. "Humphreys, Andrew Atkinson."

confirm the value of a West Point education. Therefore, they had a vested interest in defending its integrity.

During the Civil War, Humphreys received numerous promotions, reaching the rank of major general, U.S. Volunteers.[8] On August 8, 1866, sixteen months after the war's completion, he became the new chief of engineers. His illustrious record promised an exciting and productive tenure in his new position. Looking back, Abbot clearly thought that Humphreys's stewardship fulfilled this promise. "Whatever success may have attended the operations of the corps at this important period of its history," Abbot asserted, "deserves to be associated with the chief who directed its labors with a skill for which his own studies had been so admirable a preparation."[9]

How accurate was Abbot's observation? With a worldwide reputation, largely resulting from the widespread dissemination of the Mississippi Delta report, Humphreys was unquestionably in the perfect position to enhance both his reputation and that of the corps. Ironically, events took a different course. Humphreys oversaw and partly caused a decline, not an increase, in the corps's influence. He defended his report before both the political and engineering communities and came to identify attacks on the report as attacks on the corps itself. At the same time, he had to defend the corps in unrelated matters, such as the question of retaining responsibility for conducting Western explorations. Conservative by nature, possessing an ego largely untouched by failure, and convinced of the soundness of his position, Humphreys became increasingly frustrated and defensive in the face of changing political and engineering concepts. The more he was attacked, the less willing he seemed to modify his position. Tragically for the corps, it was this inflexibility that became his main legacy, rather than the scientific dedication to truth that had characterized his report on the Mississippi. An examination of his career and its effect on the Army Corps of Engineers shows how a bureaucracy can be crippled when it elevates theory to dogma and forgets that scientific research is, by definition, innovative and not stagnant.

* * *

When, in 1850, Humphreys first learned that he had been assigned to the Mississippi Delta survey, he wrote, "It is a work which I should

[8]For good overviews of Humphreys's Civil War service, see the works by Henry H. Humphreys and John Watts De Peyster cited in n. 7.

[9]Henry L. Abbot, "Memoir of Andrew Atkinson Humphreys, 1810–1883," *National Academy of Sciences, Biographical Memoirs* 2 (1886): 212.

desire, as it is one of much difficulty and of great importance."[10] The statement was characteristic of Humphreys, who eagerly accepted challenges throughout his life. What he hardly could have anticipated was that it would take him eleven years to complete the work. In hindsight, it is remarkable that the survey and final report were completed at all. Only Humphreys's tenacity insured a successful conclusion of the effort.

Humphreys was forty years old in 1850. Until then, his professional life had followed an erratic, but unremarkable, course. He was commissioned a second lieutenant of artillery in 1831, after graduating from the United States Military Academy at West Point, thirteenth in a class of thirty-one. According to one biographer, Humphreys's "frolicsome nature and consequent demerits" pulled down what would have otherwise been an outstanding academic record.[11] Clearly, he was not afraid of a scuffle, either as a cadet or later in life. One observer, Charles A. Dana, wrote that Humphreys was "very pleasant to deal with, unless you were fighting against him, and then he was not so pleasant. He was one of the loudest swearers I ever knew."[12] While fighting against the Seminole Indians in the mid-1830s, Humphreys caught a fever that nearly killed him. He returned to Philadelphia to recuperate and, eventually, to become a civilian engineer. However, in 1838 Congress approved the establishment of a separate Corps of Topographical Engineers, and Humphreys decided to join. He obtained a new commission as a lieutenant and became one of thirty-five officers (including twenty lieutenants) in the new corps.[13] For several years thereafter, he worked on various engineering projects, mainly in the Washington, D.C., area. His intelligence and industry caught the attention of Alexander D. Bache, superintendent of the United States Coast Survey, and in 1844 Bache obtained Humphreys's assignment to his office.[14]

The two men developed a deep and lasting respect for one another, and it is likely that some of Bache's scientific curiosity infected Hum-

[10]Humphreys to Alexander Dallas Bache, September 7, 1850, as cited in Humphreys, *Andrew Atkinson Humphreys* (n. 7 above), p. 57. The original of this letter is in the Andrew A. Humphreys papers, vol. 3/fol. 12, Historical Society of Pennsylvania, Philadelphia. Hereinafter the Humphreys papers will be cited as HP, followed by vol. no./fol. no.

[11]Round (n. 7 above), p. 23.

[12]Charles A. Dana, *Recollections of the Civil War*, intro. Paul M. Angle (New York, 1963), pp. 173–74.

[13]Attachment to printed report of the chief of engineers to the secretary of war for the year 1868, HP 22/31–39.

[14]Shallat (n. 7 above), p. 9. For more on Bache, see Nathan Reingold, "Alexander Dallas Bache: Science and Technology in the American Idiom," *Technology and Culture* 11 (April 1970): 163–77.

phreys. When Colonel John J. Abert, chief of the Corps of Topographical Engineers, requested Humphreys's reassignment to the
corps for the Delta survey, Bache reluctantly agreed. He noted to
Thomas Corwin, the secretary of the treasury, "The loss of the services
of so valuable an officer as Captain Humphreys will be very much felt
by the Coast Survey."[15] Yet, he admitted to Charles M. Conrad, the
secretary of war, it was "most fortunate for the success of this important work that it has been assigned to Captain Humphreys. To sound
knowledge he joins a practical turn which renders available his theoretical acquirements. He is cautious in obtaining data, energetic in using
them when obtained; is not likely on the one hand to run into unnecessary refinement nor on the other to mistake rough guesses for accurate
conclusions."[16] Bache had already attempted to procure a brevet promotion for Humphreys. Failing in that endeavor, he nevertheless took
satisfaction that the army had recognized Humphreys's abilities by
assigning him an important and challenging task.[17]

Humphreys received his appointment on October 6, 1850. Within a
month he suffered a major disappointment. Charles S. Ellet, Jr., one of
the best-known engineers of the day, who had already written about
flood problems in the Mississippi Valley,[18] applied to make the Delta
survey. Secretary of War Conrad desired that Ellet work alongside the
army engineers, but Ellet preferred to work independently.[19] Under
pressure from some congressmen and after confering with President
Millard Fillmore, Conrad relented and divided the $50,000 congres-
sional appropriation between the army survey and Ellet's. Humphreys
was infuriated. He wrote Bache, "I shall be crippled in my operations
by this division of the appropriation should I conclude to go on. If in
my power I shall give Mr. Conrad a touch from his Whig friends in
Philadelphia—who I imagine are not very favorable to Mr. Ellet."[20]

Neither Humphreys nor Conrad's Whig friends influenced the matter. Ellet spent the next half-year in New Orleans writing his report.[21]

[15]Bache to Corwin, October 3, 1850, as cited in Humphreys, *Andrew Atkinson Humphreys*
(n. 7 above), p. 58.

[16]Bache to Conrad, November 4, 1850, HP 21/48. The letter is also reprinted in
Humphreys, *Andrew Atkinson Humphreys*, pp. 59–60.

[17]Bache to Humphreys, October 3, 1850, HP 3/17. Parts of the letter are quoted in
Humphreys, *Andrew Atkinson Humphreys*, pp. 57–58.

[18]Charles Ellet, Jr., "Contributions to the Physical Geography of the United States. Part
I: Of the Physical Geography of the Mississippi Valley, with Suggestions for the Improve-
ment of the Navigation of the Ohio and Other Rivers," *Smithsonian Contributions to
Knowledge* 2 (1851): 1–58.

[19]Gene D. Lewis, *Charles Ellet, Jr.: The Engineer as Individualist* (Urbana, Ill., 1968),
p. 139.

[20]Humphreys to Bache, November 5, 1850, HP 3/24.

[21]Lewis (n. 19 above), p. 139.

Humphreys also went to New Orleans, but not before he stopped at Napoleon, Arkansas, to consult with Lieutenant Colonel Stephen H. Long, whom Abert had also appointed to the Mississippi Delta survey team. Long had gained much attention early in his career when he conducted surveys that carried him as far west as the Rocky Mountains. In 1850, the sixty-six-year-old officer was in charge of the Office of Improvements on Western Rivers and was mainly occupied with building marine hospitals such as the one then undergoing construction at Napoleon.[22] Humphreys and Long drafted a report to Abert that outlined the manner in which the survey would be executed. In light of later controversies over the effectiveness of levees, it is particularly noteworthy that both men stated their wish to avoid discussing "the propriety or impropriety of the construction of artificial levees as a means of preventing overflows" until the survey was complete.[23] Humphreys was even more adamant. He desired "to be distinctly understood as not having in view exclusively any one method of protection against the overflows of the Mississippi river."[24]

Humphreys left Napoleon for New Orleans, where he bought supplies and established the field office for the survey. Assisted by Lieutenant Gouverneur K. Warren, later a distinguished explorer and Civil War general, Humphreys organized three teams—topographical, hydrographical, and hydrometrical—and put a civilian in charge of each of them. He drafted detailed instructions for each team, specifying when, how, and under what conditions to measure river depth and velocity, rate and amount of sediment accumulation, and many other technical matters.[25] Ever the conscientious supervisor, Humphreys also took frequent excursions to the work sites to be sure his instructions were being followed.

The arduous work in the hot climate exhausted Humphreys. In the summer of 1851 he suffered a physical collapse and had to return to Philadelphia. The survey was incomplete, but Long drafted a report based on Humphreys's notes. He simply explained what had been

[22]Richard G. Wood, *Stephen Harriman Long, 1784–1864: Army Engineer, Explorer, Inventor* (Glendale, Calif., 1966), pp. 228–29. On Long's explorations, three sources that complement Wood's book are Roger L. Nichols and Patrick L. Halley, *Stephen Long and American Frontier Exploration* (Newark, Del., 1980); William H. Goetzmann, *Army Exploration in the American West, 1803–1863* (New Haven, Conn., 1959); Frank N. Schubert, *Vanguard of Expansion: Army Engineers in the Trans-Mississippi West, 1819–1879* (Washington, D.C., 1980).

[23]Long and Humphreys to Abert, December 18, 1850, Senate Executive Document 13, 31st Cong., 2d sess., January 17, 1851, p. 13.

[24]Ibid.

[25]Humphreys's instructions for "observing the rise and fall of the River," 1851, HP 3/39.

accomplished without offering any specific recommendations. Therefore, Ellet's essay became the first comprehensive study of flood control on the Mississippi. Both reports were sent to Congress in January 1852.[26] What distinguished Ellet's submission was his insistence on both the practicability and value of building reservoirs on the Mississippi's tributaries to reduce flooding. When Abert forwarded the report to Conrad, he commented, "While I willingly admit that all the speculations of a man of intellect are full of interest and deserving of careful thought, yet I cannot agree with him that these reservoirs would have any good or preventive effects upon the pernicious inundations of this river. . . ."[27]

Humphreys's convalescence was long. He spent most of 1852 in Philadelphia on sick leave. Early in 1853 his doctor recommended that he go to Europe to continue his recuperation. Nothing could have suited Humphreys more. He asked Abert for orders to be sent to Europe to investigate various rivers and compare them with the Mississippi.[28] Abert, apparently unconvinced of the benefits the corps would receive from such a trip, hesitated. He finally agreed, but only after he sent Humphreys detailed instructions about the information to be obtained on both artificial harbors and river improvements.[29] In the summer of 1853, Humphreys went abroad. He visited such rivers as the Rhône, Rhine, Vistula, and Po, taking careful notes in his diary.[30]

After an eighteen-month sojourn, Humphreys returned to the United States and was given a new assignment. His friend, Jefferson Davis, then secretary of war, directed him to supervise the Pacific railroad surveys. Assisted by Lieutenants Warren and—a new acquaintance—Abbot, Humphreys carefully analyzed the data sent him by various field parties. His final report described the major routes eventually used to girdle the Western Plains.[31]

Humphreys's mind was still on the Delta survey. It could hardly have been otherwise, since Ellet's ideas had stirred up a great deal of controversy. Herman Haupt, a West Point graduate and chief engineer for the Pennsylvania Railroad, insisted that the reservoir plan was impracticable. Not incidentally, he also argued that, no matter how waterways

[26]Senate Executive Document 49, 32d Cong., 1st sess., January 21, 1852.
[27]Abert to Conrad, January 19, 1852, in ibid.
[28]Humphreys to Abert, January 22, 1853, HP 3/85.
[29]Lt. Col. James Kearney to Humphreys, March 30, 1853, HP 3/85. Kearney, next to Abert the senior officer in the Corps of Topographical Engineers, apparently was acting on Abert's behalf in giving Humphreys his instructions.
[30]Humphreys's journal of his European trip is in the Humphreys papers.
[31]Schubert (n. 22 above), pp. 109–10; Goetzmann (n. 22 above), p. 312; Garry David Ryan, "War Department Topographical Bureau, 1831–1893: An Administrative History" (Ph.D. diss., American University, 1968), pp. 186–90.

are improved, the work should be in the hands of private companies and not governmental agencies.[32] In 1857, William Milnor Roberts, a respected civil engineer later employed by the Army Corps of Engineers, attacked Ellet's plan in an essay published by the distinguished Franklin Institute in Philadelphia. Roberts maintained that Ellet took no account of the enormous relocation expenses and underestimated the engineering and construction tasks.[33] Humphreys himself published a broadside, apparently at his own expense, in which he concluded that a survey of the Ohio River and its tributaries should be made "to determine if it be practicable by the use of reservoirs upon its tributaries to render the flow of that river more equal at all seasons, and thereby to maintain permanently a depth suitable for navigation by steamers of considerable draft and restrain the floods within harmless bounds."[34]

Soon after he completed the Pacific railroad report, Humphreys obtained the Mississippi Delta survey's records that Long had held at his headquarters in Louisville. In 1857, he received permission to reopen the Mississippi Delta survey office in the Topographic Bureau in Washington. From there, he directed Lieutenant Abbot to go to the Delta and finish the necessary studies.[35] Abbot performed his work with skill and dispatch. He and his assistants took gauge readings, determined discharges, measured cross sections, and reported on the state of various river improvements. When possible, he compared his data with those of earlier surveyors. "In a word," Abbot later said, "the finger was to be firmly placed on the pulse of the great river, and every symptom of its annual paroxysm was to be noted."[36]

In the shadow of civil war, Humphreys and Abbot finally put together their huge report. Humphreys later admitted that he "was deeply anxious for many reasons to complete the work & to put it beyond the risk of loss by having it printed. I was also desirous of taking part, at the earliest day practicable, in the military operations which indeed were likely to separate me from what I considered the work of my life while it was still in an unfinished condition. A few hours more or less under such circumstances became important. . . ."[37]

[32]Herman Haupt, *A Consideration of the Plans Proposed for the Improvement of the Ohio River* (Philadelphia, 1855), pp. 52–53.

[33]Lewis (n. 19 above), pp. 147–48. Roberts's essay took the form of articles that appeared monthly from July through December, 1857, in vol. 64 of the *Journal of the Franklin Institute*.

[34]Printed broadside signed by A. A. Humphreys, January 1857, Joshua Humphreys papers, vol. 1775–1831 (*sic*), Historical Society of Pennsylvania, Philadelphia.

[35]Ryan (n. 31 above), pp. 265–69.

[36]Abbot, "Memoir of Andrew Atkinson Humphreys" (n. 9 above), p. 212.

[37]Humphreys to Sir Charles Lyell, Washington, May 28, 1866, as reprinted in Humphreys and Abbot (n. 1 above), p. 646.

Humphreys was the report's principal author, since he had been involved with the project from its inception; however, he generously acknowledged Abbot's valuable help by making the young lieutenant his coauthor.[38] Their 500-page report arrived at the office of the chief of topographical engineers in August 1861, a few months after the firing on Fort Sumter.

* * *

Once Humphreys became the new chief of engineers in 1866, he faced several organizational, political, and technical questions, any of which would have been sufficient to test one's patience and endurance. Humphreys's reaction was uncompromising. He vigorously defended the corps's integrity and performance, particularly when it came to questions involving his own report. In the most famous and far-reaching controversy, Humphreys pitted himself against the famous civil engineer James Buchanan Eads on the question of how to make the mouth of the Mississippi permanently navigable. Appreciation of this controversy, however, requires a review of the other problems confronting Humphreys.

In regard to his own report, Humphreys dismissed its detractors as self-promoters, ignoramuses, or, at the very least, inexperienced. For instance, Daniel Farrand Henry, a corps employee working on the St. Clair River in Michigan for the Lake Survey, questioned the accuracy of the double-float method of measuring river current developed by Humphreys and Caleb Forshey for the Delta survey. It was a method employing a wooden keg, its top and bottom removed, with lead strips attached to its side to add weight. To the keg a line was attached that led to a piece of cork on top of which was mounted a small flag. The idea was to throw the keg into the water and trace its movement by carefully observing the flag that rose from the cork bobbing on the water's surface. In place of this primitive system—Abbot suggested it could be traced back to Leonardo da Vinci—Henry developed the first cup-type current meter, a cross between an anemometer and a Morse telegraph sounder, to measure river current.[39] His invention was vastly superior to the double float; yet Humphreys would not accept it. He abruptly stopped Henry's work and relieved Henry's superior, Brevet Brigadier General William F. Raynolds, of his command. Henry himself left the

---

[38]Humphreys and Abbot (n. 1 above), p. 15.

[39]Arthur H. Frazier, "Daniel Farrand Henry's Cup Type 'Telegraphic' River Current Meter," *Technology and Culture* 5 (Fall 1964): 541–65; Henry L. Abbot to Caleb Forshey, February 28, 1873 (copy), HP 25/10.

corps in 1871 and two years later published a book explaining his method. By the end of the decade, even some army engineers were using Henry's meter, and Humphreys, finally persuaded by a wealth of new data, retreated from his dogmatic position.[40]

Humphreys vigorously defended not only the credibility of the Delta survey but also his own claims to specific contributions. Consequently, in 1873, when Caleb Forshey, who had led the hydrometric party in the early 1850s, proposed that he had actually discovered the double-float method,[41] Humphreys was astounded. He wrote Abbot that the claim was "so preposterous that I hardly think he can mean what his words convey, for they prefer a claim utterly unfounded." Abbot more diplomatically wrote Forshey that the idea was in Humphreys's mind when he penned Forshey's instructions, and "the twenty-two years that have rolled over our heads since those days have confused your recollections. The official records show it most conclusively—even your own journal kept at the time."[42] It is worth clarifying this dispute only because an explanation reveals something about the petty jealousies involved in making large claims for small improvements. Forshey's journal shows that Humphreys indeed suggested the double-float system for measuring the velocity of the Mississippi. However, Humphreys recommended tying pieces of weighted wooden blocks to the cork. Forshey's improvisation was the topless-bottomless keg, which, he observed, would expose a more uniform surface to the water.[43] The mechanism then resulted from a joint effort for which neither man could rightfully claim full credit.

These minor disputes tell us something about Humphreys's character, but they naturally must be put in a larger context in order to enable us to judge the man fairly. The growth of professional engineering societies, combined with the increased congressional demand for public works projects, resulted in intensified scrutiny of the corps in the post–Civil War period. Both private engineers and congressmen questioned the corps's capabilities. Consequently, Humphreys was forced to stretch every dollar while constantly defending the corps's performance and integrity. His problems were formidable.

The first problem was management. The volume of congressionally

[40]Frazier, pp. 561–62. Henry's book is titled *Flow of Water in Rivers and Canals* (Detroit, 1873).

[41]Forshey to Abbot, February 13, 1873 (copy), HP 25/6.

[42]Humphreys to Abbot, February 25, 1873, HP 25/8; Abbot to Forshey, February 28, 1873 (copy), HP 25/10.

[43]"Report of Operations of Hydrometric Survey of Mississippi River, 1851 and 1852," diary 55, pp. 14–25, carton 1, folder 22, Caleb Goldsmith Forshey papers, the Mississippi Valley Collection, Memphis State University Library, Memphis, Tenn.

authorized rivers and harbors work increased dramatically in the post–
Civil War period. The work jumped from about $3.5 million for
forty-nine projects and twenty-six surveys in 1866 to nearly $19 million
for 371 projects and 135 surveys in 1882.[44] So much activity ought to
have delighted the army engineers, but this was not the case. To handle
the increased number of projects, Humphreys needed more officers;
but, even with the reincorporation of the topographical engineers into
the Corps of Engineers during the Civil War, the corps's authorized
strength was only 109 officers. Actually, the number of officers Humphreys commanded was smaller than that: ninety-seven, for instance,
in 1872. Humphreys himself occupied the one authorized general's
position. Even the full complement of officers, Humphreys complained, reduced the percentage of engineer officers in the army
officer ranks to the level existing in 1817, when the corps had had far
fewer responsibilities. The corps also employed about 110 civil engineer assistants, often young and inexperienced, for the many projects spread around the country.[45]

In light of the manpower shortage, Humphreys created boards that
oversaw several projects at once. He appointed his more experienced
senior officers to these boards, while younger officers became assistants, exercising local supervision and learning engineering theory and
practice. The system required senior officers to serve on numerous
boards and do much traveling in order to monitor the work of the
junior officers. According to Humphreys, in the period from 1868 to
1872 his senior officers supervised on the average not less than twenty
projects at any one time that were "scattered in almost every case, over
hundreds of miles of territory, . . . taxing the physical and mental
abilities of these officers to a degree embarrassing to the service."[46]
What he did not say—but Abbot did—was that even some of his senior
officers were inexperienced.[47] They had advanced rapidly in rank
during the Civil War but had had little or no prior experience with
rivers and harbors work. To be sure, there were a number of excellent
senior engineer officers on whom Humphreys placed an inordinately
heavy load of work. Nevertheless, just at a time when the corps needed
competent engineers, its ranks were filled with unseasoned men.

It was a time not only of increased activity for the corps but of·

[44]Edward Lawrence Pross, "A History of Rivers and Harbors Appropriations Bills,
1866–1933" (Ph.D. diss., Ohio State University, 1938), p. 44.
[45]Memorandum by Humphreys, March 1872, HP 56/9.
[46]Ibid.
[47]Abbot, "Memoir of Andrew Atkinson Humphreys," p. 212.

increased competition for engineers, and this was Humphreys's second problem. The United States Military Academy at West Point, the country's preeminent engineering school during the first half of the century, did not produce nearly so many prominent engineers after 1850. In 1866, its supervision was transferred from the corps to the secretary of war; henceforth, its superintendent could be from any branch of the army.[48] Meanwhile, other colleges began to develop engineering departments, in a short time obtaining respectability and challenging West Point for the best students, many of whom had no desire to become military officers.

Engineering societies as well as classroom professors established professional standards. The American Society of Civil Engineers, created in 1852, was reestablished in 1867, and through its professional papers and annual meetings, and the generally high quality of its leadership, exerted growing influence over the course of American civil engineering. Unfortunately for the corps, members of the ASCE, believing firmly in the superiority of civilian over military instruction, began to perceive the army engineers as inferior competitors, not colleagues.[49]

Not only professional engineers but politicians as well became critical of the broad scope of Army Corps of Engineers activities; this was the third challenge facing Humphreys. In 1873–74, the House Committee on Appropriations investigated the overlapping activities of the Army Corps of Engineers, the Smithsonian Institution, and the Department of the Interior in surveying the American West. Some sort of consolidation seemed advisable, but no one knew how to bring one about. Representative James A. Garfield, chairman of the committee, asked the various heads for their views on the subject. Humphreys justified the army's involvement by noting that corps surveys and explorations served legitimate military operations and helped in the settlement and commercial development of the territories. The corps, Humphreys claimed, should manage the surveys since it had long been interested in them and possessed officers experienced in the command and conduct of such operations.[50] Testifying before the committee, Humphreys lashed out at critics who charged that the army was no proper agency to

[48]Raymond H. Merritt, *Engineering in American Society, 1850–1875* (Lexington, Ky., 1969), p. 127.

[49]Ibid., pp. 102–5; Edwin T. Layton, *The Revolt of the Engineers* (Cleveland, 1971), pp. 29–34.

[50]Humphreys to the Hon. James A. Garfield, February 5, 1873 (copy), HP 56/23; memorandum by Humphreys, 1873, HP 56/26.

manage scientific expeditions. Army officers "are not in antagonism with the science of the country," he asserted, "but have always maintained friendly and intimate relations with it, and . . . have always been associated with the various scientific societies, and are not in an attitude of hostility toward them or the scientific institutions of the country."[51]

The committee decided against consolidation for the moment. Members thought that each agency was performing valuable work and that it would be premature to terminate any of the surveys. Nevertheless, the controversy remained. In 1878, Congress requested that the National Academy of Sciences propose how best to conduct federal surveys with "the best possible results at the least possible cost."[52] The academy responded with recommendations, among others, to terminate Corps of Engineers and Department of the Interior expeditions and to create a new United States Geological Survey Office. The report infuriated Humphreys, who resigned from the academy less than a week after all but one of the thirty-four members present (Humphreys was not) voted for it. He carried his protest to Congress, arguing that the corps could produce maps more quickly and cheaply than any other federal department. His efforts were in vain. On March 3, 1879, President Rutherford B. Hayes signed into law legislation establishing the Geological Survey within the Department of the Interior and excluding the corps from further Western exploration.[53]

Less than four months later, Humphreys and the corps suffered another legislative setback. On June 28, 1879, the president approved a bill that established the Mississippi River Commission to coordinate river improvement work on the Mississippi and to insure that both civilian advice and military advice were obtained on the subject. Humphreys, predictably, opposed the bill since he felt that the corps was entirely capable of managing the Mississippi, given sufficient manpower and money. Had not he, a corps officer, been the principal author of the basic reference book on the Mississippi? Congress, however, had become increasingly sensitive to controversies dealing with river engineering and was no longer willing to rely solely on the advice of the army engineers when it came to the Mississippi. The act stipulated that the commission president and two other members come from the corps, one from the Coast and Geodetic Survey, and three

[51]Cited in Mary C. Rabbitt, *Minerals, and Geology for the Common Defence and General Welfare, vol. 1, Before 1879* (Washington, D.C., 1979), pp. 216–17.

[52]Cited in ibid., p. 262.

[53]Ibid., pp. 275–84. On the origins of the Geological Survey, see also A. Hunter Dupree, *Science in the Federal Government: A History of Policies and Activities to 1940* (Cambridge, Mass., 1957), pp. 195–214; Thomas G. Manning, *Government in Science: The U.S. Geological Survey, 1867–1894* (Lexington, Ky., 1967), pp. 34–35, 44–47.

others, two of whom had to be civil engineers, from civilian life.[54] Two days after the bill became law, Humphreys retired from the army, having served as chief of engineers for nearly thirteen years.

* * *

Of all the blows that befell the corps during Humphreys's time as chief of engineers, none exasperated him more than the challenges to the methodology and conclusions of the Humphreys-Abbot report. His reaction to Henry's invention of the cup-type current meter was a minor, though revealing, matter. Far more consequential to the corps was the dispute between Humphreys and Eads on building jetties at the mouth of the Mississippi. Eads was convinced that jetties could provide the necessary navigational depth, while Humphreys vigorously argued against the jetties project. Old rivalries and friendships, competing egos, and changing political conditions determined the course of the dispute as much as dispassionate analyses of competing engineering theories. Congressmen, most without any engineering experience or education, debated highly technical questions in the House and Senate chambers. They called on both government and, increasingly, nongovernment experts to testify before congressional committees. Given the prominent coverage in newspapers from coast to coast, this was probably the first technical engineering dispute that became a national issue. More important for the corps, Humphreys used data and ideas first published in the Mississippi Delta report to defend his position. As a result, among professional engineers this dispute took on the character of a referendum on the report itself; and, since Humphreys had so closely tied the corps to the conclusions of the report, it also became a judgment on the corps.

Some background is necessary to understand the dispute between Humphreys and Eads. As far back as the early 18th century, there had been attempts to eliminate the bars at the mouth of the Mississippi in order to allow ocean vessels easier access to New Orleans and ports farther upstream. The Army Corps of Engineers was involved in

[54]Pross (n. 44 above), pp. 82–84; John R. Ferrell, "From Single- to Multi-Purpose Planning: The Role of the Army Engineers in River Development Policy, 1824–1930" (Washington, D.C., 1976), pp. 63–67. A revised version of this manuscript, which is a significant contribution to water resources history, will be published by the Historical Division, U.S. Army Corps of Engineers. On Humphreys's position on the establishment of the Mississippi River Commission, see his letter to the Honorable E. W. Robertson, chairman of the Committee on Levees and Improvement of the Mississippi River, House of Representatives, May 1, 1878 (print copy), HP box marked "Pamphlets. The Mississippi River."

unsuccessful attempts in 1837, 1852, and the early 1870s.[55] Pre–Civil War efforts were confined mainly to dragging iron harrows along the river's bottom to stir up the sediment, thereby allowing the current to carry the material into the Gulf. However, these efforts did not succeed for long, for the bars soon re-formed. After the Civil War, the corps attempted to dredge the passes at the mouth, but this method proved no more successful than harrowing. Months of work could be destroyed by a single night's storm. Pressed by commercial interests, in 1871 the House of Representatives requested that the corps conduct a new survey to determine whether a canal might be dug from the Mississippi to the Gulf that would circumvent the passes altogether. Although Humphreys had endorsed dredging in his Mississippi Delta report, he became an enthusiastic supporter of the canal once dredging proved unsatisfactory. He charged Captain Charles W. Howell, the New Orleans engineer officer, to conduct the survey. Not surprisingly, Howell reported favorably on a canal near Fort St. Philip which would connect the Mississippi to Breton Sound to the east. He estimated the cost at about $7.5 million, but subsequent reviews raised the figure to $10 million.[56]

Humphreys created a board of engineer officers to review Howell's report. With one significant exception, board members approved Howell's conclusions. The exception was the board's president, Brevet Major General John G. Barnard. He thought that Howell's report did not adequately address problems associated with the canal's defense, nor did he find a sufficient appreciation for the engineering difficulties involved. In response to an additional question from Humphreys on whether the canal should be constructed as an alternative or supplement to improving the mouth of the Mississippi, the board—again, with the exception of Barnard—emphatically answered that the canal was the best response to the problem. Jetties could not overcome "practical difficulties" involved in improving the passes, while dredging had already proved a failure. Barnard, however, correctly argued that Humphreys and Abbot themselves had admitted that the construction of jetties would increase "the erosive or excavating power of

[55]For the history of early efforts to eliminate the bars at the Mississippi's mouth, see the excellent dissertation by Walter M. Lowrey, "Navigational Problems at the Mouth of the Mississippi River, 1698–1880" (Ph.D. diss., Vanderbilt University, 1956). The Historical Division, U.S. Army Corps of Engineers, has acquired the literary rights to the Lowrey work and plans to publish it in the near future.

[56]Captain Charles W. Howell, Corps of Engineers, to Brig. Gen. Andrew A. Humphreys, chief of engineers, February 1873, *Annual Report of the Chief of Engineers* (1874), pp. 778–88 (hereinafter cited as *ARCE*).

the current" by increasing its velocity.[57] The Delta report simply suggested that the cost of constructing and maintaining jetties would probably be prohibitive. Yet jetties had been successful at the mouth of the Danube, like the Mississippi an alluvial river, and Barnard concluded by calling for more study on jetties before proceeding with the canal.[58]

Barnard's dispute with Humphreys—the only person who outranked him in the corps—went beyond the jetties question. Barnard had been offered the position of chief of engineers in 1864 but had turned it down because he thought Major General Richard Delafield, his senior, deserved the honor. Two years later, when Delafield retired, Humphreys, not Barnard, succeeded him. The disappointment rankled Barnard for years. He accused Humphreys of having insufficient field experience, as either a military or a civil engineer, to warrant being appointed chief of engineers.[59] Unlike Humphreys, whose work, though significant, was mainly confined to surveys before the Civil War, Barnard had actually supervised the construction of coastal fortifications and rivers and harbors projects in the 1840s and 1850s. He also planned the defenses of Tampico and surveyed the battlefields around Mexico City during the Mexican War. Later, he served for a short time as superintendent at West Point. During the Civil War, he developed the defenses for the city of Washington, became chief engineer of the Army of the Potomac, and ended the war as chief engineer "of the Armies in the field" on the staff of General Grant. Like Humphreys, he was the author of a number of significant scientific papers. For Johnson's *Universal Cyclopedia* he wrote over seventy articles, including one on jetties. In 1864, Yale College conferred on him an honorary doctor of laws.[60]

Barnard's confidant on the jetties matter was Brevet Major General Cyrus B. Comstock, his chief assistant in the early years of the Civil War. He told Comstock that Howell was an inexperienced young officer whose poorly prepared report should never have gone beyond the desk of the chief of engineers. Further, his colleagues on the

---

[57]Board of Engineers to Humphreys, January 9 and 13, 1874, and minority reports of Barnard, January 20 and 29, 1874, *ARCE* (1874), pp. 823–53. Barnard quotes from p. 489 of the Humphreys and Abbot report.

[58]Minority Report of Col. J. G. Barnard, Corps of Engineers, January 29, 1874, *ARCE* (1874), pp. 841–53.

[59]Barnard to Cyrus B. Comstock, April (?) 1874, Cyrus Ballou Comstock papers microfilm reel 2, Library of Congress, Washington, D.C. Part of this long and vituperous letter is printed in Arthur E. Morgan, *Dams and Other Disasters: A Century of the Army Corps of Engineers in Civil Works* (Boston, 1971), pp. 138–41.

[60]*Dictionary of American Biography*, s.v. "Barnard, John Gross."

examining board appointed by Humphreys were poorly qualified. As for his own report, Barnard explained, "I *did* not ask that the government should fling blindly into the jetty system. I *did* ask that it should *not* do so in the canal, and I *did* ask—setting forth dispassionate and unprejudiced reasons for [the] belief that jetties would succeed—[they] at least should be thoroughly studied."[61] Barnard was not the only army engineer who thought the jetties could succeed. However, he was the most distinguished, and his support greatly strengthened the case of Humphreys's primary antagonist, James Buchanan Eads.

Eads had first proposed jetties in May 1873, but he did not gain significant congressional support until February 1874, when he dramatically proposed that he be allowed to construct jetties at the Southwest Pass of the Mississippi at his own risk.[62] The jetties would be built to maintain a channel 28 feet deep from the Southwest Pass to the Gulf of Mexico, and they would cost no more than $10 million. The government would not pay a dollar until 20 feet had been secured, when Eads would receive $1 million. He would then receive another million dollars for each additional 2 feet secured, for a total of $5 million when the 28-foot depth had been obtained.[63]

Humphreys absolutely rejected the jetties concept. He sent Howell's recommendations and the report of the board of engineers to the secretary of war, William Belknap, with the conclusion that the jetties system "does not present either in its construction or cost superior advantages to the canal plan. One of the chief objections to the jette [*sic*] system is the unavoidable necessity of constantly extending the piers in the open sea, exposed to the full force of storms."[64] Humphreys had first raised this objection in the Mississippi Delta report. He and Abbot had argued that a "dead angle" results from "vertical eddies" which form where "the river-water meets and rises upon the sea-water . . . when the river-water begins to ascend upon the salt-water of the gulf, the rolling material is not carried with it, but is left upon the bottom in the dead angle of salt-water. A deposit is thus formed, whose surface is along or near the line upon which the fresh-water rises on the salt-water as it enters the gulf. *This action produces the bar*."[65] This fairly complicated argument can be reduced to a very simple statement: it

[61]Barnard to Comstock, April (?) 1874, Comstock papers, microfilm reel 2.
[62]Lowrey (n. 55 above), pp. 342–43, 377–78; Elmer L. Corthell, *A History of the Jetties at the Mouth of the Mississippi River* (New York, 1881), pp. 60–68.
[63]Corthell, p. 62.
[64]Humphreys to the Honorable William W. Belknap, secretary of war, February 4, 1874, *ARCE* (1874), p. 778.
[65]Humphreys and Abbot (n. 1 above), p. 478. Emphasis in the original.

was impossible to prevent the buildup of a sandbar in the Mississippi Delta. Building jetties would simply move the bar further out in the Gulf, and the jetties would have to be extended every year to keep the channel clear. The chief of engineers predicted that the jetties at Southwest Pass would have to be extended 1,200 feet each year.[66]

Eads disputed Humphreys on almost every point. There was no such thing as a dead angle; and, contrary to Humphreys's theory, there was a direct relationship between current velocity and amount of sediment suspended in the water. By constructing a system of parallel jetties which constricted the river's width, Eads maintained, he could create a sufficiently rapid river current to carry sediment out into the Gulf. The jetties would not have to be extended, nor would they fall apart as Humphreys feared. New construction techniques, successfully practiced in Holland and elsewhere, insured that jetties could withstand the onslaughts of both man and nature.[67]

It is worthwhile to inquire why Eads made his astounding proposal to do the work at his own risk. There is no conclusive evidence, but one plausible theory focuses on the reaction of the Corps of Engineers to the construction of the famous Eads bridge at St. Louis. In August 1873, in response to pleas from certain navigation interests, the secretary of war ordered Humphreys to convene a board to investigate whether the bridge being built by Eads across the Mississippi would obstruct shipping on the river. The board determined that the bridge's height posed a serious obstacle because it was too low for steamboats with high chimneys. Although the bridge was nearly finished, the army engineers recommended its abandonment and the construction of a canal spanned by a drawbridge over which trains would run. The proposal amazed Eads, who replied that steamboat chimneys could be built to smaller dimensions without impairing performance. He called in his defense thirteen riverboatmen who supported the bridge. On January 31, 1874, the board rebutted Eads in a second report reiterating its position and impugning the expertise of the riverboatmen siding with Eads. One board member, Brevet Major General Gouverneur K. Warren—a friend of Humphreys since the Delta survey days and the brother-in-law of Washington Roebling, whom Eads was then suing over alleged infringement on a caisson design—suggested that a

[66]Humphreys to Belknap, April 15, 1874, *ARCE* (1874), pp. 854–67.

[67]*Report of General Humphreys, Chief of Engineers U.S.A. (Ex. Doc. 220) Reviewed by James B. Eads, C.E.* (May 1874), and James B. Eads, *Mouth of the Mississippi Jetty System Explained* (St. Louis, 1874). Both pamphlets are in the Elmer L. Corthell papers, box 28, Brown University, Providence, R.I. Although rarely used, the Corthell collection is a very rich source of information on late 19th- and early 20th-century civil engineering developments. Corthell was Eads's resident engineer on the jetties project.

cheaper and more satisfactory bridge could have been built years ago "had not the authors of the present monster stood in the way."[68] Humphreys submitted the board's report, Eads's response, and various affidavits and supporting material to Secretary of War Belknap in March 1874, but Belknap never acted on the report. One reason, presumably, was that President Grant opposed the army engineers' interference. Another was that the bridge was nearly finished.[69]

What is most intriguing about this controversy is the timing. The bridge board submitted its second report condemning the Eads bridge only about two weeks after Barnard's board had recommended, despite Barnard's objections, that work proceed on a canal to circumvent the mouth of the Mississippi. The bridge board's adamant and unreasonable position doubtless enraged Eads, and the recommendations of the other board, directly counter to Eads's own publicly stated position, gave him the opportunity to reveal the corps's shortsightedness. The stage was set for Eads's confrontation with Humphreys and for one of the most significant experiments in the history of American river engineering.

The feud between Eads and Humphreys attracted public attention for many reasons. It pitted St. Louis interests, favorable to Eads, against New Orleans businessmen, generally supportive of the canal. The American Society of Civil Engineers provided a professional forum—its annual meeting and published *Transactions*—to debate the enterprise. Indeed, members of the ASCE visited the jetties during their annual meeting in New Orleans in 1877.[70] Midwest interests followed the debate because they were concerned with navigation improvement which would allow more tonnage on the upper Mississippi, the Missouri, and other Western rivers. Elsewhere, people wondered whether jetties could solve problems at the mouths of other

[68]"Personal statement of Maj. G. K. Warren, Corps of Engineers," January 16, 1874, appended to the supplementary report of Board of Engineers, January 31, 1874, *ARCE* (1874), p. 679. For more on Warren's involvement in the dispute between Washington Roebling, the builder of Brooklyn Bridge, and Eads, see David McCullough, *The Great Bridge* (New York, 1972), pp. 346–47, 457–61. Also see Florence L. Dorsey, *Road to the Sea: The Story of James B. Eads and the Mississippi River* (New York, 1947), p. 152; Morgan (n. 59 above), pp. 112–13.

[69]For more on Eads and the famous Eads bridge, see John A. Kouwenhoven, "The Designing of the Eads Bridge," *Technology and Culture* 23 (October 1982): 535–68; Kouwenhoven, "James Buchanan Eads: The Engineer as Entrepreneur," in *Technology in America: A History of Individuals and Ideas*, ed. Carroll W. Pursell, Jr. (Cambridge, Mass., 1981), pp. 80–91; C. M. Woodward, *A History of the St. Louis Bridge containing a Full Account of Every Step in Its Construction and Erection, and Including the Theory of the Ribbed Arch and the Tests of Materials* (St. Louis, 1881).

[70]Lowrey (n. 55 above), p. 444.

rivers. Probably the biggest reason for interest in the feud was because it matched a self-made engineer, with only a few years of formal education, against the chief of engineers, a West Point graduate who had coauthored a book on hydraulic engineering that had been honored around the world. It was David against Goliath—at least Eads thought so. The people who bet on David were right.

Congressional support for the jetties came slowly. In June 1874, the Senate forced the House, which supported Eads's plan, to agree only to the establishment of a board to investigate the various engineering reports and recommend the best approach. The board, appointed by the president, included Generals Horatio G. Wright, Barton S. Alexander, and Cyrus Comstock of the Army Corps of Engineers, Professor Henry Mitchell of the Coast Survey, and civil engineers T. E. Sickles, W. Milnor Roberts, and H. D. Whitcomb. Members visited the Mississippi Delta and also traveled to Europe to inspect river improvement projects there. They were particularly interested in the jetties at the mouth of the Danube that had been designed by Charles A. Hartley, an English engineer, in the late 1850s. In January 1875, the board submitted its report. All but General Wright agreed that jetties afforded the most likely chance of success. However, they suggested that the jetties be built at the smaller South Pass rather than the Southwest Pass favored by Eads.[71] Hartley had first suggested this alternative to Barnard the year before, and Barnard had mentioned the advantages of South Pass in his dissent to the recommendations of the board over which he had presided.[72] According to the 1875 report, jetties at South Pass could be constructed and maintained for just under $8 million, while jetties at Southwest Pass would cost twice as much. Given this fact and their judgment that South Pass was perfectly adequate for "present and prospective wants of commerce," board members saw no reason for work at Southwest Pass.[73]

In March 1875, Congress passed a bill authorizing Eads to build jetties at South Pass. Eads was to begin construction within eight months after adoption of the act and had to secure a depth of 20 feet through South Pass within thirty months. Payments would depend on Eads's reaching specified depths. By the time the required final 30-foot depth and 350-foot width had been attained, Eads would have been

[71]Board of Engineers to Belknap, January 13, 1875, *ARCE* (1875), pp. 948–59.

[72]Hartley's early contacts with Barnard are discussed in an article by his grandnephew, C. W. S. Hartley, "Sir Charles Hartley and the Mouths of the Mississippi," *Louisiana History* 24, no. 3 (Summer 1983): 261–87. C. W. S. Hartley possesses Charles Hartley's unpublished diaries and makes extensive use of them. See also Barnard's mention of Hartley in *ARCE* (1874), p. 852.

[73]Board of Engineers to Belknap, January 13, 1875, *ARCE* (1875), p. 953.

paid $4.25 million. For twenty years thereafter the government would pay Eads or his estate $100,000 annually to maintain the 30-foot depth. If the depth were maintained, Eads would receive another $.5 million at the end of ten years and another $.5 million at the end of the twenty.[74]

During the year before passage of this act, Eads and Humphreys had waged a constant war of words. At his own expense, Eads published his correspondence with various engineers—including Humphreys—congressmen, and other public officials, all of which showed his careful attention to the jetties problem. Meanwhile, Humphreys denounced the jetties to the secretary of war and congressional committees.[75] He wrote three memoranda at the beginning of 1875 and a fourth the following August attacking Eads's project, and he printed these in the 1875 *Annual Report of the Chief of Engineers.*[76] The controversy degenerated into personal attacks. Eads made unfair accusations and inaccurate claims.[77] To some degree both men were right. The barrage of charges and countercharges insured that the jetties project would continue to receive public attention. Eads also suffered from financial troubles; twice, in 1878 and 1879, Congress, encouraged by Eads's progress, agreed to modify the contract and advance funds to him.[78] "Will you, Senators," cried William Windom of Minnesota to his colleagues, "consent to bankrupt this man who has done more for the commerce of this country than any other man since DeWitt Clinton?"[79] The Senate could not withstand such eloquence; neither could the House.

Eads was careful about obtaining advice on his project from both

[74]*Law of the United States Relating to the Improvement of Rivers and Harbors from August 11, 1790 to March 4, 1913* (Washington, D.C., 1913), 1:245–50. Section 4 of the 1875 Rivers and Harbors Act pertains to Eads.

[75]*Correspondence between the Business Men of New Orleans and James B. Eads,* privately printed pamphlet (April 1874), box 89, Corthell papers; Eads, *Mouth of the Mississippi* (n. 67 above); *Report of General Humphreys* (n. 67 above); *Mouth of the Mississippi. Correspondence &c. 1874–5* (London, 1875), box 28, Corthell papers. This pamphlet was evidently published by Hartley and contains mainly letters between Hartley and Eads; Humphreys to Belknap, April 15, 1874, *ARCE* (1874), pp. 854–67; Lowrey (n. 55 above), pp. 384–86.

[76]"Improvement of the Entrance to the Mississippi River by Jetties," by A. A. Humphreys, Brig. Gen. and Chief of Engineers, *ARCE* (1875), pp. 959–75. Humphreys also printed these memoranda in the 1876 edition of *Physics and Hydraulics,* pp. 678–91. He attacked Eads one more time in his May 1, 1878, letter to E. W. Robertson opposing the establishment of the Mississippi River Commission. See n. 55 above.

[77]Lowrey (n. 55 above), pp. 384–413.

[78]Ibid., pp. 445–61.

[79]Cited in ibid., p. 461.

military and civil engineers. He persuaded Generals Barnard and Alexander to join an advisory board to review work on the jetties. Sir Charles Hartley, Professor Mitchell, and civil engineers Sickles, Roberts, and Whitcomb were also members. Comstock had refused Eads's invitation because the secretary of war had already appointed him the official government inspecting officer to monitor progress on the jetties. Eads then suggested that Comstock simply meet with his advisory board, but Comstock once more demurred and informed Humphreys of Eads's overture. Humphreys was outraged. "The effrontery of Mr. Eads," he fumed, "is on a par with that of Tweed, Gould, and Fiske [*sic*]."[80]

On July 10, 1879, ten days after General Humphreys's retirement, Captain Micah Brown of the Army Corps of Engineers certified that Eads had satisfied the requirements of his contract. A 26-foot-deep channel existed at the head of the passes and a 30-foot-deep channel between the jetties.[81] The people of New Orleans, suspicious and skeptical four years before, were exultant. "There is no achievement of mechanical genius," trumpeted the New Orleans *Daily Times*, "which compares with it in the splendor of its economies or in the magnitude of its results."[82] Eads was vindicated. Had he lived long enough, he no doubt would have enjoyed an 1899 army engineers report which recommended that jetties be built at Southwest Pass—Eads's choice all along—to form a 35-foot-deep navigable channel. The South Pass channel was no longer adequate for the increased volume of shipping going up the Mississippi.[83] Congress authorized the channel in 1902.[84]

Eads's victory culminated a series of events which considerably diminished the corps's responsibilities and reputation. First, Congress took away the corps's right to conduct scientific expeditions in the West. Then it weakened the corps's authority on the Mississippi River by creating the Mississippi River Commission. Finally, the success of the jetties showed that the corps was far from perfect in river engineering (and that the Humphreys-Abbot report was seriously flawed), thus

[80]Eads to Comstock, August 7, 1875; Eads to Comstock, August 25, 1875; and Humphreys to Comstock, August 27, 1875, all in the Comstock papers, microfilm reel 1. Eads to Comstock, telegram, August 17, 1875, Comstock papers, microfilm reel 2. Humphreys's comment is in the August 27 letter.

[81]Lowrey (n. 55 above), pp. 462–63.

[82]Cited in ibid., p. 463.

[83]"Survey Relative to the Practicability of Securing a Navigable Channel of Adequate Width and of 35 Feet Depth at Mean Low Water of the Gulf of Mexico throughout Southwest Pass, Mississippi River," printed as House Document 142, 55th Cong., 3d sess., *ARCE* (1899), pp. 1863–1951.

[84]*Laws of the United States Relating to the Improvement of the Rivers and Harbors from August 11, 1790 to June 29, 1938* (Washington, D.C., 1940), 2:61–62.

encouraging private civil engineers who wished to break the corps's monopoly on federal public works projects.

Throughout the 1880s, private engineers attacked the corps's competence. It is clear from the literature, however, that these engineers were at least as concerned about their exclusion from positions of responsibility on public works projects, and consequent loss of work, as about the alleged poor performance of military engineers. Articles against the corps appeared in *Lippincott's Magazine* (1883), *The Iron Age* (1885), *Engineering and Mining Journal* (1885 and 1892), *Engineering News and American Contract Journal* (1886), *Forum* (1887), and *The Engineering Magazine* (1892).[85] A civil engineers' convention in Cleveland in 1885 urged Congress to establish a "Civil Bureau of Public Works" to develop a "comprehensive system of public works." Presidentially appointed members would include three military engineers, three civil engineers, and one member of the legal profession.[86] The following year, again in Cleveland, twenty-one civil engineering societies and clubs formed a "Council of Engineering Societies on National Public Works" with the object of promoting "an improved system of national public works."[87] The 1886 meeting of the American Society of Civil Engineers occasioned a discussion on the South Pass jetties, and civil engineers took the opportunity of explicitly arguing that Eads's success showed the need for drawing on the talent of graduates from engineering schools other than West Point.[88]

[85]Frank D. Y. Carpenter, "Government Engineers," *Lippincott's Magazine* 32 (1883): 159–68; "A New Department of Government," *Iron Age*, February 19, 1885; series of letters exchanged between "C.U.E." and "Spectator" on "Civil vs. Military Engineers," *Engineering and Mining Journal*, March 7, 14, and 28, 1885, and April 4, 1885; "Re-Organization of the Scientific Bureaus of the General Government," *Engineering News and American Contract Journal* 15 (1886): 104–6; W. F. Smith, "A New Executive Department," *Forum* 3 (May 1887): 296–304; W. R. King, "Government Engineering Defended," *Engineering Magazine* 2 (February 1892): 664–74; George Y. Wisner, "Worthless Government Engineering," *Engineering Magazine* 2 (January 1892): 427–34, and, by the same author, "Worthless Government, A Rejoinder," *Engineering Magazine* 2 (March 1892): 743–52; "Army and Civil Engineers," *Engineering and Mining Journal* 53, no. 11 (March 12, 1892): 307; Raymond W. Rossiter, "The Civil Engineers and the Public Works of the United States," ibid., p. 298.

[86]*Proceedings of the Civil Engineers Convention*, Bulletin 2, Cleveland, December 3–5, 1885, p. 4.

[87]*Proceedings of the Council of Engineering Societies, on National Public Works, Held at Cleveland, Ohio, March 31st, April 1st and 2d, 1886*, Bulletin no. 13, p. 8. One of the best sources of old engineer society proceedings is the pamphlet collection of the Corthell papers at Brown University.

[88]"Discussion on the South Pass Jetties. —Ten Years' Practical Teachings in River and Harbor Hydraulics," *American Society of the Civil Engineers Transactions* 15, no. 325 (April 1886): 223–336. Participants included Gen. Comstock, Gen. Quincy A. Gillmore (first president of the Mississippi River Commission), and Col. William E. Merrill (in charge of Ohio River improvements), all of the Corps of Engineers, and Corthell and Eads.

A series of bills introduced into Congress reflected the engineers' growing political power. In 1885, Congressman Clifton R. Breckinridge of Arkansas proposed the creation of a department of rivers and harbors which would be supervised by a commission of rivers and harbors appointed by the president. All Army Corps of Engineers rivers and harbors work would be transferred to this new department. The following year, Congressman John F. King of Louisiana introduced a bill to create a cabinet-level department of public works that would replace the various scientific and public works agencies within the federal establishment, including the civil works functions of the Corps of Engineers.[89]

In 1888, Breckinridge joined with Senator Shelby M. Cullom of Illinois in sponsoring legislation to establish a bureau of harbors and waterways "to be officered by a corps to be known as the Corps of United States Civil Engineers." In a concession to opponents, Breckinridge and Cullom agreed that the new bureau be under the secretary of war, just as the Corps of Engineers was. Their bill had more appeal than its predecessors, and the House Committee on Expenditures in the War Department recommended passage of a substitute bill very nearly the same as the original. Nonetheless, the bill never came to a vote in either the House or the Senate.[90] Supporters, however, continued to call for reforms in the development and administration of the nation's waterways. Their demands contributed to the growth of the Progressive conservation movement at the turn of the century.[91]

Eads's accomplishment at the mouth of the Mississippi seriously undermined the credibility of Humphreys and the Army Corps of

[89]H. R. 7855, 48th Cong., 2d sess., "A bill to provide for the creation of a River and Harbor Department and for other purposes," introduced by Breckinridge; H. R. 9628, 49th Cong., 1st sess., "A bill creating a Department of Public Works," introduced by King. Copies of both bills are in the Corthell papers, box 140.

[90]The Cullom-Breckinridge bill was introduced on January 16, 1888, as S. 1448 and H. R. 4923. The bill is reprinted and discussed in a pamphlet published by the Engineers' Club of St. Louis, *The Proposed Change of Plan in the Execution of all River and Harbor Improvements. The Organization of a Civil Bureau of Harbor and Waterways. Arguments on the Cullom-Breckinridge Bill* (St. Louis, 1888?), pp. 13–19, box 140, Corthell papers; Committee on Expenditures in the War Department, House of Representatives, "Expenditures in the War Department," 50th Cong., 2d sess., Report no. 4093, February 19, 1889, box 140, Corthell papers.

[91]See Samuel P. Hays, *Conservation and the Gospel of Efficiency: The Progressive Conservation Movement, 1890–1920* (Cambridge, Mass., 1959). The Cullom-Breckinridge bill did have one lasting result. One of its provisions was to locate department engineers permanently around the country to supervise projects in their areas. Partly in response to this proposal, Brig. Gen. Richard C. Drum, the adjutant general, issued General Order No. 93 on November 8, 1888, creating five engineer divisions (not departments). Divisions have remained a part of the corps's organizational structure to the present day.

Engineers. If some corps officers felt slightly paranoid, it was under-standable. Their integrity had been questioned and their missions reduced. Still, many officers continued to hold the Humphreys-Abbot report in high esteem and defended the one conclusion that continued to enjoy widespread, if not unanimous, support: "levees only" could control floods on the Mississippi.

*   *   *

One of the common popular misconceptions about the corps, or about any large bureaucracy, is that it is a monolithic organization in which personnel unquestioningly accept official policy. This is prob-ably not true of most bureaucracies and has rarely been true of the corps. The "levees only" policy was questioned, but it remained corps philosophy until the Mississippi flood of 1927.

Humphreys thought that levees would not keep the water from rising but would, if sufficiently high, prevent flooding. Eads argued that levees would actually lower the bed of the river because they would allow floodwater to scour a deeper channel. Closing all gaps in the levees and then imposing a uniform width on the river by narrowing wide places would, Eads thought, eventually secure a deep enough depth to render levees unnecessary. In short, Eads sought to apply to the main stem of the Mississippi the same principles he had successfully applied to its mouth. Of course, the riverbanks would have to be protected through various bank stabilization devices. As a charter member of the Mississippi River Commission, Eads insisted on filing a minority report in 1882, when he thought the commission's report did not state emphatically enough the necessity of closing all outlets along the river and repairing the levees without delay. Twentieth-century research has proved both Humphreys and Eads wrong. Levees alone have never adequately confined the river, but neither have they re-sulted in a significant deepening of the channel.[92]

---

[92]Eads makes his point about levees causing channel deepening in *Physics and Hydrau-lics of the Mississippi River* (New Orleans, 1876). See also his letter to Sen. William Windom, St. Louis, March 15, 1874, reprinted by the Executive Committee, Interstate Mississippi River Improvement and Levee Association, in *The Mississippi River: Its Hydraulics, Value, and Control* (Washington, D.C., 1890), pp. 1–4. James B. Eads to Gen. Quincy A. Gillmore, president, Mississippi River Commission, April 12, 1882, part of the 1882 Mississippi River Commission Report, *ARCE* (1882), pp. 2766–80. The report was separately printed in Washington, D.C., in 1882 and distributed as a pamphlet. A copy is in the Corthell papers, box 96, II. See also Albert E. Cowdrey, *This Land, This South: An Environmental History* (Lexington, Ky., 1983), pp. 122–23; Brien R. Winkley, *Man-Made Cutoffs on the Lower Mississippi River: Conception, Construction, and River Response* (Vicks-burg, Miss., 1977), pp. 120–22.

While Barnard supported the reliance on "levees only,"[93] Comstock did not. As president of the Mississippi River Commission, he told the Senate Committee on Commerce in 1890 that the construction of levees was not sufficient. To relieve the pressure on levees during floods, water must be allowed to escape from the Mississippi down the Atchafalaya River in Louisiana and into the Gulf of Mexico.[94] Captain John Millis of the fourth Mississippi River Commission district went so far as to recommend the construction of two artificial outlets in the levees to allow the dispersion of floodwaters.[95] The question of outlets was probably the most difficult one facing engineers on the Mississippi. One suggestion which attracted much attention was the construction of a controlled outlet from the Mississippi to Lake Pontchartrain north of New Orleans. Most engineers agreed that this would reduce flood heights, but some thought, as had Humphreys and Abbot, that the sediment deposits would silt in the lake. Colonel Curtis M. Townsend, president of the Mississippi River Commission, concluded in 1914 that such an enormous amount of dredging would be required as to render the project impractical.[96] Despite periodic major flooding on the lower Mississippi and the minority views of Comstock, Millis, and others, the commission stood by the "levees only" policy.

Humphreys and Abbot were challenged not only because of their

[93]Barnard had early decided about the primacy of levees. His article, "Outlets and Levees of the Mississippi River," advocating levees, first appeared in the New Orleans *Picayune* and then was republished in the October 1859 issue of *De Bow's Review.* The article, actually a series of letters to G. W. R. Bayley, the state engineer of Louisiana, was also reprinted in pamphlet form, and a copy is now in the possession of the Historical Division, Office of the Chief of Engineers, Washington, D.C.

[94]Benjamin G. Humphreys, *Floods and Levees of the Mississippi River* (Washington, D.C., 1914), p. 239. Appendix E (pp. 221–47) reprints testimony which Comstock gave before the Senate Committee on Commerce on May 12, 1890. Humphreys was no relation to Andrew A. Humphreys. Rather, he was the son and namesake of the first elected governor of Mississippi after the Civil War.

[95]Col. John Millis to John R. Freeman, no date, but noted as received on July 5, 1922, John Ripley Freeman papers (MC 51), Institute Archives and Special Collections, Massachusetts Institute of Technology Libraries, Cambridge, Mass. Millis's letter is actually a memorandum dealing with his time as a district engineer for the Mississippi River Commission from 1890 to 1894.

[96]Col. C. McD. Townsend, "The Flow of Sediment in the Mississippi River and Its Influence on the Slope and Discharge, with Especial Reference to the Effect of Spillways in the Vicinity of New Orleans, La.," *Professional Memoirs* 7, no. 33 (May–June 1915): 357–77. The article was originally presented as a talk before the Mississippi River Commission in July 1914. *Professional Memoirs* was a bimonthly publication of the United States Army Engineer School, Washington Barracks, District of Columbia. Townsend was a staunch defender of the "levees only" policy. He clearly expressed his views on the subject in an address in Memphis, Tenn., on September 26, 1912. The address was later printed as Senate Document 1094, 62d Cong., 3d sess., February 22, 1913.

rejection of outlets. Hiram Chittenden, a corps officer who later gained fame as the developer of the Yellowstone Park road system, thought other conclusions also required modification. For instance, Humphreys and Abbot had argued that reservoirs were not practical for controlling floods in the lower Mississippi basin because of the basin's rainfall pattern and topographical peculiarities; there were not enough potential reservoir sites to merit consideration.[97] Chittenden responded that an elaborate system of reservoirs could be developed in the Mississippi basin that would significantly reduce flooding along the lower Mississippi. He identified fifty-two potential reservoir sites on tributaries of the Mississippi. However, he thought that flood control alone would never justify construction.[98] The difficulty, he wrote, "was not so much a physical as a financial one."[99] Later in his career, Chittenden also argued against Humphreys's assertion that cutoffs in the Mississippi to shorten the river were impractical.[100] It was only in the 1930s, when experiments at the corps's hydraulic laboratory in Vicksburg, Mississippi, proved the value of cutoffs, that corps policy on this matter changed.[101]

Marshall O. Leighton, chief hydrographer of the United States Geological Survey, took Chittenden's argument about reservoirs one step further. In two influential papers published in 1908, Leighton

[97]Humphreys and Abbot (n. 1 above), pp. 407–9.

[98]Captain Hiram M. Chittenden, "Examination of Reservoir Sites in Wyoming and Colorado," *ARCE* (1898), pp. 2815–2922 (see esp. pp. 2862–64); Gordon B. Dodds, *Hiram Martin Chittenden: His Public Career* (Lexington, Ky., 1973), pp. 24–41.

[99]Chittenden, p. 2863. The corps was not absolutely opposed to reservoirs. Pursuant to congressional authorization, army engineers built five small timber dams in Minnesota on the headwaters of the Mississippi between 1881 and 1895. The resulting reservoirs were justified as aids to navigation, although they may have been more beneficial to waterpower interests. See Raymond H. Merritt, *Creativity, Conflict and Controversy: A History of the St. Paul District, U.S. Army Corps of Engineers* (Washington, D.C., 1980), pp. 71–80.

[100]"Final Report of the Special Committee on Floods and Flood Prevention," *American Society of Civil Engineers Transactions* 81 (1917): 1256–57. Chittenden's comments on cutoffs are contained in his critique of the final report of the special committee.

[101]Morgan (n. 59 above), pp. 240–51; L. E. Lyon, "Changes in Length of the Lower Mississippi," *Military Engineer* 24, no. 137 (September–October 1932): 458–62; "New Plans for the Mississippi: General Review of Present Program," *Engineering News-Record* 110, no. 25 (June 22, 1933): 795–801; George R. Clemens, "Straightening the Father of Waters," *Engineering News-Record* 116, no. 8 (February 20, 1936): 269–76; Maj. Gen. J. L. Schley, "Mississippi Control Works Tested by Flood," *Engineering News-Record* 15 (April 14, 1938): 533–37; George R. Clemens, "Cutoffs Lower Flood Crests," *Engineering News-Record* 121, no. 20 (November 17, 1938): 608–14; Harley B. Ferguson, "Construction of Mississippi River Cut-Offs," *Civil Engineering* 8, no. 11 (November 1938): 725–29; Harley B. Ferguson, "Effects of Mississippi River Cut-Offs," *Civil Engineering* 8, no. 12 (December 1938): 826–29.

claimed that, at least in the case of the Ohio River, reservoirs were not only technically feasible but economically justified.[102] He specifically attempted to rebut W. Milnor Roberts's 1857 essay, which he thought had "well-nigh crystallized" government policy.[103] A supporter of multipurpose river development, Leighton identified benefits to be gained from waterpower and navigation as well as flood control. This argument did not convince the corps, whose officers believed in the primacy of navigation interests and generally opposed multipurpose river development.[104]

Many engineers agreed with the "levees only" policy. When John A. Ockerson, a respected civil engineer and a member of the Mississippi River Commission, testified before the Senate Committee on Commerce in 1914, he was asked his opinion on the best means to prevent floods. Without hesitation, he replied, "From what I have seen of other rivers and also of the Mississippi, I am convinced that there is only one way to control floods and prevent overflow, and that is by means of levees."[105] William M. Black, a retired major general and former chief of engineers, stated before the American Society of Civil Engineers in 1926, "Researches into the laws governing the flow of water in streams date back through the centuries, but it is hardly too much to claim for Humphreys and Abbot that our present knowledge of stream flow is largely based on their pioneer work in the Mississippi, carried on during the middle of the past century."[106]

The following spring the most devastating flood ever to hit the lower Mississippi killed over 250 people, flooded 16 million acres, and destroyed 200,000 buildings and homes. The flood also destroyed the credibility of the Humphreys-Abbot report. Major General Edgar

[102]M. O. Leighton, "Relation of Water Conservation to Flood Prevention and Navigation in Ohio River," *Preliminary Report of the Inland Waterways Commission,"* Senate Document 325, 60th Cong., 1st sess., February 26, 1908, pp. 451–90; M. O. Leighton and A. H. Horton, *The Relation of the Southern Appalachian Mountains to Inland Water Navigation,* U.S. Department of Agriculture Forest Service, Circular 143 (1908). For more on Leighton's influence, see Hays (n. 91 above), pp. 107–8; Ferrell (n. 54 above), pp. 127–28.

[103]Leighton, p. 452.

[104]Hays (n. 91 above), pp. 208–18.

[105]"Hearings before a Subcommittee of the Committee on Commerce, United States Senate, on S. 2, a Bill Appropriating Funds to Prevent Floods on the Mississippi River and to Improve Navigation Thereon," Senate Committee on Commerce, 63d Cong., 2d sess. (Washington, D.C., 1914), p. 140.

[106]William M. Black, "Some of the Progress of Waterways Engineering During the Past Half Century," address given at the convention of the American Society of Civil Engineers, Philadelphia, October 8, 1926. A typed draft is in the William M. Black papers, file 57 ("Correspondence and Addresses, October–December 1926"), Historical Division, Office of the Chief of Engineers, Washington, D.C.

Jadwin, the chief of engineers, devised a new plan which added flood-ways and spillways to the flood control measures to be used along the lower Mississippi. No more were levees to be the only source of security; the water had to be dispersed as well as confined.[107]

Nevertheless, Jadwin stood firmly in the tradition of Humphreys and Abbot in his opposition to reservoirs. He established a special reservoir board of engineer officers to examine the subject, and the board concluded that Jadwin's plan was "far cheaper than any method the board has been able to devise for accomplishing the same result by any combination of reservoirs."[108] Still, the idea of locating reservoirs on the lower Mississippi was far from dead. The corps's position became politically unpopular in the midst of growing unemployment resulting from the Great Depression. Public works projects, once considered uneconomical, began looking very attractive as a means of employment. Moreover, many politicians felt that flood control was essential to protect human life, regardless of cost. Mainly reacting to this political interest, the corps reversed its position on a number of flood control projects, including several in the lower Mississippi basin. Revised reports concluded that the necessity for "public-work relief" and the suffering caused by recurring floods provided grounds for construction. Engineering judgment was subordinated to political requirements.

The 1936 Flood Control Act recognized that flood control was a "proper activity of the Federal Government in cooperation with States, their political subdivisions, and localities thereof."[109] Congress gave responsibility for federal flood control projects to the Army Corps of Engineers. Pursuant to congressional authorization and appropriation, in the following years the corps built over 300 multipurpose reservoirs whose primary benefit was flood control. During high water, they retarded downstream flooding by storing water. At other times dam operators released water for navigation, irrigation, hydropower, or water supply purposes. It is, in fact, inconceivable that many of these reservoirs would have been built had flood control been the only benefit.

It is noteworthy that so many army engineers maintained their skepticism of the value of flood control reservoirs despite the windfall of work Congress gave the corps in the 1930s. Brigadier General

[107]Martin Reuss, "The Army Corps of Engineers and Flood Control Politics on the Lower Mississippi," *Louisiana History* 23, no. 2 (Spring 1982): 131–148; see 131–32.

[108]Ibid., p. 142; House Flood Control Committee Document 2, 70th Cong., 1st sess., p. 33.

[109]*Laws of the United States Relating to the Improvement of Rivers and Harbors from August 11, 1790 to January 2, 1939* (Washington, D.C., 1940), 3:2404.

Harley B. Ferguson, president of the Mississippi River Commission and a recognized expert in flood control, stated that reservoirs in the lower Mississippi basin "never were justified except for work relief."[110] Some corps engineers, both military and civilian, simply shared the skepticism of many private civil engineers who thought it difficult, if not impossible, to operate a flood control reservoir as a multipurpose project.[111] According to Gerard H. Matthes, the senior engineer with the Mississippi River Commission, even single-purpose flood control reservoirs posed significant "practical operating difficulties."[112] While such reservoirs can perform quite well in small watersheds, such as the Miami Valley in Ohio, they were ill suited in "large drainage basins, or in any flood-control system in which a large number of dams and reservoirs are required, or where the tributary system is at all complex."[113] A pamphlet, "Notes on Flood Control," circulated within the Office of the Chief of Engineers in August 1936, two months after passage of the Flood Control Act, identified four methods of flood control: levees, enlarging the discharge capacity, providing additional channels, and constructing reservoirs. The pamphlet then noted: "Of the four methods of controlling floods mentioned above, construction of levees is the most direct and surest method. . . . Works, such as reservoirs, constructed at localities distant from areas damaged by floods are not so determinate as to effects, and the benefits of reservoirs become smaller and smaller as distances from reservoir sites increase. As a consequence, a dollar spent for levee construction is more likely to be a dollar well spent than a dollar spent for other methods of flood control."[114]

Humphreys's influence waned very slowly. Only after World War II did the Engineer School at Fort Belvoir, Virginia, publish a booklet which listed reservoirs as a flood control option, without suggesting that it was necessarily the least attractive alternative.[115]

[110]Ferguson to chief of engineers, April 23, 1938 (copy), Technical Records A-2, 1928–1943, Mississippi River Commission, Vicksburg, Miss.

[111]"Final Report of the Special Committee on Floods and Flood Prevention," pp. 1224–25; Hiram M. Chittenden, "Detention Reservoirs with Spillway Outlets as an Agency in Flood Control," *American Society of Civil Engineers Transactions* 82 (1918): 1473–92.

[112]Matthes's comments on an article by George R. Clemens, "The Reservoir as a Flood-Control Structure," *American Society of Civil Engineers Transactions* 100 (1935): 921.

[113]Ibid.

[114]"Notes on Flood Control," Office of the Chief of Engineers, Washington, D.C., August 1936, pp. 1–2, Samuel D. Sturgis, Jr., papers, file labeled "Flood Control. Capt. Samuel D. Sturgis, Jr., 1936–39," Historical Division, Office of the Chief of Engineers, Washington, D.C. Sturgis was chief of engineers from 1953 to 1956.

[115]United States Army Engineer School, *Flood Control: Orientation and Methods* (Fort Belvoir, Va., 1949), pp. 15–19.

* * *

The question remains, why did the majority of army engineers accept the "levees only" policy for so long? The support some private engineers gave to the policy was surely one reason. Humphreys's own stature as a scientist and chief of engineers was another. The Humphreys-Abbot report also seemed an apt response to those civil engineers, such as John R. Freeman and Arthur S. Morgan, who argued, not without reason,[116] that civil engineering education at West Point was inferior to that given at the better nonmilitary colleges and universities.[117] Humphreys and Abbot took the classical French and Italian engineering theory they had learned as cadets in the Military Academy and applied it to a practical problem: preventing the inundation of the Mississippi Delta. In so doing, they compiled a massive amount of data and, like good scientists, searched for the underlying formula that would enlarge mankind's understanding of the physical universe. Although they failed in this attempt, they still won the respect of other engineers who used their data and insights.

Another reason for the corps's embracing the report is that the "levees only" policy was an attractive political solution. Many engineers and scientists supported it, while Delta residents naturally preferred confining floods between levees to dispersing floodwater over a wide area. Instead of rendering valuable farmland useless through the construction of floodways and diversion channels, levee construction

[116]An analysis of the development of civil engineering courses at West Point is beyond the scope of this article. It deserves a special study of its own, especially in light of the enormous influence army engineers have had on the development of this country's resources. One fact, however, can be noted here without comment. From approximately 1830 to 1930, the Military Academy used only four different civil engineering textbooks. They were written by Dennis Hart Mahan (1837), Junius B. Wheeler (1876), Gustav J. Fiebeger (1904), and William A. Mitchell (1926). Each of these textbooks had only a minimum amount of information on hydraulic engineering.

[117]Freeman, a former president of the American Society of Civil Engineers, denounced the training of army engineers partly because of the corps's opposition during most of the 1920s to the establishment of a federal hydraulics laboratory, a project in which Freeman passionately believed. In a letter to the House Committee on Rivers and Harbors, May 25, 1928, Freeman wrote, "The U.S. Army Engineers by education first of all are to be military engineers. By training, office, tradition and routine of frequent transfer they are not specially qualified for such scientific research. . . . *It is the system that fails to train for research.* The personality of the U.S. Army Engineers averages exceptionally high" (emphasis in the original). Freeman papers. Part of this letter is quoted in Morgan (n. 59 above), pp. 216–20. Morgan, organizer of the Miami Conservancy District and first chairman of the Tennessee Valley Authority, leaves Freeman far behind in pouring scorn on civil engineering education at West Point. See *Dams and Other Disasters*, pp. 2–20.

actually held out the promise of reclaiming previously unusable land. There is no evidence that Humphreys planned a report with such obvious political appeal. Quite the contrary, he insisted on a rigorous, unbiased approach to the work. However, when he finally did arrive at the levees only policy, he firmly put his reputation behind it and defended it before critics within and outside of Congress.

In 1850, Congress appropriated $50,000 "For the topographical and hydrographical survey of the Delta of the Mississippi, with such investigations as may lead to determine the most practicable plan for securing it from inundation, and the best mode of so deepening the passes at the mouth of the river as to allow ships of twenty foot draft to enter the same."[118] Few congressmen imagined that they were sponsoring a survey that would address much of what was then called the "science of hydraulics." Many, no doubt, would have been content with a short overview of theory, an explanation of methodology, and a presentation of conclusions. Instead, they received a treatise that quickly gained an international reputation. It was a triumph for Humphreys and Abbot, for the Army Corps of Engineers, and ultimately for American science. However, Humphreys stifled instead of stimulated further research in river engineering, for he defended his report against every criticism. Other engineer officers supported him, particularly when, for a number of disparate reasons, both politicians and private engineers seemed intent on reducing the corps's role. Faced with myriad questions about their competence, engineer officers used the report to validate their education and capabilities. Their very human reaction was understandable—but unfortunate. They lost sight of that commitment to truth that is the basis of all scientific research and on which the Humphreys-Abbot report rested. With rare exceptions, basic research in river behavior was left to private and academic engineers in the United States and elsewhere. Army engineers confined their attention to structural matters, such as the design of locks and dams. Only under political pressure, and in the aftermath of the 1927 flood, did the corps finally build a hydraulic laboratory at Vicksburg in 1929, and in the next decade an outstanding experimental program was established there.[119] But some sixty years had been wasted.

[118]*Laws of the United States Relating to the Improvement of Rivers and Harbors from August 11, 1790 to March 4, 1913* (Washington, D.C., 1913), 1:116.

[119]Rouse and Ince (n. 2 above), p. 227; Rouse (n. 2 above), p. 108; Morgan (n. 59 above), pp. 185–239. The Freeman papers at MIT contain a large number of documents dealing with the controversy over the establishment of a hydraulic laboratory.

Institutions

ENGINEERS AND THE NEW SOUTH CREED: THE FORMA-TION AND EARLY DEVELOPMENT OF GEORGIA TECH

JAMES E. BRITTAIN AND ROBERT C. MC MATH, JR.

In the last quarter of the 19th century fundamental changes oc-curred in the training of American engineers. At the same time major shifts were occurring in the economy of the American South. These two trends, the development of professional engineering education and the beginning of an industrialized New South, converged in the 1880s in the establishment of the Georgia School of Technology. This paper will show how changing ideas about engineering education in America influenced the beginnings of Georgia Tech, how the estab-lishment of the Georgia school supported the large strategy of indus-trializing the South, and how the early development of Georgia Tech compared with experiments in engineering education elsewhere in the Southeast.[1]

* * *

In his monograph, *The Mechanical Engineer in America, 1830–1910*, Monte A. Calvert discussed two contrasting approaches to the educa-tion of mechanical engineers that were manifestations of a "deeper struggle between two cultures—school and shop—for control of the whole process of socialization, education and professionalization."[2] According to Calvert, the shop culture leaders were a class-conscious elite with an interlocking network of family connections.[3] When new

JAMES E. BRITTAIN and ROBERT C. MCMATH are historians in the Department of Histo-ry, Technology, and Society at Georgia Tech, where Dr. McMath is currently also associate dean of the Ivan Allen College. They are among six coauthors of *Engineering the New South*. Dr. McMath, a specialist in southern agrarian history and politics, is the author of *Populist Vanguard*. Dr. Brittain, a specialist in the history of electrotechnology, edited *Turning Points in American Electrical History*, and his biography of E. F. W. Alexanderson is in press.

[1] There are two institutional histories of Georgia Tech: Marion Luther Brittain, *The Story of Georgia Tech* (Chapel Hill, N.C., 1948); and Robert B. Wallace, *Dress Her in White and Gold: A Biography of Georgia Tech and the Men Who Led Her* (Atlanta, 1969).
[2] Monte A. Calvert, *The Mechanical Engineer in America, 1830–1910* (Baltimore, 1967), p. 62.
[3] Ibid., p. 11.

This essay originally appeared in *Technology and Culture*, vol. 18, no. 2, April 1977.

college programs created by school culture advocates began opening the engineering profession to young men from lower-class and non-shop backgrounds, the proponents of the shop culture became out-spoken critics of these new programs. The conflict between school culture and shop culture reached its maximum intensity during the 1880s and persisted until after 1900.[4]

This cultural conflict produced two major types of mechanical engineering colleges. The first type, designed by leaders of the school culture, included such colleges as the Stevens Institute of Technology, Sibley College at Cornell (after 1885), and the Massachusetts Institute of Technology. Engineering programs there were similar in their stress on higher mathematics, theoretical science, and original research. The second type, defended by shop culture advocates, included the Worcester Free Institute, the Rose Polytechnic Institute, and Sibley College (before 1885). The programs of the latter group stressed practical shop work and produced graduates who were able to work as machinists or shop foremen but who were not highly trained for engineering analysis or original research.[5]

Robert H. Thurston was the chief architect of the mechanical engineering curriculum adopted in the centers of the school culture.[6] The outline of his proposed curriculum began to take shape during the late 1860s while Thurston was teaching at the Naval Academy. It was first implemented at Stevens Institute of Technology where Thurston taught from 1871 until 1885, when he was persuaded to undertake a major alteration of the Sibley College program. A significant innovation in engineering education, the mechanical laboratory, was introduced at Stevens in 1874 and at Sibley when Thurston arrived. Thurston argued that such laboratories would, by adding to the store of knowledge needed in engineering and industry, be of "incalculable benefit to mankind." The laboratories would, he predicted, bridge the cultural gap separating scientists and businessmen.[7] Whether the mechanical laboratory would teach research skills or shop skills became a focal point of the school culture/shop culture debate.

[4]Ibid., pp. 70, 281.
[5]Ibid., pp. 56–57.
[6]The intellectual leader of the new school culture was raised in a shop culture environment in Rhode Island, where his father was a partner in a machine shop. Thurston was graduated from Brown University with a degree in civil engineering in 1859. After serving in the Engineering Corps of the United States Navy during the Civil War, he taught engineering at the Naval Academy from 1866 through 1871. See William Frederick Durand, *Robert Henry Thurston: A Biography* (New York, 1929).
[7]Robert H. Thurston, "On the Necessity of a Mechanical Laboratory: Its Province and Its Methods," *Journal of the Franklin Institute* 100 (1875): 1409–18.

By 1884 Thurston was stressing the distinction between "technical" schools like Stevens and "trade" schools such as Worcester. He suggested that the graduates of the technical schools would be members of a profession that would be served by the trade school graduates, and he wrote that the mechanical engineer should be educated as a "designer of construction, not a constructor."[8] Thurston contrasted explicitly the "construction" system of education employed at Worcester with the "instruction" or "Russian" system that he had introduced at Stevens.[9] In a commencement speech delivered in 1887, Thurston extended his distinction between technical and trade schools to a more general distinction with overtones of intellectual, rather than social, elitism. He argued that there were two classes of people and that they needed to be educated differently. One class was best suited for intellectual pursuits, while the other was endowed with "constructive faculties." But admission to the engineering profession, Thurston believed, should be based on individual ability, not on social status.[10]

In an essay published in 1895 Thurston revealed one ideological basis for his "ideal system" of education: the social Darwinism propounded by that "great ex-engineer" Herbert Spencer.[11] In that essay Thurston again referred to the two classes: those "brilliant of intellect" especially suited for the professions, and those possessing the "constructive faculty" that suited them for work with their hands. He outlined a plan of instruction that would "best fit the educated citizen to his environment." Thurston suggested that evolution toward this ideal had progressed further in "authoritarian" European nations such as France and Germany than in the United States, where greater freedom had impeded the "evolution of the fittest education." The American lag was being overcome, however, and the advanced system of technical education pioneered in Europe would soon take its place as the "capstone of a magnificent edifice."[12]

[8] R. H. Thurston, "Instruction in Mechanical Engineering," *Scientific American Supplement* 17 (1884): 6904–5.

[9] R. H. Thurston, discussion of George I. Alden, "Technical Training at the Worcester Free Institute," *Transactions of the American Society of Mechanical Engineers [ASME]* 6 (1884): 528, 556–65. The term "Russian system" evidently derived from an exhibit featuring engineering laboratory apparatus at the Philadelphia Centennial Exhibition of 1876. The exhibit was from the Imperial Technical School of Moscow. See Lewis M. Haupt, "Technical Education," *Transactions of the American Institute of Mining Engineering* 5 (1876): 510–15. See also Calvert, p. 77.

[10] R. H. Thurston, "Technical Training Considered as a Part of a Complete and Generous Education," *Scientific American Supplement* 24 (1887): 9614–16.

[11] R. H. Thurston, "The Evolution of the Fittest Education," *Cassier's Magazine* 9 (1895–96): 472–76. See also "The Vogue of Spencer," in *Social Darwinism in American Thought*, rev. ed., Richard Hofstadter (New York, 1955).

[12] Thurston, "Evolution of the Fittest Education." It has been pointed out that Fred-

Unlike some southern advocates of advanced technical education, Thurston saw clearly that the polytechnic institute was a necessary but not a sufficient stimulus to regional industrial development. As early as 1878 he had formulated a comprehensive plan for the promotion of industrialization in New Jersey. His plan included a four-level system of education beginning with common or elementary schools. At the next level he proposed a system of manual training schools for those destined to work as artisans or laborers. Third, a network of trade schools would help train students to work in particular industries such as textile manufacturing. Finally, the state would need at least one polytechnic school to furnish engineers with advanced education in science and laboratory research. Thurston's grand design also envisioned direct encouragement of industry by the state through tax relief or subsidy, improved transportation systems, and creation of governmental departments for the promotion of industry. Thurston concluded that the state might still fail to achieve prosperity through industrialization without "the capacity of the people to comprehend and to take advantage of the opportunity for self-improvement thus offered them, and their inclination to avail themselves of it."[13]

* * *

The movement for engineering education that wedded scientific learning and practical experience—in whatever mix—was compatible with a contemporary intellectual movement in the South that Paul M. Gaston has labeled the New South Creed.[14] Adherents of this new creed argued that in the bitter aftermath of the War for Southern Independence the key to recovery, modernization, and restoration of southern power within the Union lay in industrialization. The states of the late Confederacy should, so the argument went, capitalize on their abundant natural resources and vast labor supply to develop manufacturing, drawing when necessary on northern capital and expertise.[15] As part of the argument that the South should embrace

---

erick W. Taylor's well-known system of scientific management also was inspired by social Darwinism and was calculated to benefit especially an elite group of mechanical engineers in industry (see Edwin T. Layton, *The Revolt of the Engineers* [Cleveland, 1971], pp. 139–41).

[13] R. H. Thurston, "Technical Education in the United States: Its Social, Industrial, and Economic Relations to Our Progress," *Transactions ASME* 14 (1893): 855–1013.

[14] Paul M. Gaston, *The New South Creed: A Study in Southern Mythmaking* (New York, 1970), chap. 1.

[15] Ibid. Manufacturing did exist in the antebellum South, but it was economically and socially subordinated to the interests of the planter class (Charles S. Sydnor, *The Development of Southern Sectionalism, 1819–1848* [Baton Rouge, 1948], pp. 25–29; Eugene

the industrialism of the conquering North, appeals for technical education in the region were being articulated as early as the 1870s.

To suggest such a plan of action at that time was to invite recrimination, for many influential southerners interpreted such notions as a rejection of the socioeconomic system of the antebellum South, a system that southerners had convinced themselves was both economically and morally superior to the wage labor system that prevailed in the North.[16] Moreover, members of the same established elite perceived demands for college-level technical education as threats to the region's classically oriented state universities and denominational colleges.[17]

By the early 1880s overt opposition to industrialization had abated somewhat in the South. The ideas of industrial boosters like Henry Grady, Daniel A. Tompkins, and Henry Watterson were becoming the new economic orthodoxy. Furthermore, the end of the depression that had gripped the nation since 1873 made possible the kind of economic expansion that could only have been talked about before.

However radically it differed from the planters' ideal of a hierarchical socioeconomic structure based on slave labor, the idea of an industrialized New South was at heart a blueprint for southern economic and cultural independence. Beneath the rhetoric about emulating the Northeast lay the conviction that by altering the patterns of manufacturing, commerce, finance, and agricultural production, the white South could strike from its shoulders the chains of mercantilistic subservience to the North.[18]

In this anticolonial crusade education played a secondary, but important, role. New South boosters and their Bourbon Democratic allies opposed large state expenditures for schools, or for anything else. Indeed, some southern industrialists, particularly in the cotton textile industry, argued that it was more efficient to import technically trained personnel and skilled operatives from the North than to train them in the South.[19] In the short run, the importation of engineers

---

D. Genovese, *The Political Economy of Slavery: Studies in the Economy and Society of the Slave South* [New York, 1961], chap. 8).

[16]William R. Taylor, *Cavalier and Yankee: The Old South and American National Character* (New York, 1961). A perceptive account of the southern affirmation of the slave system as economically and morally superior is found in Jack P. Maddex, "Pollard's *The Lost Cause Regained:* A Mask for Southern Accommodation," *Journal of Southern History* 40 (1974): 595–612.

[17]See David A. Lockmiller, *History of the North Carolina State College of Agriculture and Engineering* (Raleigh, 1939), chap. 2.

[18]Gaston, pp. 98–99.

[19]Jack Blicksilver, *Cotton Manufacturing in the Southeast: An Historical Analysis,* Studies in Business and Economic Research, no. 5 (Atlanta, 1959); Judson C. Ward, Jr., "Geor-

and skilled personnel *was* essential, for the training of local people in numbers sufficient to operate southern mills and factories would take years, even decades. Nevertheless, some of the New South advocates, including Atlantan Henry Grady, called for a new kind of *technical* education in the region that would provide southern youths with the expertise to manage the region's industrial expansion.[20]

Perhaps the first public endorsement of this kind of training in Georgia came from Benjamin H. Hill, erstwhile Confederate senator, who had acquiesced briefly in Radical Reconstruction. In 1871 Hill, back more or less in the Democratic fold, told the Alumni Association of the University of Georgia that southerners' dependence on slavery and the resultant demeaning of physical labor by whites had reduced the South's ability to compete economically with the North. The South must overcome its antipathy for commerce and industry, Hill argued, if it is to control its own destiny. Centrally important in this value revolution would be the expansion of higher education to include "practical" training of all sorts. Said Hill: "We *must* have an educated labor. We must have multiplied industries. We must have schools of agriculture, of commerce, of manufacturing, of mining, of technology, and, in short, of all polytechnics, and we must have them as sources of power, and respectability, and in all our own sons must be qualified to take the lead and point the way."[21]

Not surprisingly many heard in Hill's speech only an attack upon the sacred memory of the Old South, and they responded accordingly. But within a few years Hill's ideas were being repeated by influential Georgians. In 1881 an International Cotton Exposition in Atlanta focused attention on the possibilities for manufacturing in the state, and, in several public speeches during the exposition, appeals were made for the kind of practical education that Hill had described.[22]

gia under the Bourbon Democrats, 1872–1890" (Ph.D. diss., University of North Carolina, 1947), pp. 445–63; Gaston, pp. 105–7.

[20]*Atlanta Constitution* (July 28, 1885); Ernest Culpepper Clark, "The Response to Urbanism in Henry W. Grady's New South" (M.A. thesis, Emory University, 1968), pp. 122–23.

[21]Benjamin H. Hill, Jr., *Senator Benjamin H. Hill of Georgia: His Life, Speeches and Writings* (Atlanta, 1891), p. 343. Hill intended that all these activities should be encompassed in the University at Athens.

[22]Jack Blicksilver, "The International Cotton Exposition of 1881 and Its Impact upon the Economic Development of Georgia," *Atlanta Economic Review* 7 (June 1957): 12. Promoters of the exposition included Henry Grady and Samuel Inman, who subsequently helped secure the location of the Georgia School of Technology in Atlanta.

In March 1882, the *Macon Telegraph and Messenger* editorially called for the establishment of a state polytechnic college. The Macon paper was then managed by John F. Hanson, one of the founders of the Bibb Manufacturing Company (textiles) and future president of the Central of Georgia Railroad. In December 1882, Hanson's protégé, Nathaniel E. Harris, introduced a resolution in the state legislature leading toward the establishment of a technological school.[23] The resolution failed of adoption, but not for lack of support in the capital city. A week after Harris introduced his resolution the *Atlanta Constitution* endorsed "technical schools in which the boys of Georgia can acquire the rudiments of mining engineering, architects, contractors, mechanical engineers, chemists, foremen, builders, and all other positions embraced in material progress."[24]

The editor of the *Constitution*, Henry Woodfin Grady, was a leading proponent of southern industrialization and an important booster of technical education in Georgia. While still a cub reporter for another Atlanta paper, Grady had listened to and been impressed by Benjamin Hill's appeal for practical training. Like Hill, Grady was vague about the specific content of technical education, but he did envision a multilevel system similar to that proposed by Robert Thurston, in which trade and technical schools would be located around the state with a "high polytechnique school" to be established, probably as part of the University at Athens.[25]

If Grady was vague about the content of technical training, he was explicit about the anticipated consequences. He and other proponents repeatedly stated that a supply of technically trained personnel was the sine qua non of industrialization in the South. In 1886, the day after Atlanta had been chosen as the site of the new technological school, the *Constitution* made a pointed comparison: "A single technological school in a few years converted Worcester, Mass., from a struggling village into a city of 200,000 people, and one of the busiest manufacturing centers in New England. Our school will be quite as finely equipped and officered as the school at Worcester."[26] As a weapon in the struggle for public support of technical education such hyperbole was no doubt useful. As an analysis of the process of economic development it was overly sanguine. At its inception the

[23]*Macon Telegraph and Messenger* (March 4, 1882), quoted in Brittain, p. 6; Brittain, pp. 7, 8. Harris, who represented Bibb County (Macon) in the legislature, had made the establishment of a technical school the major issue in his campaign for election.
[24]*Atlanta Constitution* (December 14, 1882).
[25]Ibid.
[26]Ibid. (October 2, 1886).

Georgia School of Technology, like the industrial new South itself, was burdened with promises made on its behalf that were impossible to fulfill.[27]

* * *

While editorialists, industrialists, and politicians agitated for a technical school in Georgia, industrial training, of sorts, was already being offered in a rather unlikely place, Emory College, a classically oriented Methodist school located in Oxford, Georgia. Isaac Stiles Hopkins, by training a physician, by calling a Methodist minister, and by trade a professor of Latin and English at Emory, established there in 1883 a school of "Tool-Craft and Design." After Hopkins became president of Emory in 1884 he expanded the program, although it remained an adjunct of the classical curriculum, a series of elective courses designed to give college boys an appreciation of the work done by artisans and mechanics.[28] Hopkins's experiment at Emory was not intended to produce professional engineers or even skilled mechanics, but he cited as models for the program the curricula at Worcester and Stevens Institute. Hopkins saw in the new program an antidote to the rising tide of class antagonism in America. Technical training would break down the barriers of prejudice between the educated classes and working people by showing both the positive value of work. Technical education thus became a defense against "communism and revolution."[29]

Hopkins was not alone among the Georgia advocates of industrial education in his concern about class antagonism. In 1883 the legislative committee charged with investigating the need for technical education in Georgia argued bluntly that such training for working-class youths would help "stop the drift towards communism, and insure subordination to law and order in all classes of our complex population."[30] At its inception technical training in Georgia was intended to fit the needs of an interlocking social, economic, and

[27]In the summer of 1974 an enterprising features writer for the *Atlanta Journal* discovered that Tech had not single-handedly transformed Georgia into a high technology region and on that basis judged the school to be a failure (*Atlanta Journal* [August 1, 1974]; see also reply by Tech President Joseph M. Pettit, ibid. [August 9, 1974]).

[28]Isaac Stiles Hopkins, "Industrial Education: An Alumni Address, Emory College, 1883," *Emory University Publications*, ed. Goodrick Cook White, ser. 7, no. 1 (Atlanta, 1952), pp. v–ix. After Hopkins left Emory in 1888 the program was quickly scrapped.

[29]Ibid., pp. 2, 7, 15.

[30]"Report of the Committee on School of Technology," *Journal of the House of Representatives of the State of Georgia* (July 24, 1883).

political elite. Technical training would serve, its proponents claimed, as an agency of social control and a release for pent-up class tensions at the same time it was producing skilled personnel for industry.[31]

In contrast to Emory's manual training program that flourished briefly during the 1880s the University of Georgia had offered professional training in engineering since 1866. The program at Athens remained quite small and was limited to civil engineering, the oldest branch of the engineering profession and the one for which an educational program could be most easily grafted onto a classical college.[32]

In July 1882, Patrick H. Mell, chancellor of the University, recommended that a technological department be established at Athens. Mell noted in his proposal the growing public support for such an undertaking. His cognizance of the political forces at work, coupled with the vagueness of his recommendation, suggests that Mell might have been doing what administrators of state universities, including Georgia's, had done earlier to prevent the establishment of separate agricultural colleges; that is, to co-opt campaigns for such schools by establishing similar programs within the existing university.[33]

Whatever his motivation, Chancellor Mell added his voice to the chorus demanding technical education in Georgia. He corresponded with Harris, whose efforts in the legislature had led in 1883 to the creation of a committee to investigate the possibility of establishing a technical school.[34] The members of that committee, including Harris, had no preconceived notions about the curriculum such a school should provide, but they intended to do more than ratify Mell's suggestion to add a technological division to the University or to emulate the rudimentary kind of technical training then being offered at land grant colleges in the South. The prototype of the southern land grant colleges was Mississippi A & M (now Mississippi State University), which was initially an agricultural school with a course in "mechanical arts" tacked on. Georgia already had a land grant agricultural college

[31]For evidence of both elite and grass roots support for the establishment of A & M colleges in the Southeast, see "Will of Thomas G. Clemson," published in the *Clemson Agricultural College Bulletin* 21 (1925): 9; and Duncan Lyle Kinnear, *The First 100 Years: A History of Virginia Polytechnic Institute and State University* (Blacksburg, Va., 1972), p. 54.

[32]Robert Preston Brooks, *The University of Georgia under Sixteen Administrations, 1785–1955* (Athens, 1956), p. 47; University of Georgia, *Annual Announcements,* 1882, p. 31. Between 1872 and 1885 the chair of engineering was held by L. H. Carbonnier, a graduate of St. Cyr.

[33]Chancellor's Report, July 1882, in University of Georgia Trustee Minutes, pp. 320–22, Special Collections Department, University of Georgia Library. For a full account of Mell's actions that reaches different conclusions about the episode, see Brooks, pp. 76–78.

[34]Brooks, p. 77.

at the University, and for all their uncertainty about the nature of a technological curriculum, the committee agreed on the need for a school devoted exclusively to training Georgia's youths in the skills needed in an industrialized economy. Thus, they turned for models to the engineering schools of the northeast.[35] In June 1883, the committee visited Boston Tech (now the Massachusetts Institute of Technology), the Worcester Free Institute of Industrial Sciences (now Worcester Polytechnic Institute), Cooper Union, and Stevens Institute of Technology, all schools that were deeply involved in the debate over the best method of training engineers.[36]

The committee looked most carefully at Boston Tech and Worcester Free Institute, finally deciding upon Worcester as the best model for the Georgia school. In so doing they picked the institution most heavily committed to the shop culture tradition and the only one of the four they visited at which items made by students in the shop were sold to produce income for the school.[37] In addition to recommending that a shop system similar to that at Worcester be established, the committee suggested an academic curriculum that was virtually identical to the one there, with work to be offered in mechanical engineering, mining engineering, geology, building and architecture, chemistry, design, and textiles.[38]

Why Worcester? The financial success of the Washburn Shop at Worcester must have impressed the Georgia legislators, who knew that governmental support for *all* education in their state was small and that no large-scale appropriations would be forthcoming for a new institution of higher learning. The enthusiasm with which they described the financial potential of the commercial shop system suggests that economic considerations loomed large in their decision.[39] But the shop culture approach to technical education was congenial to the Georgians on intellectual as well as economic grounds. New South boosters would feel comfortable with its emphasis on practical application of existing technology. Even more importantly, the shop cul-

[35]Kinnear, p. 57; John K. Bettersworth, *People's College: A History of Mississippi State* (Tuscaloosa, Ala., 1953), pp. 133–35; Alabama A & M College, *Catalogue* (1885), p. 19.
[36]"Report of the Committee on School of Technology," pp. 231–37.
[37]Ibid., p. 234; M. P. Higgins, "Worcester Polytechnic Institute: The Washburn Shops," pamphlet (October 1904) in Worcester Polytechnic Institute (WPI) Archives, Worcester, Mass.
[38]"Report of the Committee on School of Technology," p. 234. The committee cautiously suggested that in the future the school might accept women students in certain departments, such as drawing, design, and typewriting, but stressed the fact that they were not trying to develop " 'strong mindedness,' so called, in our lovely women" (ibid., pp. 252–53).
[39]Ibid., p. 234.

ture tradition, in which education served a socializing function, comported well with what many educational reformers in Georgia and elsewhere meant by "industrial education." In the crusades that led to a new kind of college for southern whites and to the establishment of black institutions such as Hampton and Tuskegee, the term referred to that kind of education which instills the character trait of industry, or diligent attention to work.[40] For Benjamin Hill, Isaac Hopkins, Henry Grady, and Nathaniel Harris, the instilling in white southerners of the Protestant work ethic (a virtue allegedly possessed in abundance by Yankees) was a central reason for the establishment of technical education.

The legislative committee's recommendation, presented in July 1883, rehearsed the now familiar litany of New South boosterism and urged that a school of technology be established along the lines of the Worcester Free Institute. A bill to implement their recommendation was narrowly defeated, but a similar measure was adopted two years later in 1885. That bill established a school of technology, nominally as a branch of the state university, and created a commission to determine the location and character of the new school and thereafter to become its board of trustees.[41]

The commission subsequently appointed by Governor Henry D. McDaniel was chaired by Harris. The five commissioners were men of considerable talent, but only two of them had professional experience in fields related to engineering. Oliver S. Porter had established a major cotton mill in Newton County, and Samuel M. Inman controlled a multifaceted business empire in Atlanta that included manufacturing and land development.[42]

The commissioners turned first to the selection of a site. After a spirited competition Atlanta was chosen over four other locations, including Athens. The decision to locate the school in Atlanta apparently did not represent a reasoned preference for an urban-industrial setting. Initially, each of the five contending sites received one vote. Of the five, only Atlanta and Macon could be considered urban, although Athens had been something of an industrial center since the antebellum period. Not until after twenty-three ballots had been cast

[40]For an understanding of this idea we are indebted to Elizabeth Jacoway Burns, who has developed it in a paper entitled "The Industrial Education Myth" (presented at the meeting of the Southern Historical Association, November 13, 1975, in Washington, D.C.); and more fully in "An Experiment in Negro Education: The Penn School" (Ph.D. diss., University of North Carolina, 1974).

[41]Brittain, pp. 326–27.

[42]Ibid., pp. 13–15. The other two members of the commission were Edward R. Hodgson, a Clarke County businessman, and Columbus Heard, a Greene County lawyer and politician.

did a majority of the commissioners cast votes for Atlanta, which had offered $100,000 to help establish the school, an amount larger than the initial state appropriation.[43]

With a population of 65,533, Atlanta was the largest city in Georgia at the time of the 1890 census. Macon was fourth largest with a population of 22,746, while Athens had 9,334. The population growth rates of Atlanta and Macon were comparable during the decade 1880–90 with Atlanta increasing by 75.2 percent and Macon by 78.4 percent. Atlanta was also the state's leading manufacturing city in 1890 as measured by number of establishments, capitalization, number of employees, and gross value of product. Macon trailed Augusta and Savannah as well as Atlanta in manufacturing in 1890 but moved ahead of Savannah by 1900.

The commissioners adopted the legislative committee's recommendation of Worcester as a model without prolonged discussion of the matter. Milton P. Higgins, superintendent of the Washburn Shop at Worcester, and George I. Alden, professor of mechanical engineering there, were brought to Atlanta for consultation. The commission tried to hire both of them to set the school in operation. Both declined to accept permanent appointments in Georgia, but Higgins obtained a year's leave of absence from Worcester and was hired by the Georgia commission "to organize the school."[44]

[43]Commission on the School of Technology, Minutes, October 12 and 20, 1886, in Georgia Institute of Technology Archives, Atlanta (in 1888 the Commission became the Board of Trustees of the Georgia School of Technology); Brittain, pp. 16–18. S. M. Inman, one of the commissioners, stated in a speech given several years later that several leading citizens of Atlanta including Henry Grady had been instrumental in persuading the City Council to appropriate $50,000 and $2,500 per year for twenty years if Atlanta were to be selected. Citizens of Atlanta had raised an additional sum of $20,000, and 4 acres of land had been promised. The bids of the other four contenders were never revealed fully although Inman recalled that they had offered a "great many inducements" (see S. M. Inman's speech transcript "recollections of the Founding of the Georgia School of Technology," Georgia Tech Archives). Some question remains as to which of the commissioners cast the decisive vote for Atlanta. According to the official minutes, Commissioner Hodgson joined Commissioners Inman and Porter on the twenty-fourth ballot to give a majority. Another version that was included in the Brittain monograph is that it was Commissioner Heard rather than Hodgson who provided the majority and that Governor McDaniel may have persuaded Heard to end the deadlock.

[44]Commission on School of Technology, Minutes, December 8, 1886, and January 24, April 8, December 1, 1887; *Worcester* (Mass.) *Telegram* (March 1, 1887), in Scrapbook, 1882–94, WPI Archives. Some of the Worcester trustees opposed granting Higgins the leave, although they would have been willing to accept his resignation. Finally the leave was granted, on condition that Higgins continue to supervise the Washburn Shop during his year in Atlanta (Worcester Polytechnic Institute, Record of the Board

When the commissioners met with the trustees of the University of Georgia seeking authorization to hire a faculty, Henry Grady, a trustee and alumnus of the University, proposed that the commissioners "confine themselves to the chairs now in existence in the Worcester School." His resolution was not adopted, but in fact the positions authorized were virtually identical to those that composed the Worcester faculty. However, the only degree to be offered was one in mechanical engineering.[45]

Although the Worcester system was adopted initially at Georgia Tech, the school's first decade was a recapitulation of the larger struggle between shop culture and school culture in which, by the time Tech opened in 1888, school culture advocates were prevailing in most major American engineering colleges. Isaac Hopkins, president of Emory College, was named president and professor of physics in the Georgia School of Technology. His views on the value of manual training, reinforced by his avocational work in Atlanta machine shops, were quite compatible with those of Higgins, who presided over a corps of four shop supervisors at the North Avenue campus. The critically important chair of mechanical engineering remained vacant when the school opened in October 1888. Alden of Worcester had turned down an offer to come to Atlanta, and efforts to have a naval engineering officer detailed to the position proved unsuccessful.[46] The following autumn Commissioner Oliver Porter succeeded in recruiting for the position John Saylor Coon, a distinguished young graduate of Sibley College at Cornell. Coon, a founding member of the American Society of Mechanical Engineers, was a protégé of John Sweet of Cornell, a leading shop culture man.[47] The chair of mathematics went to West Point graduate Lyman Hall, under whose tutelage higher mathematics, including the calculus, became a principal component of the early curriculum, as it was at

---

of Trustees, November 12 and 23, 1887, W.P.I Archives; *Worcester Telegram* [November 20, 1887]).

[45]University of Georgia Trustees, Minutes, December 21, 1887, recorded in Commission on School of Technology Minutes. The following chairs were authorized: chemistry, mechanical engineering, physics, free-hand and mechanical drawing, architecture, mathematics, English, and geology and mineralogy. The chairs of architecture and mineralogy-geology were not filled.

[46]Commission on School of Technology Minutes, March 1 and July 5, 1888. In the 1880s naval engineers filled chairs of mechanical engineering in a number of fledgling engineering schools (Calvert [n. 2 above], pp. 50, 51).

[47]Brittain, p. 14; Trustees of the Georgia School of Technology (GST), Minutes, May 2, 1889; Calvert, pp. 90–91.

engineering colleges in the school culture tradition. The chair of chemistry was filled by W. H. Emerson, a graduate of the Naval Academy, who had earned his Ph.D. at Johns Hopkins. Upon the chair of English, filled initially by Charles Lane of Macon, devolved all nontechnical instruction.[48]

The exclusion of classical studies and the diminution of liberal arts and the nascent social sciences was intentional. The School of Technology was, theoretically, a branch of the University, and the commissioners, several of whom were alumni of the Athens school, worked with that institution's trustees to insure that existing programs in Athens were not threatened by competition in Atlanta.[49] In addition, the notion of a "pure technical institute" found favor with both New South boosters and engineering educators. Henry Grady and Robert Thurston could have applauded the reply of Lyman Hall, as president of Tech in 1898, to a request that the School of Technology join the State Oratorical Association: "The Faculty of the School of Technology will take no steps toward joining the Oratorical Association at present," Hall replied. "Our courses are exclusively scientific and we have but little time to devote to oratory."[50]

Nathaniel Harris discussed his adamant opposition to early efforts to create an A.B. curriculum at Tech in an illuminating passage in his autobiography. Basically he opposed liberal studies at Tech because they would duplicate existing programs at the University of Georgia, his alma mater. He concluded that the University had offered an engineering course before Tech was founded but stressed the differences between the two and argued that for Tech students "the machine is the principal text book." In an address made in 1888 during the opening ceremonies at Tech, Harris elaborated on the relationship between the University and the School of Technology: "The head is in Athens; the hands are here. We have here thought versus work; practice against theory; the shop against the study; the hammer against the book; the blouse against the cutaway."[51]

In contrast, the curricula of North Carolina A & M, Clemson A & M, and Alabama A & M had strong humanities and social science components from the beginning. Why the difference? Alabama A & M had evolved from a Methodist liberal arts school, the East Alabama Male College. The founders and early administrators of all three schools were for the most part identified with the old collegiate tradi-

[48]GST, *Catalogue, 1888–89*, pp. 29–31, and addendum, p. 7.

[49]University of Georgia Trustees, Minutes, December 21, 1887.

[50]Lyman Hall to John Roach Stratton, October 27, 1898, correspondence of Lyman Hall, Georgia Institute of Technology Archives.

[51]Nathaniel E. Harris, *Autobiography: The Story of an Old Man's Life with Reminiscences of Seventy-five Years* (Macon, 1925), p. 224; *Atlanta Constitution* (October 6, 1888).

tion in American higher education or the new scientific education that had emerged in the nation's leading colleges since 1850, rather than with the shop culture. Unlike the shop culture advocates, men like A. Q. Holladay and H. A. Strode, first presidents of North Carolina A & M and Clemson, respectively, perceived liberal studies to be complementary to the practical instruction to be offered in their new schools.[52]

* * *

At its inception, then, the Georgia School of Technology was within the shop culture tradition in its emphasis on manual training and its near exclusion of nontechnical studies. The single-minded adherence to technological studies on the part of Tech's leaders precluded the school's development into a multipurpose university. But the addition to the faculty of men who were grounded in the school culture tradition and were aware of national trends in technical and scientific education insured that Tech would not remain a carbon copy of Worcester as it existed in the 1880s.

Within a decade the initial design of the curriculum was altered significantly. In part these changes reflected the difficulty of transferring northeastern systems of technical education to the South. The schools whose programs Tech had copied could draw students from a well-developed system of public high schools. The high school systems in Georgia and other southern states were still embryonic in the 1880s, and Georgia made no immediate effort to implement Robert Thurston's "pyramid" plan to feed students into Tech from lower grade trade schools. Consequently Georgia Tech resorted to a practice that state universities and land grant colleges outside the South had abandoned in the 1870s, the establishment of a preparatory class as part of the college. In 1891 Tech added a "subapprentice" (subfreshman) class as one means of dealing with the generally inadequate preparation of potential students.[53] In 1896, on recommendation of the faculty, the trustees formally lowered some of the school's academic standards, an action that probably ratified existing practices.[54]

[52]Ralph B. Draughon, *Alabama Polytechnic Institute* (New York, 1954), pp. 8–16; North Carolina A & M, *Catalogue* (1891), pp. 26–28; Clemson College Trustees, Minutes, March 4, 1891, February 1, 1893, typescript in Special Collection, Clemson University Library, Clemson, S.C.; Lockmiller (n. 17 above), chap. 3; biographical sketch of H. A. Strode in Presidents' Papers, Clemson University.

[53]Frederick Rudolph, *The American College and University: A History* (New York, 1962), pp. 283–84.

[54]Trustees of GST, Minutes, January 7, 1891, December 30, 1896. Specifically, standards in mathematics were eased and the studies of geography and history were removed from the examination of the apprentice (freshman) class.

For its first eight years of operation Tech adhered rigorously to the construction shop system as employed at Worcester. When Higgins returned to Worcester after one year at Tech he was replaced as shop superintendent by another Worcester man, William F. Cole. In 1890 President Hopkins recommended that time spent by students in the shop be almost doubled, to bring Tech's practices into line with those of Worcester.[55] But from the outset the construction shop system at Tech was on shaky ground. Local manufacturers objected to the competition from state subsidized student-made goods. After Higgins left, the shop never came close to matching the financial success of the Washburn Shop at Worcester. Indeed, Worcester was the only American school in which the system had been commercially successful, and by the mid-1890s academic opposition forced even the Washburn Shop to divest itself of its commercial interests.[56] In 1892 the Georgia Tech shop building was gutted by fire. The facilities were rebuilt, but in the interim emphasis was shifting to laboratory work and classroom instruction.[57] A precipitous decline in enrollment following the fire was not immediately overcome when the shop was repaired, a circumstance that added urgency to the task of rethinking the school's method of operation.

The controversy between the constructional and the instructional shop systems reached a climax in 1896, when the trustees decided that the shop should be changed "to a purely educational institution as far as possible," and that Professor John S. Coon should be made superintendent of the shop as well as professor of mechanical engineering. At the same meeting of the trustees Isaac Hopkins, still a supporter of the Worcester system, resigned the presidency and left the faculty. The board appointed Lyman Hall to succeed him as president.[58]

[55]Trustees of GST, Minutes, May 2, 1889, February 6, 1890.

[56]Trustees of WPI, Minutes, June 1, 1895, January 18, 1896. The Tech shop did enjoy a modest success at least during its first year under the direction of Higgins. Higgins, in fact, used the Tech shop during its first year in a paper defending the Worcester approach to engineering education. He reported that the shop had sold almost $10,000 worth of manufactured products and had operated at a net cost of only $1,865, excluding his own salary. He stated that it had taken the Washburn Shop five years to do as well (M. P. Higgins, "Education of Machinists, Foremen and Mechanical Engineers," *Transactions ASME* 21 [1899–1900]: 66 ff.). On local opposition to the commerical shop at Tech see "Dedication to John Saylor Coon," by W. H. Emerson, in *Tech Alumnus*, vol. (1923), in the John Saylor Coon file, Georgia Institute of Technology Archives.

[57]Trustees of GST, Minutes, April 29, 1892. After the fire some students worked in machine shops in Atlanta and visited industrial facilities elsewhere in the state.

[58]Ibid., January 3, 1896. One indication of the shift away from the construction shop can be seen in the illustrations of the school's annual catalogs. Prior to 1896 they depict

The abolition of the construction shop and the resignation of Hopkins were only part of the shift that occurred in the late 1890s. In 1896 the school began offering degrees in electrical and civil engineering and added to the faculty R. M. Quick, an assistant professor at Cornell, as professor of electrical engineering and physics.[59] Three years later a department of textile engineering was added. One side effect of these curricular changes was to dilute further the shop tradition at Tech, for that tradition had been primarily associated with mechanical engineering. Coon, the influential professor of mechanical engineering, had been educated in the shop culture tradition, and he retained an affinity for shop training, but he was aware of the prevailing trends in engineering education and was flexible enough to adjust to the changes.[60]

The curricular and administrative changes, although substantial, did not mark a radical transformation of the school's mission or method of operation. Well into the 20th century Tech's leaders prided themselves on the school's commitment to "practical" training. Required shop work persisted, alongside of modern laboratory instruction and advanced scientific and mathematical courses. Indeed, the last vestiges of the shop system in the mechanical and electrical engineering curricula did not disappear until the 1950s. And the School of Textile Engineering still operates a mill, Tex-Tech, where "students design, develop, manufacture and market textile products while exposed to all the real life textile business."[61]

The addition of a textile department was clearly intended to help

---

students manufacturing items, presumably for sale, whereas after that date they show students conducting laboratory experiments.

[59]GST, *Catalogue, 1897*, pp. 23–31; Trustees of GST, Minutes, January 3, 1896; updated memorandum (ca. November 1898), Hall Correspondence. Georgia Tech lagged behind other engineering schools in the region in diversifying its curriculum. Between 1892 and 1895 North Carolina A & M, Clemson, and Alabama A & M all transformed their vaguely defined and poorly staffed departments of "mechanical arts" into professional programs in mechanical, electrical, and civil engineering and hired graduates of leading engineering schools to staff them (N.C. A & M, *Catalogue, 1895–96*, pp. 3, 25–32; Alabama A & M, *Catalogue, 1894–95*, pp. 73–76; Clemson College, *Catalogue, 1895*).

[60]Brittain, pp. 39–40. Lyman Hall's efforts to establish the textile school are documented in his correspondence during 1898. Textile schools were also established at N.C. A & M in 1900 and at Clemson in 1898. There was considerable competition between the three institutions for northern financial assistance in establishing the schools and in recruiting instructors (see J. H. M. Beatty to Henry S. Hertzog, March 4, 1899, in President's Papers, Special Collection, Clemson University Library).

[61]Stephen Krebs, "Tex-Tech Institutes Textile Labor for Business," *Technique* (February 7, 1975), p. 11.

the school meet its original objective of promoting southern industrialization. Lyman Hall, who as president spearheaded the drive to establish the textile department, explained its significance this way to the president of the University of Georgia: "The establishment of this department is the most important step in education ever taken in Georgia. It means the education of our sons for those industries which derive the greatest profit from the cotton plant, and the successful operation of the Textile school will bring a wealth to Georgia almost inestimable."[62]

In discussing the matter with Henry Grady's successor as editor of the *Atlanta Constitution*, Hall was even more explicit in his use of the New South rhetoric: "When the first brick is layed in the textile department of the Georgia School of Technology the South declares war against New England; a war not of secession but of aggression, a war against slavery, and we are the slaves who shall be free."[63]

* * *

Thus, over a decade after the Georgia school opened its doors, the rhetoric used to justify its existence and expansion was that of the New South creed. As Paul Gaston has pointed out, by the end of the century the New South advocates for all their high-flown rhetoric had not transformed the region into an industrial paradise, but the proponents of industrialization nevertheless began boasting that the golden age was at hand: the New South creed became the New South myth.[64] Was the persistence of booster rhetoric exhibited by Georgia Tech's leaders yet another attempt to obscure the fact that an economic revolution had not occurred in Georgia? To put the question another way, what impact *did* Georgia Tech have on the city of Atlanta, the state of Georgia, and the southern region?

[62] Lyman Hall to William E. Boggs, November 3, 1898. Much of Hall's energy as president was devoted to expanding the student body and soliciting external funds to expand the school's physical plant. He envisioned the textile school as a means of attracting large numbers of students, including women, to Tech. Enrollment did increase sharply in the three years after the textile school was opened, but leveled off after that. In an effort to attract young men to Tech from outside the state Hall persuaded school teachers in neighboring South Carolina to canvass prospective students in Charleston and Columbia, promising to pay the teachers $5.00 for each student recruited in this manner (Hall to Mrs. Robert E. Park, December 4, 1898; Hall to J. A. Finger, August 4, 1898; Hall to E. C. Long, August 27, 1898; all in Hall correspondence; GSY *Catalogues, 1900–1910;* Brittain, p. 45).

[63] Hall to Clark Howell, March 7, 1898, Hall Correspondence. Hall failed to win financial support from textile manufacturers in Georgia for the new school. Ironically, money was contributed to build the textile building by Aaron French, a Philadelphia manufacturer, and much of the equipment was donated by northeastern machine manufacturers.

[64] Gaston, p. 214.

One is tempted merely to quote the response of Daniel Coit Gilman, first president of Johns Hopkins, to a similar question: "The university," he said, "renders services to the community which no demon of statistics can ever estimate. . . ."[65] The school's impact on the way Georgians thought about industrial, and economic matters, and even its direct impact on the state's economy, defy precise measurement. Obviously a school that had produced only about one hundred graduates by 1900 could not have single-handedly transformed the economy of Georgia or Atlanta, despite the prophecies of men like Henry Grady and Lyman Hall. The impact of the school on the industrial work force of the South was of course greater than the number of graduates would suggest. More than four times as many freshmen were enrolled than graduated during the first decade of the school's existence. Presumably many of the dropouts used their shop experience in work situations.[66]

In a study of manufacturing in Georgia between 1850 and 1900 that was published in the Twelfth Census, an unidentified analyst noted the accelerated growth rate during the decade 1890–1900 (despite the nation's worst depression to date) that increased manufactured product valuation from $68.9 million to $106.6 million. He asked rhetorically whether it could be a coincidence that this was the first decade of the Georgia School of Technology. The 1900 census also indicated that Atlanta was the state's leading manufacturing city, both in number of employees and value added, followed by Augusta, Macon, and Savannah. Atlanta had also shown the most rapid population growth during the decade 1890–1900 with a 21.8 percent increase. It is difficult, if not impossible, to determine precisely how much of this growth might be attributed to Georgia Tech and how much to other factors such as Atlanta's transportation advantages, its aggressive business leadership, and the various industrial expositions held in the city in the 1880s and 1890s.[67]

It is possible to learn something about the school's impact by identifying the backgrounds and career patterns of some of the young men who first attended and graduated from the new school. For purposes of analyzing students' socioeconomic background and the

---

[65]Rudolph, p. 272.

[66]The point was made during a heated legislative debate over the appropriations bill for Tech late in 1888. Representative Martin Calvin, a staunch supporter of the new school, used the example of two graduates of a six-month night school of technology in Atlanta. He claimed that their value to the state had been increased five times by this training, from $1,000 to $5,000 per year (*Atlanta Constitution* [December 12, 1888]; see also "Industrial School at Atlanta," *American Machinist* 9 [1886]: p. 3).

[67]*Twelfth Census of the United States Taken in the Year 1900*, vol. 8, *Manufactures* (Washington, 1902), pp. 131–35.

demographic distribution of the student body this study deals only with evidence concerning the students enrolled in 1898–99, the tenth year of the school's operation and the only year prior to 1900 for which even limited information on the occupations of students' parents is available. To get some ideas about career patterns we have traced for a ten-year period the job changes of all those who graduated from Tech before 1899.[68]

Based on data from 1898–99, it appears that the school was only moderately successful in making college education available to groups that had not traditionally received such training, although its record in this regard is probably better than that of the old line state universities in the region. One-third of the students were the sons of farmers, laborers, or clerks, while slightly over half were the sons of professional men, government officials, merchants, or craftsmen.[69]

Although the legislature attempted to bring technological education within the reach of people from all parts of the state by providing up to eight tuition scholarships to students from each county, the rural areas of the state were underrepresented. Fifty-six percent of the students came from towns with populations over 2,500, although less than 16 percent of the State's population lived in such towns. Three out of ten students were from Fulton County (Atlanta). No doubt the overrepresentation of cities and towns in the student body reflected the inequality of college preparatory schools in the state. Only in the principal towns was one likely to find high schools that would prepare students for work in a technical institute.[70]

In examining the career patterns of graduates (and no attempt was made to follow the students who did not graduate), we sought to identify the kind of work the graduates did and their locations. Did graduates actually apply the skills they had acquired or did they follow the pattern set by most graduates of agricultural colleges in the late 19th century and move into jobs unrelated to their training? In 1899, 73 percent of the Tech alumni from the classes of 1890–98 were engaged in manufacturing, construction, or other technical pursuits, although it is impossible to tell from the evidence how many were

[68]The following generalizations are based on tables 1–9.

[69]Data from 1905–9 indicate that about 20 percent of the students' parents in that period were farmers or laborers.

[70]Of the 109 graduates through 1900, 45 (42.2 percent) were from only four counties: Fulton (Atlanta), Richmond (Augusta), DeKalb (adjoining Fulton County), and Floyd (Rome). Twenty-six (23.8 percent) were from Fulton County. Of 286 Graduates through 1909, seven counties had 175 (45.3 percent): Fulton, Chatham (Savannah), DeKalb, Bibb (Macon), Floyd, Richmond, and Cobb (adjoining Fulton). Fulton alone had 94, or 24.3 percent.

TABLE 1

OCCUPATIONS OF STUDENTS' PARENTS:
GEORGIA SCHOOL OF TECHNOLOGY, 1898–99

Parent's Occupation	No. of Students
Farmer	38
Laborer/clerk	21
Drummer	7
Contractor	7
Merchant/businessman	37
Craftsman/artisan	6
Bookkeeper	5
Railroad business	9
Local official/government employee	7
U.S. Army	1
Editor	1
Minister	9
Engineer	4
Banker	6
Lawyer	15
Physician	6
Manufacturer	8
Congressman	1
Total	177

SOURCE.—Lyman Hall to Clarence Knowles, 24 November 1898, Lyman Hall Correspondence, Georgia Institute of Technology Archives, Atlanta. Unfortunately, this source does not provide explicit information about the economic status of students' fathers. In the absence of a list of the fathers' names it was impossible to collect that information from other sources. However, it is probably safe to assume that most of those identified as "farmers" were in fact planters or relatively well-to-do farm owners.

TABLE 2

HOMETOWNS OF GST STUDENTS, 1898–99,
WHO WERE GEORGIA RESIDENTS

Hometown	% of Georgia Students
Georgia towns over 2,500 population	56.4
Fulton County (Atlanta)	28.0
Georgia cities other than Atlanta (Augusta, Columbus, Macon, Savannah)	12.9

SOURCE.—GST *Catalogues, 1898–99.* Less than 6 percent of the students in 1898–99 were from out of state.

TABLE 3

LOCATION OF ALL GST GRADUATES, 1890–98, IN 1899

Location	No. of Graduates (%)
Atlanta	33 (37.9)
Georgia cities other than Atlanta	9 (10.0)
Elsewhere in Georgia	19 (21.8)
Other southern states	12 (13.8)
U.S., outside the South	13 (14.9)
Deceased	1
Total	87

SOURCE.—GST *Catalogue, 1899.*

TABLE 4

LOCATION OF ALL NORTH CAROLINA A & M
ENGINEERING GRADUATES, 1893–99, IN SIXTH
OR SEVENTH YEAR AFTER GRADUATION

Location	No. of Graduates (%)
North Carolina	33 (56.9)
Other southern states	20 (34.5)
U.S., outside the South	4 (06.9)
Outside the U.S.	1 (01.7)
Total	58

SOURCE.—North Carolina A & M *Catalogues, 1893–1905.*

TABLE 5

LOCATION OF ALL CLEMSON COLLEGE ENGINEERING
GRADUATES, 1896–1900, IN 1901

Location	No. of Graduates (%)
South Carolina	51 (66.2)
Other southern states	7 (09.1)
U.S., outside the South	19 (24.7)
Total	77

SOURCE.—*List of Graduates of Clemson College, S.C., and Occupations Represented, 1896–1901* (n.p., n.d.).

TABLE 6

LOCATION OF IDENTIFIABLE ALABAMA A & M
ENGINEERING GRADUATES TO 1900, IN 1906

Location	No. of Graduates (%)
Alabama.......................	48 (48.5)
Other southern states	31 (31.3)
U.S., outside the South	15 (15.1)
Outside the U.S.	5 (05.0)
Total.......................	99

SOURCE.—*Catalogue of the Officers and Alumni of the Alabama Polytechnic Institute, Auburn, Alabama, 1872–1906* (Opelika, Ala., 1906). The early Alabama A & M catalogs did not distinguish between engineering and nonengineering graduates. Only graduates in identifiable technical jobs or associated with technically oriented firms are included in this table.

actually practicing engineers. This figure is slightly higher than comparable statistics for engineering graduates of North Carolina A & M and Clemson College, but it is considerably lower than the 1904 data on graduates of Sibley College at Cornell, according to which 86 percent of the graduates were engaged in technical careers.[71]

In underdeveloped regions there is typically a serious outmigration of technically trained people. Information on the location of all Tech graduates in 1899 suggests that no such pattern emerged among alumni of the Georgia school, or among graduates of North Carolina A & M, Clemson, and Alabama A & M. Three-fourths of the Tech graduates were still living in Georgia in 1899, and only 15 percent were living outside the South. Graduates living in Georgia worked primarily in cities and the principal towns (38 percent were in Atlanta), but the distribution of graduates was roughly equal to the distribution of students' home counties. After the turn of the century, however, there was an increased outmigration of graduates from Georgia. By 1909, 93 percent of Tech's students were Georgia residents, but only 53 percent of the graduates were working in the state. Another 24 percent were working elsewhere in the South. Slightly over one-fourth of the graduates were in Atlanta. Thus Atlanta continued to do well in retaining Tech graduates, but the state as a whole suffered a substantial loss of trained personnel during the school's second decade.[72]

[71]Calvert, p. 150.

[72]Through the class of 1909, 98.8 percent of the graduates were from the South and 92.5 percent were from Georgia. Seventy-seven percent of these were employed in the

TABLE 7

OCCUPATIONS OR BUSINESS AFFILIATIONS OF GST GRADUATES, 1890–98,
IN FIFTH AND TENTH YEARS AFTER GRADUATION

Occupation or Affiliation	Fifth Year (%)	Tenth Year (%)
Engineer, undesignated	6	8
Mechanical engineer	5	8
Electrical engineer	4	1
Civil engineer	2	2
Consulting engineer	1	2
Draftsman	7	4
Member or principal officer, manufacturing firm	3	13
Member or principal officer, construction firm	2	1
Manager or superintendent, manufacturing firm	11	10
Manager or superintendent, operating firm	3	7
Inspector	4	0
Foreman	1	1
Manufacturing firm, unspecified job	5	2
Operating firm, unspecified job	3	1
Engineering professor or superintendent of college shop	4	4
Total technical jobs	61 (70.0)	64 (73.6)
Commercial firm, unspecified job	6	3
Attorney	0	1
Farmer	4	3
Clerk	1	1
Student	3	0
Army/naval officer	1	3
Federal employee, nontechnical job	2	1
Schoolteacher/principal	2	1
Total nontechnical jobs	19 (21.8)	13 (14.9)
No information	4	4
Deceased	3	6
Total	87	87

SOURCE.—GST *Catalogues, 1895–1910.* To compensate for gaps in the collection of catalogs, some data are from the year closest to the fifth or tenth year for which a catalog was available.

We can conclude from this evidence that Georgia Tech added rather significantly to the pool of technically trained people needed to build the industrial New South. It is equally clear that the school could

South and 53 percent in Georgia. In 1909, 26 percent of Tech's graduates were working in Atlanta. Georgia's other major cities did poorly in attracting Tech graduates. Except for Atlanta with ninety-nine, only Savannah had as many as ten (eleven) in 1909. Of course these figures do not indicate how many graduates of other engineering colleges were moving *into* Georgia during this period.

TABLE 8

OCCUPATIONS OR BUSINESS AFFILIATIONS OF N.C. A & M ENGINEERING GRADUATES,
1893–99, IN THIRD OR FOURTH AND SIXTH OR SEVENTH
YEARS AFTER GRADUATION

Occupation or Affiliation	Third or Fourth Year (%)	Sixth or Seventh Year (%)
Electrical engineer	4	3
Civil engineer	5	2
Mechanical engineer	0	3
Engineer, unspecified	1	1
Locomotive engineer	1	2
Federal employee, technical job	3	5
Manufacturing (unspecified job):		
Iron	3	2
Cotton textiles	3	5
Other	3	2
Manufacturing superintendent or foreman	2	3
Machinist	3	2
Textile designer	1	0
Architect	4	4
Draftsman	1	2
Chemist	1	1
Instructor at N.C. A & M	5	2
Total technical jobs	40 (68.9)	39 (67.2)
Farmer	3	4
Schoolteacher/college instructor	2	2
Commercial firm	11	12
Army	0	1
Total nontechnical jobs	15 (25.9)	19 (32.8)
No information	2	0
Total	58	58

SOURCE.—N.C. A & M *Catalogues, 1893–1905*. Totals for N.C. A & M include only living graduates.

not socialize and control the working classes in the manner that its founders had intended. Technological education remained beyond the reach of most of Georgia's poor and rural youths.

The decision of Tech's founders to adopt the Worcester Free Institute as a model and to locate the institution in the state's leading urban center made the Georgia School of Technology a unique educational experiment in the South. Tech was the only southern engineering school that was not an adjunct of an agricultural college and the only one to adopt the commercial shop system. The shop system was an experiment based on both economic considerations and

TABLE 9

OCCUPATIONS OR BUSINESS AFFILIATIONS OF CLEMSON COLLEGE
ENGINEERING GRADUATES, 1896–1900, IN 1901

Occupation or Affiliation	No. of Graduates (%)
Cotton textiles	11
Electrical industry	17*
Other manufacturing	5
Federal employee, technical job	8
Chief draftsman	3
Machinist	2
Street railway company	1
Miller	1
Telephone Company	1
Total technical	49 (63.6)
Farmer	5
College professor	6
School teacher	2
Merchant	3
Army officer	1
Lawyer	1
Physician	1
Salesman	2
Federal employee, nontechnical job	2
Bookkeeper	2
Total nontechnical jobs	28 (36.3)
Total	77

SOURCE.—*List of Graduates of Clemson College, S.C., and Occupations Represented, 1896–1901*
(n.p., n.d.).
*Nine of the seventeen were with General Electric.

ideological preferences. The experiment failed by the mid-1890s for
a number of reasons: Higgins's successors as shop superintendents
were less capable than he; the shop fire temporarily forced the use of
alternative methods of education; local manufacturers objected to
tax-subsidized competition from Tech; new faculty joined the school
who were not enamored of the construction shop approach; and enrollment fell to a low of eighty-five students in 1894, making changes
seem necessary if the school were to survive.

The diversification of the School's curriculum in 1896 and succeeding years brought Georgia Tech closer to the national mainstream in
engineering education and, ironically, made the school more like the
engineering departments of the principal A & M colleges of the
Southeast, all of which had adopted diversified professional engineer-

ing programs by 1895. Yet Tech remained unique in the region with its singular focus on technological and scientific education, a focus that remains largely unchanged today.[73] In this respect, at least, the New South of the 1880s and the Newer South of the 1970s are strikingly similar.

[73]A fascinating sequel to our story occurred during World War I, in which the non–land grant status of the Georgia School of Technology was significant on a national stage. Phineas V. Stephens, a lobbyist for Georgia Tech, was instrumental in blocking passage of a bill introduced by Senator Francis G. Newlands of Nevada that would have provided federal funding of engineering experiment stations at land grant colleges. The Smith-Howard bill supported by Stephens that would have made Georgia Tech and other non–land grant institutions eligible for the prospective funds also failed to pass when the various factions could not agree (see Daniel J. Kevles, "Federal Legislation for Engineering Experiment Stations: The Episode of World War I," *Technology and Culture* 12 [1971]: 182–89).

AT THE TURN OF A SCREW: WILLIAM SELLERS, THE FRANKLIN INSTITUTE, AND A STANDARD AMERICAN THREAD

BRUCE SINCLAIR

At a time when the superb drama of exploiting a new continent filled the minds of most Americans, nineteenth-century technical arguments about the shape and number of threads on a screw often have a remote, somewhat comic quality—a bit like Jonathan Swift's mock epic struggle between the Little Enders and the Big Enders. But the issue was not a matter of satirical trivia. Industrial development on a national scale demanded that nuts and bolts of the same diameter be interchangeable. Interchangeability, in turn, required that manufacturers conform to a standard system which fixed the contour of screw threads and established for each diameter the number of threads per inch. For America, in 1860, no such standard existed. Since that era also marked the emergence of this country into the arena of international industrial competition, the search for a standard cast reflections which illuminate such related considerations as a national style of engineering, American industrial practice, and the role of government in technological change.

The most prevalent system—where system was used at all—was that which had first been proposed in 1841 by England's Sir Joseph Whitworth.[1] Whitworth's standard was a synthesis of the best English practice and answered the general case well enough so that it was widely employed. American usage was disuniform, however, varying from manufacturer to manufacturer and from locality to locality. Some firms developed systems to fit the particular requirements of their own processes. Others purposely used special threads to prevent outside repairs on their own machinery, in the same vein as the Erie Railroad's ill-fated use of wide-gauge track. By the 1860's, the appalling lack of national uniformity clearly called for reform. "If there is any one thing in the transac-

BRUCE SINCLAIR is the Melvin Kranzberg Professor of History of Technology at Georgia Institute of Technology.

[1] Sir Joseph Whitworth, "On a Uniform System of Screw Threads," *Papers on Mechanical Subjects* (London, 1885), pp. 17–26.

This essay originally appeared in *Technology and Culture*, vol. 10, no. 1, January 1969.

tions of the machine shop more incomprehensible than another," the editor of *Scientific American* claimed in 1863, "it is the want of some settled size or number for screw threads."[2] To eliminate the anarchy, the magazine called for some agreement among the country's principal manufacturers or, failing that, governmental action. Whatever the standard, Whitworth or any other, the adoption of some uniform national system of screw threads was of vital necessity.

But the development of a standard in the first instance posed certain difficulties. As Whitworth had pointed out, a system depended on compromise, not on theory or experimentation.[3] There was simply no way in which all the factors involved in screw-thread design could be stated as a precise rule, applicable to all cases. Where principle provided no final answer, practice determined the issue. That opened the door to non-technical factors. A further difficulty in America was that no single agency seemed capable of dealing with the problem. Editorial injunctions notwithstanding, the federal government had no ready means either to develop a system or to regulate its usage, not even to speak of an inclination to do so. Nor did any professional engineering society exist to propose a standard and advance its adoption. It was in this apparent vacuum that William Sellers presented a paper before a meeting of the Franklin Institute at Philadelphia in April 1864 outlining a uniform system for American screw threads.

In his paper, Sellers addressed himself immediately to the central question: Why should there be yet another system of threads? Why should not Americans adopt the Whitworth standard and by its consistent usage rationalize current practice? The English system was the result of several years of study by an outstanding mechanician, who had carefully analyzed the three main factors of screw-thread design—pitch, or the number of threads per inch, thread depth, and thread form. To arrive at his system of pitches, Whitworth had collected bolt samples from all of the principal manufacturers in England, averaged their characteristics, and developed a standard table for pitch and diameter. The form of his thread, which comprised flat sides at an angle of fifty-five degrees, with rounded tops and bottoms, was also arrived at by the same averaging technique. It was a compromise, but on the basis of the best English screw-thread practice. Any alternative scheme, Sellers suggested, should clearly "demonstrate its practicability and its superiority."[4]

The central difference between Whitworth's system and the one

2 *Scientific American,* IX (October 10, 1863), 233.

3 Whitworth, p. 18.

4 *Journal of the Franklin Institute,* XLVII (May 1864), 344.

proposed by Sellers was in the form of thread. Sellers leveled three objections to the English form. First, the fifty-five degree angle was difficult to gauge with consistent accuracy. Second, in ordinary practice the rounded tops and bottoms of English threads did not fit the corresponding bottoms and tops of nuts, and the thread's wearing surface was therefore reduced only to its sides. The point was not that a fit was impossible, but that in normal usage it was difficult to achieve contact with that form of thread. Finally, Sellers objected to Whitworth's thread form on the grounds that it was more complicated, and therefore more costly to manufacture. According to Sellers, obtaining the rounded top required "three kinds of cutters and two lathes to perform what with our practice requires but one cutter and one lathe."[5] The core of Sellers' opposition to the Whitworth standard was that it was complicated, expensive, and difficult to produce with consistent accuracy.

In place of the Whitworth form, Sellers proposed a contour then already in use in several Philadelphia machine shops, including his own. The thread took the shape of an equilateral triangle, sides inclined at an angle of sixty degrees, with the top and bottom flattened one-eighth of the thread depth (see Fig. 1). It had, Sellers claimed, precisely those virtues which were defects in the Whitworth form; it could be made with greater ease and less machinery, and could be verified with greater accuracy. The table of pitches which Sellers proposed varied only slightly from the Whitworth standard, but was notable primarily because those variations allowed for the use of a relatively simple formula to calculate pitch for any diameter. To his thread form and table of pitches, Sellers also proposed a uniform range of sizes for bolt heads and nuts and a standard thread gauge, thus offering for consideration a complete and standardized system of screw threads, nuts, and bolts.

Since screw threads were not a matter of objective precision, as Whitworth had pointed out, the determination of a system obviously lent itself to bias. Even particular elements of a screw thread reveal the possibilities. For example, assuming a generally V-shaped contour, pitch is the most significant factor in uniformity. It is important that, for any given diameter, the number of threads per inch is the same. The close similarity between the pitches of Sellers' system and Whitworth's suggests that on this issue there was little to separate American from English practice. But the form of Sellers' screw thread highlights the differences in American engineering and industrial attitudes. Nowhere was that bias more clearly revealed than in the subsequent competition between the two thread systems.

One of the keynotes in the battle was sounded almost immediately.

[5] *Ibid.*, p. 346.

William Bement, head of the Philadelphia machinery firm of Bement and Dougherty and chairman of the Franklin Institute committee to investigate and report on Sellers' system, set the tone when he advised the editor of *Scientific American* that the committee would be pleased to have the opinions of "all good *practical* mechanics."[6] The emphasis of the word was Bement's. The magazine reiterated the idea, congratulating American mechanics that their interests had not fallen "into the toils of schemers and theorists who would have confused instead of

[6] *Scientific American*, XI (October 29, 1864), 278.

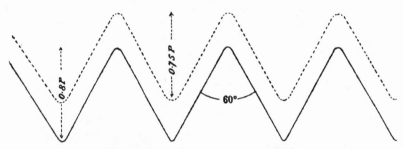

Fig. II. Usual Form of Threads best Workshop practice.

Fig. III. Form of Thread introduced Mr. Whitworth.

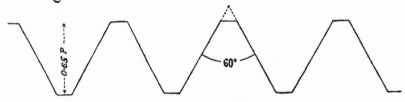

Fig. IV. Form of Thread proposed by Mr. Sellers.

Fig. 1.–Figure II above shows normal screw-thread practice when no particular standard was employed. Figures III and IV outline the characteristics of the Whitworth and Sellers threads. From Robert Briggs, "A Uniform System of Screw Threads," *Journal of the Franklin Institute*, LXXIX (February 1865), 124.

making the subject plain and practical."⁷ What was meant by "practical" was made clear in the Institute's report. The Sellers thread form was based on an angle "more readily obtained than any other." It was a thread contour which any ordinary workman could make with accuracy. The committee's conclusion was that a uniform American thread system should embody a shape which would enable "any intelligent mechanic to construct it without any special tools."⁸ There was nothing theoretical or abstruse about it. It sounded democratic and it saved money.

All subsequent consideration of the Whitworth and Sellers thread systems reiterated the basic difference in thread form. The U.S. Navy's Bureau of Steam Engineering conducted a fairly extensive analysis of screw-thread systems in 1868, with a view toward the adoption of a standard for the Navy. "We find but two *systems* in general use," the Board noted, "one known as Whitworth's, deduced many years ago from the general practice of English mechanics, and the other known as that of Mr. William Sellers, deduced more recently from the practice of American mechanics."⁹ In regard to pitch ranges, strength, and durability, the Board of Naval Engineers found no appreciable difference between the two systems. According to their report, the form of the Sellers thread was one of the major factors which led the Board to recommend his system. The Whitworth thread required "such skill on the part of the workmen" that uniformity was out of the question. Conversely, the Sellers thread enjoyed the "very important advantage of ease of production."¹⁰

In England, uniform screw threads depended on a high level of craft skill. And to Englishmen, such skill seemed a perfectly appropriate foundation. As one editorialist succinctly put it: "With good tools and good men to use them, there is no difficulty whatever in producing the Whitworth cross-section of thread."¹¹ Where labor costs were lower and craft traditions stronger, the system worked. But it was for precisely those reasons that Whitworth's system was rejected here. In America, interchangeability depended on reducing a job to the point where an ordinary workman could produce the desired uniformity. It was not Whitworth's intention to conserve labor. The fifty-five degree angle of his thread was an arithmetic mean; it was mathematically indifferent

⁷ *Ibid.*, XII (March 4, 1865), 151.

⁸ *Journal of the Franklin Institute*, LXXIX (January 1865), 54–55.

⁹ *Report of the Board To Recommend a Standard Gauge for Bolts, Nuts and Screws Threads for the United States Navy, May 1868* (Washington, D.C., 1880), p. 5.

¹⁰ *Ibid.*, pp. 22–23.

¹¹ *Engineering*, XLII (September 10, 1886), 266.

to production costs.[12] American engineering practice had labor costs as one of its basic considerations, and Sellers' system reflected the difference. The genius of his standard, however, was that it was not only easier and cheaper to construct, but easier as well to maintain with accuracy.

Because his thread perpetuated his name, Sellers is probably more widely known for his system of threads than for any other facet of his career. But its acceptance depended in part on his position in American industrial and engineering circles. William Sellers was Philadelphia's leading machine-tool builder when that city was the country's pre-eminent tool-building center. "Probably no one," according to Joseph Roe, "has had a greater influence on machine tools in America than William Sellers."[13] At a time when it seemed to some that the screw-thread problem lacked a solution simply for want of any leadership, Sellers provided it. Although he had just celebrated his fortieth birthday when he presented his paper on a uniform thread system, Sellers had already established a reputation as one of Philadelphia's leading machinists. He came to his trade early. It was one of the characteristics of the city's industrial elite to go directly to business, not college.[14] After a basic education at a private school maintained by his family, Sellers was apprenticed to his uncle's machine shop in Wilmington, Delaware. Following his apprenticeship, he superintended the family-connected machine shop of Fairbanks, Bancroft and Company in Providence, Rhode Island, for three years and then, in 1848, returned to Philadelphia where he established a partnership with Bancroft in the manufacture of machine tools and mill gearing. So successful was the firm, a contemporary observer remarked, that within less than a decade it was the city's leading machine tool company.[15]

Sellers was frequently styled the "American Whitworth," and there is no reason to believe he shunned the comparison. In fact, there is a remarkable parallelism in the accomplishments for which the two men were remembered. Both conducted large and successful industrial enterprises. Both advocated high standards of precision in machine-shop practice, and to that end both produced machines whose form depended

[12] As H. J. Habakkuk has noted, Whitworth's initial concern was in saving equipment (*British and American Technology in the Nineteenth Century* [Cambridge, 1962], p. 119).

[13] Joseph Wickham Roe, *English and American Tool Builders* (New York, 1926), p. 247.

[14] E. Digby Baltzell, *Philadelphia Gentlemen: The Making of a National Upper Class* (Glencoe, Ill., 1958), p. 109.

[15] Edwin T. Freedley, *Leading Pursuits and Leading Men* (Philadelphia, 1856), p. 331. The best biographical sketches are in Roe, pp. 247–51, and an obituary published in the *Journal of the Franklin Institute*, CLIX (May 1905), 365–81.

on function rather than on artificial architectural embellishment. But there were also important differences between the two. Whitworth's passion was accuracy. For him, "a *true plane* and *power of measurement*" were the outstanding elements of mechanical engineering, and the major thrust of his career was to upgrade English practice along those lines.[16] Sellers was by no means unconcerned with precision, but it would be fair to say that his work showed greater originality. In the sense that Whitworth's interest was with the maintenance and extension of precision workmanship, he was a conservator; by comparison, Sellers was an innovator.[17]

From the beginning, Sellers' machine tools were dominated by the same thinking which characterized his screw-thread system. The catalogues of his firm explicitly stated that position:

> It has been said that good workmen can do good work with poor tools. Skill and ingenuity may indeed accomplish great results, but the problem of the day is not only how to secure more good workmen, but how to enable such workmen as are at our command to do good work, and how to enable the many really skillful mechanics to accomplish more and better work than heretofore; in other words, the attention of engineers is constantly directed to so perfect machine tools as to utilize unskilled labor.[18]

Sellers' machines embodied that view. His 1857 bolt machine marked the advent of commercially interchangeable nuts and bolts. The gear-cutting machine, which he showed at the Paris Exhibition of 1867, was one of the earliest to be automatically operated. And he produced an increasing number of special-purpose lathes on the principle that they would allow "less skillful workmen" to produce more work of better quality.[19]

Sellers was also a grand-scale industrialist. He organized the Edge Moor Iron Company in 1868, which supplied all the structural iron work for Philadelphia's Centennial Exhibition and all structural materials except the cables for the Brooklyn Bridge. Significantly, at Edge Moor bridge-making was reduced to a standardized manufacturing process. In 1873 Sellers also became president of Midvale Steel Company, where he supported Frederick W. Taylor's long series of experiments in metal cutting. By middle age Sellers was already the dean of Philadelphia's

[16] Roe, p. 99.

[17] For that aspect of Sellers' career, see W. Paul Strassmann, *Risk and Technological Innovation: American Manufacturing Methods during the Nineteenth Century* (Ithaca, N.Y., 1959), pp. 130–32.

[18] William Sellers & Co., *A Treatise on Machine-Tools, etc. as Made by William Sellers & Co.* (4th ed.; Philadelphia, 1877), p. 113.

[19] *Ibid.*

machine industry; at his death, he was recognized as "the greatest tool builder of his day and generation."[20] But it was Whitworth himself who bestowed the highest accolade when he described Sellers as "the greatest mechanical engineer of the world."[21]

By talent and reputation, Sellers was probably better qualified than any other American machinist to effect a substantial change in the country's screw-thread practice. Proposing a system, however, was only half of the battle. The utility of a standard lay equally in its widespread acceptance. Whitworth had recognized the dual nature of the problem. He had presented his own system at meetings of the Institution of Civil Engineers, and he was aware of the importance of engineering opinion for the system's adoption. Not only was it difficult to achieve agreement on a system, inconvenience in the change-over and the inertia of traditional usage would also inhibit the spread of a standard. These were "obvious" reasons, Whitworth claimed, why engineers and their societies should push for the adoption of a uniform system.[22] Sellers was faced with similar impediments, but he was closely connected with major elements of private industry and was president of the only technical society in the country which enjoyed a national reputation—the Franklin Institute of the State of Pennsylvania. Therein lay the means for adoption.

A general society in the pre-professional era of American science and technology, the Franklin Institute was founded in 1824 as one of the earliest American mechanics' institutes, with the aim of educating young apprentices in science and its applications.[23] In the decades which followed, a series of technical studies projected the organization into national prominence. Its experiments into the motive force of water provided an intensive analysis of hydraulic power.[24] The Institute's investigation into the causes of steam boiler explosions was the most exhaustive and detailed study conducted in this country and remained so for another half-century.[25] The organization acted as technical advisor to

20 *Journal of the Franklin Institute*, CLXXIX (May 1905), 375.

21 *Ibid.*, p. 381.

22 Whitworth, "On a Uniform System of Screw Threads," p. 25.

23 For a study of the Institute's early years, see Bruce Sinclair, "Science with Practice; Practice with Science: A History of the Franklin Institute, 1824–1837" (unpublished Ph.D. dissertation, Case Institute of Technology, 1966).

24 The results of those experiments were published in the *Journal of the Franklin Institute* throughout 1831 and 1832 and in the issues from March to July 1841.

25 That research is described in Bruce Sinclair, *Early Research at the Franklin Institute: The Investigation into the Causes of Steam Boiler Explosions, 1830–1837* (Philadelphia, 1966). John G. Burke's article, "Bursting Boilers and the Federal Power," *Technology and Culture*, VII (Winter 1966), 1–23, documents the emergence of federal regulation to cope with the problem.

the state of Pennsylvania on a standard of weights and measures and to the city of Philadelphia on the best method of paving streets. It participated in the investigation into the U.S.S. *Princeton* disaster and, along with the National Academy of Sciences, co-operated in a Navy Department study of steam expansion in 1863. These activities gave the Franklin Institute a reputation for informed judgment, and that prominence was a significa~t factor in the adoption of Sellers' system.

Equally important, there was a solid link between the Institute and Philadelphia's leading machinery firms. The committee which was appointed to consider the subject of screw threads was mainly composed of representatives from those firms. Chairman of the committee was William Bement, of Bement and Dougherty, a firm second only to William Sellers and Company in the city's machine-tool hierarchy. Baldwin's Locomotive Works had two men on the committee. The Southwark Foundry and Merrick and Sons, sprawling industrial firms established by Samuel V. Merrick, one of the founders of the Franklin Institute, were also represented, as were the firms of Morris, Towne and Company, the Pencoyd Iron Works, and of course, William Sellers and Company. The Institute's Board of Managers bore the same stamp. When William Sellers was elected president of the organization in 1864, most of the same firms were represented on the Board. Most of them were also using Sellers' system in their own practices. In their combined interests, these men and their firms wielded enormous influence in the American mechanical community.[26] Promoters of rival thread systems may have felt cause to complain, as one did, that the committee simply confirmed Sellers' conclusions rather than conduct a thorough investigation of the whole problem.[27]

From the very beginning, the Institute zealously pushed for widespread adoption of the Sellers, or, as it was soon also called, the Franklin Institute system. A special committee was appointed to consider the subject raised by Sellers' paper, to the end that the Institute would recommend a system "for the general adoption of American engineers." At the same time, copies of Sellers' paper were sent to other mechanics' institutes, "with a view to promote the introduction into general use of the system advocated."[28] When the special committee made its report, it was unanimously in favor of Sellers' plan. The Institute then moved to

[26] For an analysis of the power of this group, see Monte A. Calvert, *The Mechanical Engineer in America, 1830–1910: Professional Cultures in Conflict* (Baltimore, 1967), pp. 10, 176, 229–30.

[27] Robert Briggs, "A Uniform System of Screw Threads," *Journal of the Franklin Institute*, LXXIX (February 1865), 111.

[28] *Journal of the Franklin Institute*, XLVII (May 1864), 351.

160   *Bruce Sinclair*

secure general adoption by American engineers. Formal resolutions were
sent to the quartermaster general of the Army, chief of the Navy's
Bureau of Steam Engineering, the chiefs of ordnance of both the Army
and Navy, the chiefs of the engineer and military railroad corps, and to
the superintendents and master mechanics of railroad companies. The
resolutions called for the adoption of the system "by requiring all
builders under any new contracts to conform to the proportions recommended."[29] Similar resolutions were sent to all mechanics' institutes and
to the major machine shops of the country.

Governmental approval was the most obvious and direct means of
obtaining conformity, and the Institute's resolutions were heavily
weighted in that direction. The Navy was the first to respond. In March
1868, the secretary of the Navy authorized the Bureau of Steam Engineering to recommend a standard screw-thread system for the service.
Benjamin Isherwood, chief of the Bureau, appointed Theodore Zeller,
chief engineer at the Philadelphia Navy Yard, to head a board of naval
engineers for that purpose. After visiting the principal machinery establishments in the country and after formal meetings in Philadelphia, the
Board in its report of May 1868 "unhesitatingly" recommended adoption
of the Sellers system. By direction of the secretary of the Navy, it was
immediately authorized as the Navy standard.[30]

Since the same Philadelphia machinery interests which backed Sellers'
system were then involved in a nasty dispute with Isherwood and Zeller,
the Bureau's quick accord is perhaps surprising. The quarrel involved
used machinery which Zeller had purchased from a New York dealer
for use at the Philadelphia Navy Yard. Local machinery manufacturers
resented the loss of business to an outsider and raised serious questions
about the transaction. The issue led to a congressional investigation
which reinforced an already bubbling intraservice squabble, and Isherwood ultimately lost his job.[31] But the Bureau's action suggests the importance of the thread system's identification with the Franklin Institute.
The *Journal of the Franklin Institute* had long been a favorite vehicle of
professional expression for ambitious young Navy engineers. And when
both the Navy and the National Academy of Sciences declined to publish Isherwood's *Experimental Researches in Steam-Engineering*, the
Franklin Institute did so.[32] It is also true that the Navy advocated

29 *Ibid.*, LXXIX (January 1865), 56.

30 *Report of the Board To Recommend a Standard Gauge*, p. 3.

31 The whole affair is described in Edward William Sloan III, *Benjamin Franklin
Isherwood: Naval Engineer* (Annapolis, Md., 1965), pp. 213–32. See also Leonard A.
Swann, Jr., *John Roach: Maritime Entrepreneur* (Annapolis, Md., 1965), pp. 26–32.

32 Sloan, pp. 91–93, 95.

Sellers' system in the belief that it would probably become the established practice in private industry. As the Board stated in its report, "We were naturally desirous to select a system which, while meeting all the essential requirements of a system, would be most likely to be generally acquiesced in and adopted."[33] In other words, the Navy's concern was with Navy practice; it had no concept of leading a movement for the Sellers standard. Private interests were important only to the extent that they had already reached a consensus in favor of the system.

The Board's conclusions effectively threw the burden of standards enforcement back onto private shoulders, and in the last analysis private interests were the most significant element in the adoption of the Sellers system. Two factors shaped the response of industry. First, the standard reinforced technical advances in machine tools; it meant maximum utilization of self-acting and automatic machinery. The laborious process of thread cutting by hand had long since given way to the use of self-acting lathes, which employed change gears to obtain different pitches.[34] And after 1860, machine shops had automatic bolt cutters at their disposal, further to speed and standardize the thread-cutting process. Improvements in machinery made it all the easier to manufacture a standard screw. The adoption of a system rationalized those technical advances.

A standard system also had particular benefits for certain segments of industry, and it was from those areas that the most immediate and effective support came. Interchangeability was critical on railroads, for example. It was obviously important that locomotives and rolling stock be manufactured according to standards, since repairs were frequently required at points distant from a central shop.[35] Steam engines and the equipment of heavy industry fell into the same category, and acceptance

<hr />

[33] *Report of the Board To Recommend a Standard Gauge*, p. 30.

[34] One of the best descriptions of early nineteenth-century thread-cutting methods is given in Eugene S. Ferguson (ed.), *Early Engineering Reminiscences [1815–40] of George Escol Sellers* (Washington, D.C., 1965), pp. 31–34. The eminent English machinist Charles Holtzapffel argued for a rational system of screw threads on the basis that lathes with change gears had come into general use, solving the mechanical aspect of the problem (*Turning and Mechanical Manipulation*, II [London, 1856, 667]). For a thorough study of lathe developments to mid-century, see Robert S. Woodbury, *History of the Lathe to 1850* (Cleveland, 1961).

[35] As the standards committee appointed by the Master Car-Builders' Association reported, "Your committee were at a loss how to proceed, as no two roads were found where all threads used were the same, the varieties in use being much greater than they had supposed possible. Such a state of things must evidently be prejudicial to the best interest of all railway companies, and felt most keenly by master car-builders in their daily repairs of foreign cars" (*History and Early Reports of the Master Car-Builders' Association* [New York, 1885], p. 80).

came first from these sources. By 1868, Sellers' system was used in most Philadelphia firms and had also been widely adopted by New England machine shops and by several manufacturers of machinery in the mid-Atlantic states.[36] The Pennsylvania Railroad took up the system in 1869. It was adopted by the Master Car-Builders' Association and the Master Mechanics' Association (both railroad organizations) shortly afterward and, within the next decade, by most of the country's railroads. Standard gauging tools for the system, developed by Pratt and Whitney and by Browne and Sharp, only hastened the process of adoption.[37]

But discussion and controversy did not end. General American acceptance of the system raised the debate to an international level, where it became entangled in the metric system controversy and in engineering competition between nations. For example, when German engineers proposed the adoption of the Sellers thread form for use in a metric system to replace the Whitworth standard, English technical response was sharp. The effort, according to one editor, was an eccentricity on the part of "a handful of obscure scientific men."[38] It ran counter to the enlightened practice of thirty years and would bring ruin to Germany's export trade. Furthermore, it was claimed that after a brief period of experimentation in America, the Sellers thread contour proved to be a conclusive failure, and like errant children come home, machine makers in the United States were returning to the Whitworth pattern.

Faced with increasing engineering competition from America and from a Continent potentially united by the metric system, English technical opinion was confused and divided. The Verein Deutscher Ingenieure (Society of German Engineers), also perplexed by the notion that there might not be an American standard thread system, addressed a communication of inquiry to the Franklin Institute, with particular reference to thread form.[39] The Institute's secretary, Dr. William H. Wahl, adopted a novel form of response. He directed a circular letter to the officers of the major American railroads and to a number of manu-

[36] A listing of most of the country's principal machinery firms, with the thread system employed, is printed in *Report of the Board To Recommend a Standard Gauge*, pp. 7–13.

[37] For screw-thread gauge developments, see George M. Bond, "A Standard Gauge System," *American Society of Mechanical Engineers, Transactions*, III (1882), 132–41, and John A. Brashear, "The Evolution of Standards of Measurement," *Cassiers' Magazine*, XX (May–October, 1901), 410–19, also presents some information.

[38] *Engineering*, XLII (September 10, 1886), 266.

[39] The inquiry and subsequent correspondence was published under the title "The Sellers or Franklin Institute System of Screw Threads," *Journal of the Franklin Institute*, CXXIII (April 1887), 261–77.

facturing firms, asking whether they employed the Sellers standard and, if so, how successfully the thread form was maintained. The replies, together with a counter-blast from *Railway Gazette*, were all forwarded to the German society.

It is difficult to say which offended the Americans most—English condescension, aspersions on their engineering practice, or technical inaccuracy. *Railway Gazette*'s editor retorted hotly: "This is not the first time that facts and figures as to American practice have been 'evolved from the inner consciousness' of writers across the water, to suit the occasion, but it is not often that such a complete perversion and reversal of the facts is given currency in a journal of standing."[40] According to George Bond, of Pratt and Whitney, 90 per cent of the orders they received for taps and dies were for the Sellers system. Furthermore, Bond claimed that experimentation and two decades of experience had proved the demonstrable superiority of its thread form. There was an American standard, Dr. Wahl informed German engineers. It was called the Sellers, or Franklin Institute, standard and was used "throughout the United States, to the exclusion of any other."[41] The exchange reflected, once again, the importance of non-technical elements in the problem of screw-thread standardization. International industrial competition and engineering nationalism were the issues.

The establishment of professional engineering societies in America did not change the equation. Debate continued along much the same lines, though the forum was shifted to specialized groups. At the first annual meeting of the American Society of Mechanical Engineers, George R. Stetson, of the Morse Twist Drill and Machine Company in New Bedford, pointed out that one of the functions of the new society should be to push the acceptance of the Franklin Institute standard so that "American machinery, by its uniformity in approved design and construction, may be entitled to the most favorable consideration in foreign as well as home markets, and the field of our usefulness and profits correspondingly increased."[42] In fact, the A.S.M.E. failed to reach agreement on any standard, primarily on the principle that such questions were best left for solution in the marketplace. As Monte Calvert has noted, it was not that the society was opposed to standards, but that it felt they should be determined by businessmen and business methods.[43]

[40] *Ibid.*, p. 276.   [41] *Ibid.*, p. 277.

[42] George R. Stetson, "Standard Sizes of Screw Threads," *A.S.M.E., Transactions*, I (1880), 125.

[43] Calvert, pp. 177–78.

Trade associations, business firms, and engineers working individually
—not government—established national screw-thread practice in nineteenth-century America. The Navy's adoption of Sellers' system was
important insofar as it set a precedent and created a general climate of
accord. Writers on the subject always established the government's
acceptance of the system to suggest that it enjoyed a posture of authority. But there was no reality of governmental enforcement. Implementation was always a job for private hands. In England, the subject was a
matter of debate in the House of Commons. In France, it became settled
governmental policy. But in America, private interests, more than any
other, determined the outcome. It was by the efforts of entrepreneurial-
ly minded technical groups like the Franklin Institute and the "untiring
devotion" of organizations such as the Master Car-Builders Association
that uniformity was advanced in America.[44]

To many, it was right that such questions as screw-thread systems
should be resolved by practical men rather than by politicians. Coleman
Sellers spoke to that point of view when he remarked: "The government
of France has always been in the habit of interfering with the private
affairs of people." Conversely, he noted, the American concept of governmental function was "that it should do and enforce justice, and that
Liberty in all things innocent, is the birthright of the citizen."[45] In time,
that attitude changed; simple answers no longer fit the problem. As
specialized societies examined the technical requirements of their own
groups more closely, the weaknesses of a general system of screw
threads became more apparent (see Fig. 2). That called, in turn, for a
review of the whole issue, suggesting some governmental action. With
the establishment of the Bureau of Standards in 1901, the government
also had a means of responding to the problem. Subsequent efforts involved combinations of government and technical societies.[46]

[44] "Uniformity," the Association argued, "can only be secured by the general
adoption of a *correct* standard. The Master Car-Builders' Association has, therefore, instructed its committee to urge all railroad companies to adopt the Sellers'
or Franklin Institute system for all new work" (*American Machinist,* V [September
9, 1882], 3).

[45] *Report to the Franklin Institute of the State of Pennsylvania, for the Promotion
of the Mechanic Arts, Relative to the Metric System of Weights and Measures*
(Philadelphia, 1876), p. 5.

[46] For examples, see Charles C. Tyler, "A Proposed Standard for Machine Screw
Thread Sizes," *A.S.M.E., Transactions,* XXIII (1902), 603–31, and "Standard Pro-
portions for Machine Screws," *A.S.M.E., Transactions,* XXIX (1907), 99–119; and
Joseph Wickham Roe, *James Hartness: A Representative of the Machine Age at
Its Best* (New York, 1937), pp. 67–68.

Nothing better illustrates the nineteenth-century attitude than George Bond's frequent comment that, among screw-thread systems, it was "survival of the fittest."[47] William Sellers' uniform system was fittest in the sense that it answered contemporary engineering requirements. Sellers understood the economic as well as the technical aspects of the problem and appreciated their interrelationship as a prerequisite for adoption. The Franklin Institute's indorsement and proselytizing activities created a favorable environment for survival. At a time when no

THREADS.

V. Thread.   Eng. Stand.   U. S. Stand.   Bastard.

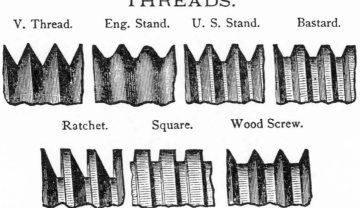

Ratchet.   Square.   Wood Screw.

FIG. 2.—This illustration shows the most commonly used screw threads of the nineteenth century. Whitworth's thread form is shown as the English standard; Sellers' thread is labeled as the U.S. standard. The V thread and bastard thread are variants from Whitworth and Sellers models. The remainder are special-purpose threads. From Cornell University Machine Shop, *A Chapter of Advice on Bolts and Nuts with Cards of Reference, Prepared for the Students in the Machine Shop of the Sibley College* (Ithaca, N.Y., 1875), pp. 9–10.

similar association existed to perform the function, the organization gave Sellers an institutional framework for his system, providing a platform, a mechanism for its advancement, and an aura of objectivity. The man and the institution were well mated, and it was the combination which produced the Sellers, or Franklin Institute, standard of uniform screw threads.

[47] *Engineering*, LXI (March 6, 1896), 325.

Part II

The Engineer and Engineering in 20th-Century America

Overview

THE ENGINEER IN 20TH-CENTURY AMERICA

TERRY S. REYNOLDS

The history of the American engineering profession in the 20th century revolves around the accommodations it made to two turn-of-the-century developments: the vastly expanded scale of organizations, particularly corporate enterprises, and, secondarily, the growing utility and prestige of science.

The Changing Environment of Engineering: Large Organizations

Late in the 19th century the environment in which American engineering was typically practiced began to change. The transportation, communications, and manufacturing systems that American engineers had designed, constructed, and sometimes managed in the 19th century had removed the ceilings limiting the size of commercial enterprises. The result was rapid growth in scale. Small enterprises with modest capital and a handful of employees serving local or, at best, regional markets were steadily displaced by much larger, capital-intensive enterprises serving national or even international markets. To secure competitive advantages, these firms sought to institutionalize and control the inventive process through industrial research laboratories, and they developed and applied steadily more sophisticated and expensive technologies. These large organizations changed the setting in which American engineering was practiced.[1]

Thomas Hughes's essay, "The Electrification of America: The System Builders," provides an excellent analysis of how technological and organizational systems in the United States grew in size, scale, and complexity around the turn of the century. Hughes traces the expansion of the utility industry by looking at three key systems builders: Thomas Edison, Samuel Insull, and S. Z. Mitchell.

Hughes argues that the first stage in the evolution of a new technological system is dominated by inventor-entrepreneurs, like Thomas Edison. Hughes describes Edison's development of the local, urban

---

[1]For additional information on the dramatic growth in the size of organizations in America in the late 19th and early 20th centuries see Alfred D. Chandler, Jr., *The Visible Hand: The Managerial Revolution in American Business* (Cambridge, Mass., 1977).

electrical power distribution system in the 1870s and 1880s, pointing out that, at this stage of system growth, technical problems are paramount and scale is small.

By the 1890s, however, engineers like William Stanley and Nicola Tesla had solved the technological problems that had limited the size, scale, and complexity of utility systems. With the emergence of the transformer, three-phase alternating current, the induction motor, the rotary converter, and the frequency changer, electrical utilities could grow beyond small, local power stations, and the industry entered a second stage dominated by the manager-entrepreneur. Hughes's archetype for this stage is Samuel Insull. Insull, an Edison protégé, left a position managing electrical manufacturing facilities to take the presidency of the Chicago Edison company in 1892. Chicago, at that time, was supplied with electric lights by twenty small, local concerns. Insull believed that further growth in the industry required lower rates, a more diversified customer base, and the more efficient use of equipment. To meet these challenges, Hughes explains, Insull developed or applied new managerial concepts, like load factor and load diversity, and sought out new technologies, like the rotary converter and the steam turbine. By 1912 Insull's company, now called Commonwealth Edison, had absorbed through competition, purchase, or merger all rival companies and served Chicago through a few very large and very efficient electrical generating facilities, selling power to a diversified customer base at much lower rates.

As the management techniques and technologies applied by Insull spread, electrical utility enterprises grew in size and scale. By the 1920s the industry had reached the point where the key problems were no longer managerial but financial: How did one finance expansion beyond regional boundaries and insure that the heavy investment required was not lost? Hughes uses S. Z. Mitchell's career to illustrate the third stage in the growth of technological systems—the development of regional systems and the primacy of financial and organizational issues. Mitchell was the key figure in the development of the utility holding company—a financial and organizational device that had revolutionized the America utility industry by the 1920s. Its key advantage was that it reduced, through diversification of ownership and provision of centralized engineering and managerial services, the risks of investment in electrical utilities. This, in turn, encouraged the inflow of money and the creation of still-larger technological systems. Mitchell's work permitted electrical utility systems to grow beyond the municipal or county level to regional or even national levels.

The rapid growth in scale and complexity of the electrical utility industry that is analyzed and described by Hughes was paralleled by similar growth in other technological systems. By the early 20th century these systems and the organizations controlling them had become the locus of most engineering employment.

The Changing Environment of Engineering: Scientization

Paralleling the change in the primary locus of engineering employment came a change in how engineering was practiced as engineers began to adopt many more of the methods and approaches of the natural sciences. Edwin T. Layton, Jr.'s essay, "Mirror-Image Twins: The Communities of Science and Technology in 19th-Century America," provides a classic overview of the early development of "scientific" methods within the engineering community.

As Layton points out, early attempts to apply science to technology largely failed. There is little evidence, for example, that most of the technologies developed in the 18th and early 19th centuries were applications of Newtonian mechanics. Instead of "applying" science, Layton argues, engineers borrowed the methods of science—systematic, quantitative experimentation and mathematical theory—to build their own unique "engineering" sciences that were based on existing craft practice rather than the study of natural phenomena.

Layton traces the emergence of "engineering science" in the 19th century, pointing out, in particular, how the bodies of scientific knowledge developed by engineers differed from those developed by scientists. For example, engineering theorists working on strength of materials tended to focus not on fundamental entities like atoms, molecules, and forces but on bundles of fibers. Similarly, engineering theorists in the late 19th century turned away from analytical solutions, relying much more on approximate methods using graphical analysis. By 1900, according to Layton, engineers had created complex systems of knowledge and the associated means of diffusing and advancing those systems (professional societies and professional journals, e.g.) that closely paralleled the knowledge systems and professional institutions of the sciences.

There were, however, differences. Layton maintains, for example, that, while the bodies of knowledge developed by scientists and engineers were similar in their dependence on mathematics, quantitative data, and theory, their approaches had so far diverged that intermediate figures (scientist-engineers or engineer-scientists) were needed to make scientific work understandable to engineers and vice versa.[2]

---

[2]Other scholars have supported Layton on this issue. For example, Hugh G. J. Aitken, *Syntony and Spark: The Origins of Radio* (New York, 1976), pp. 19–20, 311, argues that the invention of radio required the "translation" of scientific into technological knowledge rather than simply the application of science.

Layton also maintains that by the 20th century the systems of knowledge, institutions, and values developed by the engineering community had become "mirror-image twins" of the parallel systems of the sciences. He points out, for example, that in science highest status was accorded to mathematical theorists working on universal laws and lowest status to scientists working on practical matters like improved experimental apparatus. In engineering, on the other hand, the mirror-image reverse was true. Highest status went to the practical designer and builder; lowest status to the "mere" academic theorist.

Science and the systems of "scientific" knowledge developed by engineers in the 19th century became steadily more important in the 20th century, both because of the more complicated technical problems posed by the growing size of technological systems and because of the status concerns created for engineers by the organizations that employed them.

Engineering Practice in the 20th Century

The nature of engineering practice in the 20th century was largely shaped by the two developments reviewed above—the vastly expanded scale of organizations and the growing use of "scientific" methods.

The development that had the greatest impact on the engineering profession was the emergence of large organizations. Two types of large organizations provided employment to the bulk of engineers in 20th-century America: the corporation and government agencies.

The Corporation

Throughout most of the 19th century the independent consultant, an engineer who operated in a manner similar to a physician but carried technological rather than medicinal cures from project to project, dominated the engineering profession. The independent consultant shared domination with the proprietor-engineer—an engineer who simultaneously owned and operated his own machine shop or mine or construction firm. Engineers, of course, already worked as salaried employees of organizations in the 19th century, but even these positions provided considerable autonomy and high status because of their close proximity to the highest levels in organizations of limited size and complexity. For example, Ellis Sylvester Chesbrough, although technically an employee of the Chicago Board of Sewerage Commissioners, was not overshadowed by his employer.

The rise of the large corporation dramatically changed this situation. In the 20th century, engineers enmeshed in corporate hierarchies replaced independent consultants and proprietor-engineers as the archetypes of the profession. Several interrelated factors fueled the shift. First, because of the technical nature of the systems they controlled, large corporations needed large numbers of lower- and midlevel management people familiar with technology, that is, engineers. Second, by the 1890s the growing size of technological systems meant that capital costs for cutting-edge technologies were rapidly approaching a point that made it almost inconceivable for engineers to work independently. In the rapidly growing mechanical, electrical, and chemical industries, for example, the high capital costs of initiating work on increasingly complex products, processes, and systems almost insured that practice in those fields would largely be dependent on corporate employment. Corporate institutionalization of research and invention through industrial research laboratories further reinforced this trend. Third, because of the magnitude of the capital investment required for large and complex technological systems, corporations not unnaturally sought to control all aspects of the environment in which they operated. This meant corporate managers preferred to depend on engineers employed within the corporate hierarchy, *not* on outside consulting engineers. Thus, early in the 20th century, organizational hierarchies (usually corporate) became the typical place of employment for American engineers. By 1925, for example, only around 25 percent of all American engineers were proprietors or consultants—the ideals of the previous century; 75 percent were hired employees of corporate or government organizations.[3] By 1965 only around 5 percent of American engineers were self-employed.[4]

Designing, building, maintaining, operating, and managing the large and complex technical systems of 20th-century enterprises required more technical expertise than ever before. So, not surprisingly, engineers found employment in steadily increasing numbers. There were only around 40,000 engineers in America in 1900, but by 1950 there were well over 500,000 and by 1990 around 2 million.[5] Simultaneously the proportion of the work force made up of engineers increased, from 13 per 10,000 workers in 1900 to 128 per 10,000 workers by 1960.[6] By midcentury engineering had become the largest single

[3]Society for the Promotion of Engineering Education, *Report of the Investigation of Engineering Education, 1923–1929*, vol. 1 (Pittsburgh, 1930), p. 232.

[4]"What Happens to Engineers?" *Chemical Engineering Progress* 61, no. 8 (August 1965): p. 35.

[5]U.S. Department of Commerce, *Historical Statistics of the United States: Colonial Times to 1970* (Washington, D.C., 1975), p. 140; Alan Eck, "Adaptability of the Engineering Work Force: Information Available from the Bureau of Labor Statistics," in *Fostering Flexibility in the Engineering Work Force* (Washington, D.C., 1990), pp. 87–88.

[6]Edward Goss, "Change in Technological and Scientific Developments and Its Impact upon the Occupational Structure," in *The Engineers and the Social System*, ed. Robert Perrucci and Joel E. Gerstl (New York, 1969), p. 18; U.S. Bureau of the Census, *Statistical Abstract of the United States, 1979* (Washington, D.C., 1979), p. 416.

174   Terry S. Reynolds

occupation pursued by American males and the second largest of the occupations claiming professional status.[7]

Bruce Sinclair's essay, "Local History and National Culture: Notions on Engineering Professionalism in America," illustrates how large organizations, as they became the typical environment in which engineering was practiced, changed the shape of professional life. Sinclair draws on a skit prepared by the Engineers' Club of St. Louis in 1930 to get at the concerns and problems faced by rank-and-file engineers working in large organizations. The skit, titled "Every Engineer: An Immorality Play," follows the career of a young engineer, and Sinclair suggests that the skit "mirrors . . . the actual circumstances of St. Louis engineers during the 1930s." The protagonist, after graduating from a university full of self-confidence and ideals, finds it difficult to adapt to the real world. He discovers that both corporate and public employers are corrupt and not interested in independent attitudes. After securing a position in a corporation and demonstrating his worth by developing several ideas for increasing profits, he finds that his corporate employers still treat him as a "hireling" and freely steal his ideas for their own ends. As Sinclair notes, the play seems to express the individual engineer's sense of impotency in the face of the power that large organizations wielded.

Engineers, however, gradually accommodated themselves to the emergence of the large corporation as the locus of engineering work in 20th-century America. They did this by increasingly aligning themselves with the aims and aspirations of their companies. By midcentury, professional standing for many engineers had become identical with corporate standing. The approval of one's superiors in corporate or governmental hierarchies became more important than the approval of one's technical peers, contrary to the values of the traditional professions of law, medicine, and the clergy.[8]

James Brittain's essay on "The Introduction of the Loading Coil: George A. Campbell and Michael I. Pupin" illustrates the subordination of professional to corporate goals, as well as the increasing dependence of engineering on scientific concepts and methods. Brittain reviews the claims of Campbell and Pupin to priority in the development of the loading coil, the most important development in

[7]Gross, "Change in Technological and Scientific Developments," p. 18. Public school teachers outnumber engineers.

[8]For example, engineers surveyed in the 1960s held the opinions of their superiors in the organizations in which they worked to be more important than the opinions of fellow engineers and far more important that the opinions of the leaders of their professional societies. See Robert Perrucci and Joel E. Gerstl, *Profession without Community: Engineers in American Society* (New York, 1969), pp. 116–18.

telephony between Alexander Graham Bell's pioneering work and the introduction in the 1920s of electronic amplification. One of the key figures in the invention of the loading coil, George Campbell, could be considered a prototype for the engineer of the 20th century. He was a corporate employee, a member of the engineering staff of the American Bell Telephone Company in Boston, and he was a corporate loyalist to whom reputation in the company was more important than external honors and recognition. As Brittain points out, although Campbell was AT&T's leading electrical theorist in the early 20th century, he remained "almost unknown outside the telephone company" and did not actively participate in professional engineering–society activities.

Around the turn of the century AT&T's chief engineer assigned Campbell one of the key problems of long-distance telephonic transmission—elimination of voice distortion on long-distance lines. This problem had placed a ceiling on the growth of American Bell Telephone's communications system, limiting it to circuits of no more than 1,200 miles in length. As Brittain points out, Campbell's work, and the parallel work of Michael Pupin, "was symbolic in marking the appearance of a new type of 'invention' and a new type of 'inventor.'" Both Pupin and Campbell had trained in graduate-level mathematical physics, and the resulting "invention," the loading coil, was something that required the application of mathematical physics for success. In brief, Brittain's essay is a good account of an early example of the application of scientific concepts to technology.

Campbell and Pupin developed very similar solutions to the voice-distortion problem. Brittain demonstrates quite convincingly, however, that Campbell's work on the loading coil was qualitatively and quantitatively superior to Pupin's. Campbell began his experiments earlier than Pupin, carried out much more extensive experiments, and was clearly the first to use loading coils to reduce distortion on actual telephone circuits. In addition, the mathematical equations that Campbell derived to guide loading-coil design and spacing were much more exact than Pupin's and formed the basis for later work in the area.

However, Pupin filed for his patent earlier. Pupin was an academic who did not have to worry about the impact of premature disclosure like Campbell, a corporate researcher. When Campbell, shortly after, filed his claims, the patent office declared an interference. Interference hearings began in early 1903, and in December the patent examiner awarded the patent to Pupin. Brittain clearly believes the judgment was in error. He maintains that the notoriety Pupin gained from the 1901 sale of his patent claim and the technical complexity of the invention distorted the outcome. Ironically, Pupin's patent claim had been purchased before the interference hearings had begun by American Bell, Campbell's employer, to insure control over the concept. After the patent office ruled in favor of Pupin, American Bell did not pursue the appeals process to its conclusion because it did not wish a "cloud upon the title of the successful contestant." It chose, instead, to give Pupin all of the public credit for the invention of the loading coil and to downplay the contributions of its own employee in spite of his superior work. American Bell Telephone hoped this would discourage others from challenging the loading-coil patent. Significantly, Campbell apparently did not protest this decision publicly, despite being robbed of public recognition for one of the outstanding technological contributions to the history of communications. He continued to loyally serve AT&T for the remainder of his career. Corporate rewards for such acquiescence were apparently more appealing than professional or public recognition.

The subordination of professional to corporate interests was clearly one of the impacts that the organizational setting of 20th-century engineering had on engineering practice. But there were others. Most engineers in the 19th century spent their entire career as "bench engineers," that is, as engineers who focused on the technical elements of engineering like research, design, improvement of technical detail, supervision of technical work, and so on. In the 20th century many engineers began to spend only a fraction of their careers in "bench engineering." Virtually every survey of engineers made in this century has revealed that most engineers leave "bench engineering" for corporate management responsibilities within a decade or two of graduation. For example, Wickenden's study of engineering in the 1920s revealed that over 60 percent of older engineers were involved primarily in administrative work.[9] To some extent this development flowed naturally from the growing size and complexity of the technological systems controlled by American corporations. Because these systems were technological, engineers became prime candidates for managing them. Their technical expertise gave them an inside track in competing for managerial positions. In 20th-century America, an engineering position became increasingly only a stepping stone to a managerial career, not a lifelong commitment, a hallmark of the traditional professions.

American engineers became enchanted with corporate management as a long-term career for several reasons. One was certainly the greater financial rewards. Another was the fear of technological obsolescence created by the accelerating pace of technological change. A third was

[9]Society for the Promotion of Engineering Education, pp. 231–32; See also Engineers Joint Council, *A Profile of the Engineering Profession* (Washington, D.C., 1971), p. 11.

status. As engineers began to be incorporated into corporate hier-archies it was not clear at first where they fit. The engineers could have been treated as a type of highly skilled labor, the same as carpenters, machin-ists, or pipe fitters, and essentially demoted to the working class. Fearing this, engineers sought to differentiate themselves from lower-status occupations, by emphasizing their "scientific" training (discussed later) and their management skills. Frederick Taylor's promotion of "scientific management" early in the century was, in a sense, an attempt by the engineering community to create for itself a niche in industry that would accord it a position above that of the working classes.[10] In the long run, the willingness of most American corporations to view their engineering staffs as part of the "management team" and as reser-voirs for drawing potential managers provided a means of class differentiation satisfactory to most engineers and more acceptable than Taylor's scheme to corporate employers.

Mitchell, one of the utility executives discussed in Thomas Hughes's essay on systems builders, provides an example of an engineer who jumped from technical to managerial responsibilities. After graduat-ing with training as an engineer from the U.S. Naval Academy, Mitchell found a position at the Edison Company in New York. His ini-tial responsibilities were technical. He installed lights on the USS *Trenton,* worked at the Edison Machine Works on dynamos, and helped wire the distribution system of the Edison Electric Illuminating Com-pany in New York. He then became the exclusive agent for Edison products in the northwestern states. The managerial, organizational, and financial experience he gained promoting Edison's products in the Northwest enabled Mitchell in 1905 to create the Electric Bond & Share Company. By the mid-1920s Electric Bond & Share controlled at least 10 percent of the electrical energy generated in the United States.

Stuart W. Leslie's essay, "Charles F. Kettering and the Copper-cooled Engine," furnishes an additional example of an engineer who left "bench engineering" for management, as well as an illustration of the occasional problems that corporations had in integrating engineering staffs into corporate organizations. After a successful career as an in-ventor and manager of his own company, Kettering merged that company, Delco, into General Motors (GM) in 1916. By 1921 he had clearly left "bench engineering" for management; he was vice presi-dent in charge of research and a member of GM's board of directors.

---

[10]See the chapter on scientific management in Edwin T. Layton, Jr., *The Revolt of the Engineers: Social Responsibility and the American Engineering Profession* (Baltimore, 1986), pp. 134–53.

Kettering, like most 20th-century engineers, was an organization man. He shaped his work as an engineer to the needs of the corporation he worked for. Recognizing the problems that the Chevrolet division of GM was having competing with Ford's low-cost Model T, Kettering had his research staff investigate air cooling as an alternative to conventional water cooling for automobile engines. He proposed using a system in which air was blown over highly heat-conductive copper fins bonded to the engine block in place of the usual water jacket, water pump, and radiator. The copper cooling system showed promise of producing a lighter, cheaper, more efficient engine.

Ultimately, the copper-cooled engine failed commercially because of a variety of problems, including changing American tastes. But, Leslie demonstrates, a key factor in the failure was organizational. The incorporation of engineers into large and complex organizations did not occur without problems. Leslie shows how personal factors, like pride and jealousy between different engineering staffs, helped sabotage development of the new engine. He also shows how organizational shortcomings, particularly the lack of coordination between research engineers and production engineers within GM's management hierarchy, strongly contributed to the failure of an engine that GM spent four years attempting to develop.

By the late 20th century, around 75 percent of all American engineers were employed, like Campbell, Mitchell, and Kettering, in corporate hierarchies, many of them with predominantly managerial rather than technological responsibilities. But the impact of organizational employment went even beyond this. Most of the 25 percent of American engineers not employed in business and industry were employed in another type of hierarchical organization—the government agency.[11]

The Government

At the turn of the century, outside of the Army Corps of Engineers, the federal government employed few engineers. State and local government rarely employed them either. Several factors, however, promoted a sharp increase in the number of engineers employed by government. At the local government level, the growing belief in the use of "unbiased" experts in place of corrupt and incompetent patronage appointees in America's growing urban centers, a problem referred to in the skit described in Sinclair's essay, prompted increased employment of engineers. The increased dependence of urban areas

[11]Engineering Manpower Commission, *Engineering Manpower: A National Problem or a National Resource?* (New York, 1975), p. 17.

on sophisticated technological systems for water supply, sewerage disposal, transportation, and other elements of urban life reinforced this trend. By the mid-20th century most medium to large cities employed engineers.

Another factor contributing to the increase in the number of government-employed engineers was the growth of big business. Growing corporate power stimulated the growth of big government as a counterforce to regulate big business. And because much of the growth of big business occurred in highly technical areas, the agencies created by state and federal governments to oversee those industries drew heavily on engineering personnel.

The Depression of the 1930s made the federal government an even more important employer of engineers. Several of the massive public works projects undertaken to relieve unemployment and stimulate economic revival, such as the Tennessee Valley Authority, employed large numbers of engineers as design and construction supervisors during building and as managers and in research and development functions afterward.

World War II and the Cold War further increased federal employment of engineers. Postwar competition with the Soviet Union in the fields of atomic energy and space exploration, for example, led to the creation of federal agencies like the Atomic Energy Commission and the National Aeronautics and Space Administration in which engineers were employed in large numbers.

By the late 20th century around 15 percent of all engineers were directly employed in governmental organizations: 8.9 percent by the federal government and its military and 6.1 percent by state and local governments.[12] However, the nearly 9 percent of all engineers directly employed by the federal government in either a civilian or military capacity were just the tip of the federal iceberg. In the second half of the 20th century federal dollars supported many more engineers *indirectly* through contract funding.

The emergence of indirect federal employment of engineers through contract projects carried out in private corporations was largely a post–World War II phenomenon. World War II made clear the close link between technological prowess and military strength. It also made clear that new technologies had eliminated or would soon eliminate the buffer in time and space that had permitted the United States to react to external threats by converting its civilian industries into an "arsenal of democracy." This recognition and the post-1945 Cold War with the Soviet Union insured that the large government re-

---

[12]National Research Council, *Engineering in Society* (Washington, D.C., 1985), p. 48.

sources plowed into defense-related contract research during World War II would continue. And, since the Cold War also involved competition with the Soviet Union in nonmilitary areas, such as scientific prestige and aid projects to nonaligned countries, government contracting activities soon involved university and corporate engineers in nondefense contract employment on a large scale as well.

At the level of prime contractor the federal government may support as many as 24 percent of all American engineers working in private firms, and subcontracting may add another 8 percent.[13] In other words, both directly and indirectly, the federal government now supports the employment of around 40 percent of all engineers.

The emergence of the federal government as a major employer of engineers, directly or indirectly, had an important impact on the engineering profession. Massive government funding for nuclear, aircraft, and missile development contributed significantly to the emergence of new engineering specialties, notably aeronautical, aerospace, and nuclear engineering. Similarly, government support of military electronics and computing contributed to the rapid rise in the number of electrical and electronics engineers in postwar America and to the emergence of electrical and electronics engineering as the largest of the engineering disciplines. Finally, the volatile nature of federal funding led to several engineering recessions, the most notable being the engineering "glut" of the early 1970s, produced by massive layoffs in the aerospace and defense industries due to federal funding cutbacks.

The Impact of Science

The rising status and utility of science in the 20th century also had an impact on the engineering profession, although perhaps not as great an impact as the emergence of large organizations. In the 19th century, as noted previously, engineering training and engineering practice were largely empirical and practical. On-the-job training was preferred to academic training. Empirical and cut-and-try methods were preferred to dependence on mathematical analysis and theory.

In the 20th century, engineering practice began to change sharply. The new engineering fields emerging at the end of the 19th and beginning of the 20th century, like electrical and chemical engineering, were often built more heavily on basic science and often required more theoretical approaches. As Brittain noted in his essay on the loading coil (discussed above), only with mathematical physics and a knowledge of the Maxwell-Heaviside electromagnetic theory was it possible to work out loading-coil spacing.

[13]Ibid.

Bruce E. Seely's essay, "The Scientific Mystique in Engineering: Highway Research at the Bureau of Public Roads, 1918–1940," also illustrates the shift toward more "scientific" methods in American engineering. Seely notes that most historians have implicitly assumed that engineering automatically benefited from the adoption of scientific approaches and methods. This, he points out, was not always the case, but, because in the 20th century science has carried with it status associations, scientific techniques were adopted anyway.

The engineers in Seely's story are engineers employed by a government agency—the Bureau of Public Roads. Seely shows how before World War I federal highway engineers used mainly traditional empirical engineering practices to identify materials and construction techniques that worked well in actual service. The introduction of heavy trucks for transportation in World War I, however, destroyed roads never designed for such vehicles and forced the bureau to launch a new series of investigations.

In 1918 the bureau radically altered its methods, abandoning empirical tests of components and full-scale field studies to focus on the "fundamentals" of road design. This involved adopting more "scientific" methods, that is, developing experiments that would yield exact quantitative data and seeking to develop a broad theory of road construction. Seely describes in some detail the investigations that followed. He points out that the emphasis on exact quantitative data and the desire to produce a theory increased the complexity of experiments and delayed results. The investigations involved simulations that were often very ingenious but did not adequately parallel reality. Moreover, these investigations usually yielded calls for further research, as investigations in the sciences often do, not practical and immediately applicable design suggestions.

Why did federal engineers adopt "scientific" methods and retain them in spite of their limited utility for solving real problems? Seely convincingly argues that engineers hoped they could tap into the status and esteem gained by scientists as a result of their contributions to World War I and into the growing general respect being accorded to anything "scientific." Copying scientific techniques and making engineering more "scientific" offered an avenue for doing this.

Moreover, as Seely points out, practically every field of engineering followed the path that highway researchers took. In the first half of the 20th century engineers generally "scrambled to adopt the powerful problem-solving tools of science and to emulate the successes gained by electrical and chemical engineering." There is no doubt, as Seely concludes, that the adoption of scientific methods proved essential to many advances in engineering, but the Bureau of Public Roads experience demonstrates that the adoption of science came about for more than utilitarian reasons and did not invariably lead to improvements in practice.[14]

Engineering educators often led the way in the adoption of scientific methodology and provide another example of the impact of science on engineering in the 20th century. Before World War II, engineering education, while incorporating elements from the basic sciences and the engineering sciences, still retained a heavy practical focus. Engineering students were still expected to take a number of practically oriented courses, typically a year of graphics, a year of machine shop, and a host of design courses and courses emphasizing current technological practice. World War II, however, did to engineering education what World War I did to the Bureau of Public Roads. It sharply tilted engineering education toward science, mathematics, and theory.

This shift was partly stimulated by the success of directed, applied scientific research during World War II, research that led to jet aircraft, radar, and the atomic bomb. But it was also stimulated by status concerns and governmental policy. With few exceptions, academic engineers in World War II were relegated to supporting roles on large-scale government projects because they had little experience with cutting-edge research. Many returned to campus after the war hoping to continue to be involved in research and encouraged by the federal government's continuing largess in supporting scientific research as part of the Cold War competition with the Soviet Union.[15]

To effectively tap into federal research monies available for "scientific" research, engineering faculty required graduate students as research aides, and this required promoting graduate education in engineering and modifying undergraduate curricula to prepare undergraduates for graduate school and cutting-edge research instead of ordinary engineering practice. Thus, in the 1950s and 1960s American engineering schools shifted their curricula heavily toward the sciences (both basic and engineering) and toward mathematics. Courses focused on teaching practical skills were either eliminated or sharply reduced in number. By 1970 engineering education had become thoroughly scientized, closely paralleling academic science education in organization and emphasis.

Science, the engineer's incorporation in large organizational hier-

[14]Eugene Ferguson, "The Mind's Eye: Nonverbal Thought in Technology," *Science* 197 (August 26, 1977): 827–36, makes the same point in another context.

[15]Lawrence P. Grayson, "A Brief History of Engineering Education in the United States," *Engineering Education* 67 (December 1977): 260; and Frederick E. Terman, "A Brief History of Electrical Engineering Education," *Proceedings of the IEEE* 64 (1976): 1404–5.

archies, and the status concerns this engendered also had an important impact on engineering's professional institutions.

Institutional Accommodation

The growing ties between engineering careers and positions in large, bureaucratic organizations naturally had an effect on the professional institutions that engineers had created in the 19th century. These organizations had been founded and led by elites of independent consultants and proprietor-engineers with often very restrictive membership criteria. The changing nature of the engineering work force altered this. Between 1900 and 1930 virtually all of the major professional engineering societies in America loosened membership requirements to attract engineers employed by corporations and government agencies. In the process, they lost some of their cohesiveness and increasingly restricted their activities only to areas where consensus could be achieved, usually organizing meetings and publishing journals.[16]

Despite the efforts of turn-of-the-century professional societies to adapt to the changing environment and to prevent further professional fragmentation by modifying membership requirements to include corporate-employed engineers and by avoiding divisive political issues, fragmentation continued. Already at the turn of the century there were, according to one analysis, nine "major" technical societies. By 1950 there were twenty-seven.[17] New professional organizations continued to emerge both because of the continued growth of engineering specialities and because of disagreements within societies over membership requirements and scope of activities.

Terry Reynolds's essay, "Defining Professional Boundaries: Chemical Engineering in the Early 20th Century," provides a case study of how one segment of the engineering profession adapted its institutions to the changing environment of 20th-century American engineering.

In the late 19th and early 20th centuries, chemical engineers were professionally affiliated with the American Chemical Society (ACS). But in 1908 chemical engineers left ACS to found the American Institute of Chemical Engineers (AIChE). According to Reynolds, one of the primary factors behind the creation of the new professional society was "status anxiety." The rapid pace of organizational, technical, and scientific development in the chemical industry had deprofessionalized

[16]The best account of the history of the major engineering professional societies is Layton's *Revolt of the Engineers* (see n. 10).

[17]William G. Rothstein, "Engineers and the Functionalist Model of Professions," in *The Engineers and the Social System*, ed. Robert Perrucci and Joel E. Gerstl (New York, 1969), p. 76.

or threatened to deprofessionalize the most numerous type of chemist employed by industry—the analytical chemist. Instead of standing next to the proprietor or being the proprietor, as they had been in the smaller chemical plants of the 19th century, analytical chemists were increasingly isolated from the centers of power by intermediate layers of bureaucracy and were having their work routinized by new scientific and technical developments. Production chemists, serving in midlevel management positions in chemical plants, feared that remaining in a professional society dominated in numbers by analytical chemists and in leadership by academic chemists would lead to the decline of their "profession" to "craft or trade status." To prevent this from happening, production chemists splintered the institutional unity of the chemical profession by leaving the ACS and founding the AIChE. Almost simultaneously, the ACS responded to the increased importance of industrial employment by decentralizing through the creation of autonomous divisions. The first ACS division was created to serve ACS members who worked in industrial organizations.

Because status anxiety had prompted the split from the ACS, the founders of the AIChE used their new organization to try to reinforce their status in the corporate environment. As Reynolds notes, the membership standards adopted by AIChE were elitist, designed to differentiate its membership from lower-status occupations. As part of their accommodation to the corporate world, chemical engineers embraced management as "an essential part of what it meant to be a chemical engineer," and the AIChE incorporated managerial responsibilities in its membership criteria. Almost from the first, chemical engineers with corporate supervisory or managerial responsibilities dominated the society.

Engineering institutions also had to adapt to the increasing acceptance of science as a means of legitimizing professional standing. As Reynolds points out, chemical engineers in the AIChE sought to justify their separation from chemistry and differentiate themselves from practically trained chemical plant foremen by identifying the distinct body of "scientific" knowledge covered by their field. They found the key to defining chemical engineering in the concept of "unit operations" and used this concept to create a distinct chemical engineering science.

The institutions by which engineers were trained also faced the problem of adapting to the pervasive influences of corporate organization and scientific methods. Through most of the 19th century, engineering education programs, although stimulated by the rapid growth of the American economy, had evolved without developing close or continuing ties to American industries. In the 20th century this changed. As the corporation became the employer of the bulk of engineering students and as corporate need for engineers grew, the interests of the two became increasingly intertwined. Universities needed an outlet for their product; industry needed steadily more engineers.

W. Bernard Carlson's essay, "Academic Entrepreneurship and Engineering Education: Dugald C. Jackson and the MIT-GE Cooperative Engineering Course, 1907–1932," provides a case study of the interaction between the corporation and the engineering school in the early 20th century as the two attempted to adjust supply and demand. Carlson argues that this adjustment was not one-sided. Corporations did not simply dictate to engineering schools how their future employees and potential successors should be trained. Instead the integration of educational institutions into the American business system "occurred only gradually, marked by the clash of goals, values, and professional cultures." Educators had ideas they attempted to push on corporations, just as corporations had needs they wished educational institutions to accommodate.

To demonstrate this point Carlson analyses the interaction between Dugald C. Jackson, head of the electrical engineering department at MIT, and GE in their efforts to create a cooperative engineering course. Jackson believed that engineers should become the future leaders of industry. The cooperative course he pushed on GE tilted heavily toward training engineers for higher executive positions, instead of simply technical functions. Carlson argues that GE took a passive role in the initiation and development of the cooperative course. Jackson initiated the contacts, pursuing an educational vision that "complemented but did not match the needs or expectations of industry." As a result, the program had to be modified in ways that represented a compromise between Jackson's educational vision and corporate needs.

Jackson's program was a limited one, seeking to adjust one school to the corporate environment. Other educational institutions adopted other methods. Several groups, however, attempted to meld the interests of corporate and educational organizations at the national level during World War I and after. The National Industrial Conference Board, for example, cooperated with the Society for the Promotion of Engineering Education (SPEE) in the latter's comprehensive study of engineering education in the 1920s, particularly the portion of that study dealing with university-industry relationships. In 1932 these organizations spearheaded the formation of the Engineers' Council for Professional Development (ECPD; now the Accreditation Board for Engineering and Technology). Working with the major national engineering societies, ECPD became the agency charged with accrediting engineering schools. The ECPD provided a national forum for engineering educators and industrial representatives to interact by providing for significant industrial representation on the organization's ruling council. In subsequent years ECPD became an effective institutional conduit for blending the desires of the industrial and educational communities.[18]

Although most American engineers and most American engineering institutions eventually accommodated themselves to corporate power, not all engineers or engineering institutions readily accepted the idea of the subordination of the profession to corporate industry, particularly early in the 20th century. Peter Meiksins, in "The 'Revolt of the Engineers' Reconsidered," looks at the different forms opposition to the emerging probusiness orthodoxy took and maintains that the relations among these forms "must be the *focus* of an analysis of engineering professionalism and the engineers' role in American society."

Meiksins's essay identifies three broad factions in American engineering between 1910 and 1930, each with somewhat differing ideas about what the relationship between the engineering profession and the corporate world should be. One group was the "engineering establishment," made up of wealthy consultants and engineers in executive management. They dominated most professional engineering organizations and generally resisted the idea of a unified, activist, independent engineering profession. They saw no conflict between the interests of business and those of engineers, feeling that engineers operating as *individuals,* not as a group, should pursue careers in corporate management and espouse political ideals that would favor corporate interests.

A second major group Meiksins classifies as "patrician reformers." Its membership came from several sources, including "scientific management" consultants and some government-employed engineers. This group sought a unified, independent, activist engineering profession that would play a leadership role in American society. They saw engineering as a profession that could serve as the "loyal opposition" to business, even while recognizing business's importance and central role in engineering practice. The "patrician reformers" actively sought to unify the profession so it could speak with a united voice and stand against the growing power of corporate organizations. The idea of engineering subordinated to business bothered them.

---

[18]David F. Noble, *America by Design: Science, Technology, and the Rise of Corporate Capitalism* (New York, 1977), pp. 234–56, discusses, from a Marxist perspective, university-industry relations in the 20th century.

The third faction in American engineering in the early 20th century, according to Meiksins, was "rank-and-file" engineers. Many of these were young employees of large organizations suffering from low wages, poor working conditions, and status anxieties. This group did not challenge business's right to rule as the "patrician reformers" did. Its concerns centered on economic issues, particularly pay, working conditions, and promotion opportunities.

Meiksins argues that the best way to understand the history of the engineering profession in the early 20th century is to look at the interactions between these three factions. A large segment of his essay deals with the American Association of Engineers, an engineering organization that flourished between 1918 and 1920 and briefly threatened the dominant position of the older, discipline-based engineering societies. He sees the short-lived success of this organization as due to a temporary alliance between the "patrician reformers" and the "rank-and-file."

This alliance fell apart, however, as "establishment engineers" yielded to the coalition on minor issues, while simultaneously driving a wedge between the "patrician reformers" and the "rank-and-file." For example, the engineering establishment went along, for a period, with the creation of a broad, inclusive, public-service oriented society to unite all engineers—the Federated American Engineering Societies. This federation of societies blunted the drive toward creation of a unity organization based on individual membership. Simultaneously, the engineering establishment drove a wedge between the "patrician reformers" and the "rank-and-file" by accusing the American Association of Engineers of engaging in unionism.

In the long run, the form of engineering professionalism pushed by the "patrician reformers" died "as engineers became accustomed to bureaucratic employment." "In its place," Meiksins notes, "there developed a pattern of often muted but occasionally overt conflict between engineers and their employers over material issues."

Since the mid-1920s most of the major engineering professional societies have been firmly under the control of Meiksins's "engineering establishment" or at least neutralized for fear of antagonizing that group. The one society that today best represents Meiksins's third group (the "rank and file") is the National Society of Professional Engineers (NSPE). Founded in 1934, it has focused more than other engineering institutions on economic issues affecting engineers. Despite success in securing passage of laws in all states requiring licensing for certain restricted types of engineering practice, however, it has never attracted more than around 5 percent of all engineers to its membership.

By midcentury American engineering institutions had generally adjusted to corporate employment. The vision of an independent profession advocated by Meiksins's "patrician reformers" had died.

Changing Demographics

Throughout the 19th century, because of prevailing social attitudes and institutional restrictions, engineering was overwhelmingly a white, male occupation. The growing dominance of corporate hierarchies as places of engineering employment in the 20th century did not immediately change matters. Successful corporate engineering careers required working long hours and placing career and organizational concerns above family concerns. In a society where traditions and mores made women the keystone of family life, corporate engineering initially opened no more gates than the more individualistic engineering of the earlier century. Although records are sketchy, one study found only eight women graduates in engineering before 1900, and only forty-five before 1920.[19]

The institutional barriers preventing women from entering the field began to disappear rapidly in the first half of the 20th century. By 1940 at least seventy engineering schools had admitted and graduated at least one woman engineer. Engineering "manpower" shortages in World War II stimulated a burst in the number of women engineering graduates, with 710 graduating between 1940 and 1949, compared with only 156 the decade before.[20] But the entrance of males into the engineering work force in very large numbers after World War II kept the proportion of female engineering students, despite increases in absolute numbers, at less than 1 percent a year well into the 1960s.[21]

Several factors combined in the 1970s and 1980s to begin diversifying the demographic pool from which American engineers were recruited. One factor was the Cold War competition with the Soviet Union. Following Soviet successes in space in the late 1950s and early 1960s, American educational, political, and industrial leaders became concerned about the much larger number of engineers being produced by the Soviet Union and about the future supply of American engineers.[22] Concern about engineering supply was further exacer-

[19]Edna May Turner, "Education of Women for Engineering in the United States, 1885–1952," Ph.D. dissertation, New York University, 1954, pp. 186, 188.

[20]Ibid.

[21]Stanley S. Robin, "The Female in Engineering," in *Engineers and the Social System*, ed. Robert Perrucci and Joel E. Gerstl (New York, 1969), p. 207; and Martha M. Trescott, "Women Engineers in History: Profiles in Holism and Persistence," in *Women in Scientific and Engineering Professions*, ed. Violet B. Haas and Carolyn C. Perrucci (Ann Arbor, Mich., 1984), pp. 181–204, discuss elements of the early history of women in engineering.

[22]Robin, p. 203.

bated in the 1970s and 1980s as the pool of available white males declined because of lower white birth rates and the end of postwar baby boom children entering the workforce. Since white males were not entering the field in sufficient numbers, industrial organizations began to see females and minorities as reservoirs for drawing out the steadily larger number of engineers they needed to design, operate, and manage their technological systems. The civil rights and women's liberation movements of the 1960s reinforced this by eliminating additional institutional barriers and muting some of the mores and traditions that had long inhibited minorities and women from entering engineering.

As a result, the near monopoly that white males had on the engineering profession began to break in the 1970s and 1980s. Between 1972 and 1979, for example, women as a percentage of full-time undergraduates in engineering increased from 2.3 to 12.4 percent.[23] Between 1973 and 1988 the number of minority engineering graduates, excluding Asian-Americans, increased by a factor of ten from less than 1 percent to around 6 percent.[24]

Growing corporate need for engineers, coupled with political and social changes, had begun to have a significant impact on the demographics of American engineering by the late 1980s. In 1963 only 2.1 percent of all American engineers were nonwhite, only 0.7 percent were female. In 1988 the proportion of nonwhites in the engineering work force had increased by a factor of five, to 10.2 percent, and the proportion of females by a factor of ten, to 7.3 percent.[25]

Conclusion

By the late 20th century the American engineering profession had largely completed its adjustment to the requirements of large and complex organizations and the growing importance and status of science.

In the case of science, American engineers had sharply increased the emphasis placed on basic science, on mathematics, and on theory in engineering training. Further, they increasingly used possession of a "scientific" corpus of knowledge as a means to distinguish themselves from what they saw as lower-status personnel, such as skilled craftsmen and technicians. And, finally, engineering research techniques, for better or worse, had come to more closely resemble those of the sciences. By the late 20th century the boundary between scientific work and engineering work had become very difficult to distinguish.

[23]Jane Z. Daniels and William K. LeBold, "Women in Engineering: A Dynamic Approach," in *Women and Minorities in Science: Strategies for Increasing Participation,* ed. Sheila M. Humphreys (Boulder, Colo., 1982), p. 140.

[24]Edmund W. Gordon, "Educating More Minority Engineers," *Technology Review* (July 1988): 69–70.

[25]Eck, p. 92.

By the mid-20th century American engineers had also largely accommodated themselves to employment in large organizations and subordination to business. Instead of seeking wealth, status, autonomy, and influence as independent consultants or proprietor-engineers, they now sought these as corporate managers. Contributing to the success of the organizations in which they worked had become more important than the abstract advancement of engineering.[26]

The accommodation, however, had not been all one-sided. In return for loyalty and subordination to organizational interests, corporations provided employment to steadily larger numbers of engineers and began to recruit these engineers beyond the traditional demographic bounds. They gave engineers more autonomy than other occupational categories.[27] And, just as important, these enterprises accepted engineers as part of the management team, recruiting many of their midlevel management people from their engineering staffs and providing engineers with the possibility of rising to the top. In fact, by 1950, around one out of every five presidents and board chairmen in large American corporations had an engineering background.[28] One historian has even asserted that modern management "was the product of engineers functioning as managers."[29] Although it cost the American engineering profession the independence and individualistic values that it had prized in the 19th century, association with management preserved for the profession at least some of the autonomy and much of the status it had so feared losing at the beginning of the 20th century.

[26]Robert Zussman, *Mechanics of the Middle Class: Work and Politics among American Engineers* (Berkeley and Los Angeles, 1985), pp. 223–24.
[27]Ibid., pp. 110–11.
[28]Mabel Newcomer, *The Big Business Executive* (New York, 1955), p. 90.
[29]Noble, p. 263.

Background

THE ELECTRIFICATION OF AMERICA: THE SYSTEM BUILDERS

THOMAS P. HUGHES

   Isaiah Berlin in *The Hedgehog and the Fox* quoted the Greek poet Archilochus, who wrote, "The fox knows many things, but the hedgehog knows one big thing." This essay on the "Electrification of America" is about hedgehogs. Sir Isaiah describes them as those "who relate everything to a single central vision, one system less or more coherent or articulate." Foxes, in contrast, pursue many ends, "often unrelated and even contradictory." Berlin categorizes Dante, Plato, Lucretius, Pascal, Hegel, Dostoyevsky, Nietzsche, Ibsen, and Proust among the hedgehogs.[1] I want to add Thomas Edison, Samuel Insull, and S. Z. Mitchell.
   Edison invented systems, Insull managed systems, and Mitchell financed their expansion. These systems were electric light and power, now usually called utilities. Edison invented the system that took form as the Pearl Street generating station of the New York Edison Illuminating Company, now Consolidated Edison Company; Insull managed electric light and power companies that consolidated into Chicago's Commonwealth Edison Company; and Mitchell provided for the growth of large regional power systems. The three men focused upon one level of the process of technological change, such as invention, management, or finance, but in order to relate everything to a single central vision they had to reach out beyond their special competences: Mitchell managed, Insull financed, and Edison knew management and finance, as well. For this reason, Edison should be called an inventor-entrepreneur, Insull a manager-entrepreneur, and Mitchell a financier-entrepreneur—"entrepreneur" indicating the or-

THOMAS P. HUGHES is Mellon Professor of the History and Sociology of Science at the University of Pennsylvania. Recently he has published *American Genesis* (Penguin, 1990) and cowrote with Agatha Hughes *Lewis Mumford: Public Intellectual* (Oxford University Press, 1990).

[1] Isaiah Berlin, *The Hedgehog and the Fox: An Essay on Tolstoy's View of History* (New York, 1953), p. 1.

This essay originally appeared in *Technology and Culture*, vol. 20, no. 1, January 1979.

ganizational, system-building drive of the three men.[2] One hesitates to speak of inventor-hedgehog, manager-hedgehog, or financier-hedgehog.

Edison, Insull, and Mitchell were strong holistic conceptualizers and determined solvers of the problems frustrating the growth of systems. This essay, therefore, is also a history of ideas and a study of problem solving. Their strong concepts resulted from the need to find organizing principles powerful enough to integrate and give purposeful direction to diverse factors and components. The problems emerged as the system builders strove to fulfill their ultimate visions. Not one of them was satisfied to solve a part of the problem, simply to invent, manage, or finance, for each believed that the invention would not become an innovation, the managerial structure would not evolve, and the financial means would not bring growth unless electric light and power were viewed as a coherent system.

Besides focusing upon systems, directing attention to the men who presided over their growth, this essay identifies stages in the history of electric light and power. Around 1880 when Edison flourished, electric light and power were clearly in the inventive stage, and he is representative of many other leading inventors like Elihu Thomson, William Stanley, and Nikola Tesla; Samuel Insull rose to prominence about one-quarter century later, after the technology had been shaped and managing large utilities was an even greater challenge. As a result, the names of utility heads like John Lieb, Alex Dow, and Insull dominate the industry. In the twenties, invention and management remained important, but regional systems financed, organized, and managed by holding companies dominated the scene, and men like Mitchell, Charles Stone and Edward Webster, and, again, Insull were preeminent.

Edison: Inventor-Entrepreneur

Edison was not a simple tinkerer hunting and trying his way to new inventions. He said that he was no genius of heroic proportions; invention—he explained—was 99 percent perspiration and 1 percent inspiration.[3] His more scrupulous and better informed biographers[4] portray him as more than an inventor; they describe his engineering

---

[2] I have discussed my concept of an entrepreneur as one who presides over invention, development, and innovation in *Elmer Sperry, Inventor and Engineer* (Baltimore, 1971), pp. 63–70, 241, and 290–95.

[3] Frank L. Dyer and T. C. Martin, *Edison: His Life and Inventions*, 2 vols. (New York, 1930), 2:607.

[4] Matthew Josephson, *Edison, a Biography* (New York, 1959); Francis Jehl, *Menlo Park Reminiscences*, 3 vols. (Dearborn, Mich., 1937–41); and Dyer and Martin.

activities as he developed his inventions and his promotional efforts as he brought them into use. His notebooks give evidence that his concepts were bold and encompassing. Edison's activities covered the broad spectrum from invention to innovation; he approached problem solving systematically, and his inventive method synthesized the technological, economic, and scientific.[5]

In his early days, Edison was content to invent a quadruplex telegraph, a telephone transmitter (the receiver was a necessary afterthought for reasons of competition), or some other component of a technological system. Someone else, not Edison, integrated the components into a commercial system ready for the ultimate consumer. After he moved to Menlo Park to establish his research laboratory in 1876 and when he decided to introduce a system of electric lighting in 1878, his reach was far more extended and sweeping—he was ready to preside over the introduction, onto the market, of a complete system of technology synthesizing components of his own invention. As an inventor-entrepreneur, he coordinated a team of electricians, mechanics, and scientists and cooperated with associates concerned about the financial, political, and business problems affecting the technological system.

After conceiving in general and sweeping terms of a system of incandescent lighting in the fall of 1878, Edison announced his brainchild with fanfare in the *New York Sun* on October 20, 1878. Always good newspaper copy, he told reporters of plans for underground distribution in mains from centrally located generators in the great cities; predicted that his electric light would be brought into private houses and simply substituted for the gas burners at a lower cost; and confidently asserted that his central station would furnish "light to all houses within a circle of half a mile." He spoke not only of his incandescent lamp but of other envisaged components of his system, such as meters, dynamos, and distribution mains. A month earlier he had written privately of his concept: "have struck a bonanza in Electric Light—indefinite subdivision of light."[6] He was, in essence, sharing his moment of inspiration with associates and the readers of the *Sun;* he had no generator, no promising incandescent lamp, much less

---

[5]There are at least 200 laboratory notebooks for the period 1878–80 when Edison was inventing his electric light system. These, along with many more Edison notebooks, are housed at Edison National Historic Site (National Park Service), West Orange, N.J. For this paper, I have drawn especially upon notebooks for November and December 1878. I am grateful to Arthur Abel, archivist, for guidance in using the archives at West Orange.

[6]Telegram from Edison to Theodore Puskas, September 22, 1878. All Edison telegrams and letters cited are in Archives, Edison National Historic Site, West Orange, N.J., unless specified otherwise.

194   *Thomas P. Hughes*

a system of distribution—these were at least a year away. Edison, however, had the concept: "I have the right principle," he wrote, "and am on the right track, but time, hard work and some good luck are necessary too. It has been just so in all of my inventions. The first step is an intuition, and comes with a burst, then difficulties arise—this thing gives out and then that 'Bugs'—as such little faults and difficulties are called—show themselves and months of intense watching, study and labor are requisite before commercial success or failure is certainly reached."[7] But he had "the right principle."

Others also report that Edison had a general concept of his system in the fall of 1878. Francis Jehl, who joined Edison as a laboratory assistant in November and who later published reminiscences of the Menlo Park days, recalled that in October 1878—twelve months before the construction of a practical incandescent lamp and the announcement of his basic generator design—"Edison had his plans figured out, as a great general figures out his battle strategy before the first cannon is fired."[8] The secret, according to Jehl, of his accomplishments "lay in his early vision, far in advance of realization."[9]

Edison conceptualized so audaciously and embarked upon the invention of an entire system because he had a laboratory and staff to draw upon. He integrated the men and facilities with his concept just as he did the technical components. At Menlo Park there was a hierarchy of systems. His notebooks show that he assigned to his Menlo Park electricians, mechanics, and scientists problems associated with the various components (various parts of the general problem) of the system. The broad concepts were generally his; the men experimented and calculated within his guidelines. Among those to whom he turned often in the first two years of work on the electric light system were Francis R. Upton, Francis Jehl, Charles Batchelor, and John Kreusi. An analysis of the first 200 of the laboratory notebooks, which begin in November 1878 and cover the years 1879 and 1880, indicates that Francis Upton figured most often in the experimentation and calculations.[10]

Francis Upton did a literature search for Edison in the fall of 1878 in New York City before he joined him at Menlo Park. Just before taking up residence there in December, Upton asked if Edison wanted him to continue the search in Boston, Massachusetts, because the "Berlin summary of Progress in Physics since 1857 . . . and an

[7]Edison to Puskas, November 13, 1878.
[8]Jehl, 1:216.
[9]Ibid., 1:217.
[10]"Memorandum of Contents of Notebooks from Edison Laboratory," Edison Archives, West Orange, N.J.

index to Poggendorff's Annalen" were there.[11] Edison knew his aspirations to invent a system in a field of technology cultivated by scientists—as well as electricians—could only be fulfilled if he drew upon science; Upton reinforced and supplemented Edison in this regard.[12] Edison's systematic approach ignored disciplinary boundaries; today we would say that he was problem, not discipline, oriented.

Upton had come a long way to Menlo Park. Characterized as a scholar and gentleman by his plainer Menlo Park companions, he had studied at Phillips Academy Andover, Bowdoin College, Princeton University, and under Hermann von Helmholtz at Berlin University. Grosvenor P. Lowrey, Edison's counsel, business, and financial adviser, recommended him to Edison knowing of his need for a physicist and mathematician. Francis Jehl said that whatever Upton did and worked on "was executed in a purely mathematical manner and any wrangler at Oxford would have been delighted to see him juggle with integral and differential equations. . . ."[13] Upton often concentrated upon the development of a dynamo for the system.

Jehl appears frequently in the notebooks in connection with lamp filament investigations. He also came to Edison in November or December of 1878 on the recommendation of Lowrey. As a boy, he had read "every scientific paper I could find," and as a young man, he became a great admirer of Edison's.[14] Lowrey, who was general counsel for Western Union, employed Jehl as an office boy and arranged for him to take an apprentice course in the Western Union repair shops. Jehl also attended Cooper Union evenings studying chemistry, physics, and algebra.

Another member of the electric lighting team was Charles Batchelor. He, too, filled out the Edison system, for he was an ingenious master craftsman, dexterous and sharp-eyed, whose wide-ranging experimental techniques and mechanical aptitude kept him at Edi-

[11]Upton to G. Lowrey, December 12, 1878.
[12]The periodicals in Edison's library at Menlo Park show him to have been a regular recipient of the leading science and engineering periodicals, U.S. and foreign. Dr. Otto Moses, fluent in German and French, was in charge of the library (Bryon Vanderbilt, *Thomas Edison, Chemist* [Washington, D.C., 1971], p. 40). The library holdings have been reassembled in the restored library and science building at Greenfield Village, Dearborn, Mich. I am grateful to the Edison curator there, Robert G. Koolakian, for showing me these. Also Upton's 1878 abstracts of the scientific and technical periodicals survive at Edison Archives, West Orange, N.J. Edison's library preserved at his West Orange Laboratory also has foreign and U.S. scientific and technical periodicals of the 19th century.
[13]Jehl, 2:619.
[14]Ibid., 1:15.

son's right hand. Batchelor was so intimately involved with Edison in all of his work "that his absence from the laboratory is invariably a signal for Mr. Edison to suspend labor."[15] John Kreusi, who was in charge of the Menlo Park machine shop, also played a major role. Trained in Switzerland as a fine mechanic, he could adeptly construct Edison's various designs with nothing more than rough sketches and cryptic instructions. He, like Batchelor, had been with Edison in Newark, New Jersey, before the establishment of the Menlo Park laboratory.[16]

Many others at Menlo Park were assigned to work on various components of the evolving electric light system. Dr. Herman Claudius, a former officer in the Austrian telegraph corps, built simulations of the system with batteries for generators, fine wires for the distribution system, and resistors for the load. Jehl reported that Claudius had Kirchhoff's laws of conductor networks at his fingertips.[17] The names of some other Edison pioneers who made it possible for him to invent and develop an entire system include John "Basic" Lawson, J. F. Ott, Dr. A. "Doc" Haid, William J. Hammer, Edward H. Johnson, Stockton Griffin, George and William Carman, Martin Force, and Ludwig Boehm.

The availability of these varied talents helps explain the encompassing character of Edison's concept of a system. Furthermore, they were supported by a broad array of expensive machine tools, chemical apparatus, library resources, scientific instruments, and electrical equipment in the Menlo Park laboratory complex.[18] A major reason for the establishment of the Edison Electric Light Company in October 1878 was to acquire funds for additional laboratory equipment and new workers like Upton and Jehl. Obviously, the common characteristics of Edison, his men, and the laboratory were shaped by a systematic, demanding endeavor.

At Menlo Park there was more than a system, there was a community as well. Edison chose Menlo Park because the isolated rural setting insulated the staff from the distractions of an urban environment like Newark. Edison and other married members of the community bought or rented farmhouses in the vicinity; Upton brought his bride to a comfortable house provided with the new Edison light. Others lived at Mrs. Jordan's cozy, nicely appointed boardinghouse located a short walk from the laboratory compound. The meals were undoubtedly country and hearty, and the environment was well ordered.

[15]*New York Herald,* December 21, 1879, quoted in Jehl, 1:393.
[16]Jehl, 1:54.
[17]Ibid., 2:545.
[18]Jehl describes the scientific instruments in his *Reminiscences,* see esp. 1:257–70.

There are scores of anecdotes about the character of life in the laboratory, including accounts of late-hour breaks after especially arduous days. On these occasions the pipe organ at the end of the lab's second floor added to the festive consumption of food and drink. Since the working day sometimes extended nearly twenty-four hours, it can be assumed that Edison was willing to charge the expenses to business.

The system, the community, and the style of invention were essentially Edisonian. Few witnesses or historians challenge the conclusion that the organizing genius was Edison's. Yet there was one man, Grosvenor Lowrey, who during the early years of the electric light project appears to have closely advised Edison on financial and political matters. Edison laid down the guidelines for Batchelor, Kreusi, and Upton in the laboratory, but Lowrey often guided Edison when the problems involved Wall Street or New York City politicians. Edison, however, did not step back, immerse himself in technological and scientific problems and leave the "politics" to Lowrey; the correspondence shows that Edison always had a prominent role in the financial and political scenarios.

Because of his knowledge of the world of legal, business, and financial affairs, Lowrey's strengths complemented Edison's. Born in Massachusetts, Lowrey took up the practice of law in New York City and rose to prominence. He acted as counsel to the U.S. Express Company, Wells Fargo & Company, and the Baltimore & Ohio Railroad. He was also legal adviser to the financial entrepreneur Henry Villard. In 1866 he became general counsel of the Western Union Telegraph Company, a position that brought Edison and him together in connection with telegraph patent litigation. Lowrey was one of those who persuaded Edison to turn to electric lighting.[19] Having observed the sensational publicity given to the introduction of the Jablochkoff arc light in Paris in 1878, Lowrey urged Edison to enter the field and offered to raise the money Edison needed to expand Menlo Park. Not only did he advise Edison, he often encouraged the inventor. Lowrey promised in 1878 that the income from electric lighting patents would be enough to fulfill an Edison dream: "to set you up forever . . . to enable you . . . to build and formally endow a working laboratory such as the world needs and has never seen."[20] (At the time the only buildings in the Menlo Park group were the laboratory building, the carpenter shop, and the carbon shed—there were no machine shop, library, or office buildings.) Shortly afterward, Edison gave Lowrey a free hand for this purpose in negotiating the sale of forthcoming

[19]Payson Jones, *A Power History of the Consolidated Edison System, 1878–1900* (New York, 1940), p. 27; see also Jones, p. 161, on Lowrey.

[20]Lowrey to Edison, October 10, 1878.

198   Thomas P. Hughes

electric lighting patents and establishing business associations and enterprises at home and abroad: "Go ahead. I shall agree to nothing,
promise nothing and say nothing to any person leaving the whole
matter to you. All I want at present is to be provided with funds to push
the light rapidly."[21]

Lowrey had close contacts with the New York financial and political
world. His law offices were on the third floor of the Drexel
Building—Drexel, Morgan, and Company had the first floor. Working closely with his long-time friend, Egisto P. Fabbri, "an Italian
financial genius"[22] and partner of J. Pierpont Morgan, he obtained
the funds for Edison from Drexel, Morgan, and Company. His skill
and effectiveness in dealing with politicians and political problems is
conveyed by a Menlo Park episode. In December 1879, Lowrey arranged a lobbying extravaganza. The objective was to obtain a franchise allowing the Edison Illuminating Company to lay the distribution system for the first commercial Edison lighting system in New
York City. Behind the opposition of some New York City aldermen
lay gaslight interests and even lamplighters who might be thrown out
of work by the new incandescent light. A special train brought the
mayor and alderman to Menlo Park. In the dusk they saw the tiny
lamps glowing inside and outside the laboratory buildings. After a
tour and demonstration by Edison and his staff, someone pointedly
complained of being thirsty, which was a signal for the group to be led
up to a darkened second floor of the laboratory. Lights suddenly went
on to disclose a lavish "spread" from famous Delmonico's. Lowrey
presented Edison and the Edison case after dinner; in due time the
franchise was granted.[23] The franchise was as necessary for commercial success as a well-working dynamo.

The organization and early management of the companies formed
by Lowrey and Edison in connection with the electric light system
have been well told elsewhere.[24] Here it is important to stress that the
pristine character of the companies manifested Edison's determination to create a coherent system and his willingness to preside over the
broad spectrum of technological change. The first company
formed—the Edison Electric Light Company—was essentially a
means of funding Edison's inventive activity and obtaining a return of
investment by sale or licensing of patents on the system throughout
the world. The Edison Electric Illuminating Company of New York

[21]Edison to Lowrey, October 2, 1878.
[22]Lewis Corey, The House of Morgan, p. 23, quoted in Jones, p. 162.
[23]Jehl, 2:778-85.
[24]Harold C. Passer, The Electrical Manufacturers, 1875-1900 (Cambridge, Mass.,
1953).

was a licensee of the parent Edison Electric Light Company. The EEIC built the first commercial Edison system with its central generating station on Pearl Street in New York City, which was started in September 1882. Because Edison invented and developed all major components for the integrated system—except the boilers and steam engines—he had also to establish the Edison Machine Works to build dynamos, the Edison Electric Tube Company to make the underground conductors, and the Edison Lamp Works to turn out incandescent lamps in quantity. He entered into a partnership with Sigmund Bergmann, a former Edison employee, in a company to produce various accessories.[25] Not only was Edison the pivotal figure in the companies during the early years, he personally supervised the construction of Pearl Street Station. In these companies, Edison was an engineer and a manager, but the focus and the commitment for him remained invention.

Supplemented and complemented by his laboratory staff and by the particular resources of Lowrey, Edison solved problems associated with technological change on various levels and in a systematic integrated way. His systematic approach to problem solving was most clearly demonstrated, however, in the invention of incandescent light technology. Edison could not conceive of technology as distinct from economics—at least, when engaged with the electric light system. After initiating the project he read extensively and deeply about gaslighting from central stations, especially the economics of it. Also he canvassed the potential lighting market in the Wall Street district in New York where he intended to locate his first central station.[26] His notebooks show that he analyzed the cost of operating the Gramme and the Wallace arc light generators that he had acquired for test purposes.[27] From available literature, he and Upton also determined the cost of operating a Jablochkoff arc-lighting system. Laboratory notes reveal that he was especially concerned about the cost of copper and hoped to reduce it in generator and distribution wiring.[28] As early as December 1878, he estimated that the physical plant needed for one incandescent lamp in his sytem would require capitalization of $11. At an interest rate of 10 percent on this investment and assuming

[25] For a chart showing the various Edison companies and their relationships, see, Jones, p. 13.

[26] Jehl, 1:215, 2:731–32.

[27] Menlo Park Notebook no. 6 (December 4, 1878–January 30, 1879), pp. 22–30.

[28] See Menlo Park Notebook no. 1 (November 28, 1878–July 24, 1879), section on wire calculations, and Notebook no. 12 (December 20, 1878), pp. 174–75, 232–33. On cost of operating Jablochkoff candle, see Notebook no. 6 (December 4, 1878), p. 57. (If only one date is given, it is for first dated entry in notebook; if two dates, the second is for last dated entry. All notebooks are at Edison Archives, West Orange, N.J.).

lamp use of 300 hours a year, the charge per lamp would have to be more than 3.66 mills per hour.[29] Edison was clearly thinking within the context of a capitalistic system.

Perusal of Edison's notebooks should lay to rest the myth that he was a simple inventor tinkering with gadgets. There on page after page are concepts, ingenious experimentation, careful and sustained reasoning, and close economic calculation. Notebook number 120 (probable date, 1880), for example, has thirty pages of calculations (probably Upton's under Edison's instructions) about the costs and income of a central station supplying 10,000 lights. These were probably in anticipation of the Pearl Street system to be built in New York City. By the time the calculations were made Edison and Upton knew enough from experimentation and literature searches to assume that a 1-h.p. steam engine and dynamo could supply eight 16-c.p. incandescent lamps. Therefore they needed about 1,200 h.p. for the 10,000 lamp system. To house this power plant, they estimated an iron structure, or building, that would cost $8,500 (table 1). Using a Babcock and Wilcox estimate, they figured $30,180 for boilers and auxiliaries. Kreusi predicted that the steam engines and dynamos would cost $50,000. After extensive calculation, they anticipated $57,000 for conductors for the system and $5,000 for meters. It followed that on an annual basis using appropriate rates of depreciation, depreciation charges on the building, boilers, engines, dynamos, meters, and conductors would amount to $6,058. Daily labor charges were taken as: chief engineer ($5.00); engineer ($3.00); wiper ($1.50); principal fireman ($2.25); assistant fireman ($1.75); a chief voltage regulator ($2.25); an assistant voltage regulator ($1.75); and two laborers ($3.00). The total labor for the day would be $20.50, or $7,482 annually (the duty day was twelve hours). For "executive" help the annual cost would be $4,000; rent, insurance, and taxes were estimated at $7,000. (The rent is not explained.) Finally, coal cost was calculated as $8,212 annually ($2.80 per ton and 3 lb/h.p.) and oil, waste, and water taken as one-third of coal, or $2,737. Since the central station would furnish the lamps, the estimate was for 30,000 a year at 35¢ each, or $10,500 annually. The total cost then was $45,989 annually. To estimate income, Edison and Upton assumed that the 10,000 lights could be sold for five hours daily, or 18,250,000 hours annually. They had learned that 10,000 equivalent (15 c.p.) gaslights used five hours a day took 250,000 ft³ of gas each day, or 91,250,000 ft³ annually. Since the gas companies charged customers $1.50 for

[29]Notebook no. 6, p. 177; Notebook no. 120 (November–December 1880, approximate date), pp. 71–101.

each 1,000 ft³, income for the gaslight company from supplying light equivalent to Edison's planned central station would be $136,875. Having decided that the price to the consumers would be the same as gas, the calculations showed for the Edison central station an excess of $90,886 in receipts over expenses. The notation beside the excess was simply "to pay for patent rights and interest."[30] The extended analysis continues by calling for the company to capitalize at twice the cost of the plant (2 × $150,680 = $301,360), which meant that the receipts

TABLE 1

EDISON ESTIMATE FOR TEN THOUSAND LAMP CENTRAL STATION

Capital Investment:			Depreciation	
Power plant building		$ 8,500	2%	$ 170
Boilers and auxiliary equipment		30,180	10%	3,018
Steam engines and dynamos		48,000	3%	1,440
Auxiliary electrical equipment		2,000	2%	40
Conductors		57,000	2%	1,140
Meters		5,000	5%	250
Total		$150,680		$6,058
Operating and Other Expenses:				
Labor (Daily):				
Chief engineer	$ 5.00			
Assistant engineer	3.00			
Wiper	1.50			
Principal fireman	2.25			
Assistant fireman	1.75			
Chief voltage regulator	2.25			
Assistant voltage regulator	1.75			
Two laborers	3.00			
Total	$20.50			
Labor (Annual)		$ 7,482		
Other:				
Executive wages (annual)		$ 4,000		
Rent, insurance and taxes		7,000		
Depreciation		6,058		
Coal (annual)		8,212		
($2.80/ton; 3#/h.p. hour;				
5 hours daily; 1,200 h.p.)				
Oil, waste, and water		2,737		
Lamps (30,000 at 35c each)		10,500		
Total		$45,989		
Estimated Minimum Income from 10,000 Installed Lamps			$136,875	
Expenses			− 45,989	
			$ 90,886	

SOURCE:—Edison Menlo Park Notebook no. 120 (1880).

[30]Notebook no. 120, p. 99.

would allow a dividend of 30 percent on investment. Apparently the holders of the patent rights and interest (the Edison Electric Light Company) took the excess payable in dividends as the payment on patent rights and interest.

Calculations like these were as much a part of Edison's invention and development of an electric lighting system as his overly publicized and well-remembered endeavors to find the lamp filament. As a matter of fact the search for the lamp filament was conditioned by cost analyses like the above. It is known that Edison was determined to discover a high-resistance lamp filament in contrast to the low-resistance one generally tried before him by inventors of incandescent lamps; it is not widely realized that his determination was a logical deduction from cost analysis. To explain this, we must consider the cost analysis once more and also introduce science. In doing so we shall demonstrate that Edison's method of invention and development in the case of the electric light system was a blend of economics, technology (especially experimentation), and science. In his notebooks pages of economic calculation are mixed with pages reporting experimental data, and among these one encounters reasoned explication and hypothesis formulation based on science—the web is seamless. His originality and impact lie as much in this synthesis as in his exploitation of the research facilities at Menlo Park.

Among costs listed above, the $57,000 for conductors was the highest capital item. Of this, $27,000 was for copper conductors, $25,000 for the pipes containing them, and $5,000 for insulation. Early in his project, Edison saw that copper cost, especially that dependent upon the cross-sectional area and the length of his conductors, was a major variable in the cost equation. Large and long conductors might have raised the price of electric light above gas. To keep length down, he sought a densely populated consumer area; to keep the cross-sectional area small, he had to reason further using the scientific laws of Ohm and Joule. The notebooks show that Edison and Upton used Joule's Law (heat or energy = current^2 × resistance = voltage × current) to calculate energy expended in the incandescent filaments.[31] They also used an adaptation of it to show energy loss in conductors. Energy loss was taken as proportional to the current^2 × the length of the con-

[31]Notebook no. 3 (November 21, 1878), p. 107; and Notebook no. 9 (December 15, 1878–March 10, 1879), p. 41, have some of many early entries using Ohm's Law. Notebook no. 6, pp. 11 ff., shows use of Joule's Law. Since the last entry in no. 6 is January 30, 1879 (the first was December 4, 1878, see above, n. 27), Edison was using the law early in the electric light project. Notebook no. 10, p. 13, has the word "Joule" jotted down where $H = C^2R$ is being used. (This notebook has entries from December 1878 to January 1879.)

ductor × a constant dependent upon the quality of the copper used, all divided by the cross-sectional area of the conductor (energy loss proportional to C^2La/S).[32] The formula posed an enigma, for if Edison increased the cross section of the copper conductors to reduce loss in distribution, then he would increase copper costs which was to be avoided. Obviously, a trade-off—to use the jargon of the engineering profession—was in order. There was, however, another variable, the current, to consider. If current could be reduced, then the cross-sectional area of the conductors need not be so large. But current was needed to light the incandescents, so how was one to reduce it?

To solve the dilemma, Edison reasoned as follows. Wanting to reduce the current in order to lower conductor losses, he realized that he could compensate and maintain the level of energy transfer to the lamps by raising the voltage proportionately ($H = C \times V$). Then he brought Ohm's Law into play (resistance = voltage divided by current). It was the eureka moment, for he realized that by increasing the resistance of the incandescent lamp filament he raised the voltage in relationship to the current. (Resistance was the value of the ratio.)[33] Hence his time-consuming search for a high-resistance filament—but the notable invention was the logical deduction; the filament was a hunt-and-try affair.

While the essence of Edison's reasoning seems clear from the available evidence, I have yet to find in his notebooks or elsewhere the date when he realized that a high-resistance filament would allow him to

[32] Notebook no. 12, pp. 174–176.

[33] Jehl, 1:362–63, 2:852–54 stresses Edison's reliance upon Ohm's Law. This led me to look to it as an important clue to Edison's use of science and his reason. In a reference often used at Menlo Park—the essay "Electricity" in the *Encyclopaedia Britannica*, 9th ed. (1878)—Ohm's Law is stated as $R = E/C$ (resistance equals electromotive force divided by current), p. 41. In a misleadingly titled article Harold Passer argues cogently that Edison's reasoning was as suggested by Jehl ("Electrical Science and the Early Development of the Electrical Manufacturing Industry in the United States," *Annals of Science* 7 [1951]: 382–92). Passer offers no evidence, however, from the notebooks and other original Edison sources. Passer offers a similar discussion of Edison's reasoning and concepts in *The Electrical Manufacturers, 1875–1900* (Cambridge, Mass., 1953), pp. 82, 84, and 89. Dyer and Martin, 1:244–60, parallels the memorandum attributed to Edison in 1926 (see n. 4 above). Josephson (n. 4 above), pp. 193–204, 211–220, stresses Edison's use of Ohm's Law but does not note the importance of Edison's having used Ohm's Law in conjunction with Joule's in order to conceptualize his system. Josephson also dates Edison's first high-resistance lamp as January 1879 (a platinum filament), but gives no source for the statement (p. 199); nor does he provide a source for the statement that Edison came to the idea for high resistance in a "flash of inspiration" after September 8, 1878 (p. 194). A. A. Bright, *The Electric-Lamp Industry* (New York, 1949), does not offer any additional information on the way in which Edison conceived of his system. Jehl remains the most helpful published source, despite the book's lack of organization, see 1:214–15; 243–45; 255–56; and 2:820–21; 852–54.

achieve the energy consumption desired in the lamp and at the same time keep low the level of energy loss in the conductors and economically small the amount of copper in the conductors. In an essay attributed to Edison and sent to Henry Ford at his request in 1926, Edison stated that in the fall of 1878 he had experimented with carbon filaments but that the major problem with these was their low resistance. He observed that "in a lighting system the current required to light them in great numbers would necessitate such large copper conductors for mains, etc., that the investment would be prohibitive and absolutely uncommercial. In other words, an apparently remote consideration (the amount of copper used for conductors), was really the commercial crux of the problem."[34] He provided better evidence about the time of origins of his high-resistance concept in stating that "about December 1878 I engaged as my mathematician a young man named Francis R. Upton. . . . Our figures proved that an electric lamp must have at least 100 ohms resistance to compete commercially with gas."[35] Edison then said that he turned from carbon to various metals in order to obtain a filament of high resistance, continuing along these lines until about April 1879 when he had a platinum of great promise because the occluded gases had been driven out of it, thereby increasing its infusibility. Edison, then, established a search for a high-resistance filament between December 1878 and April 1879. Jehl in his *Reminiscences* maintains that Edison wanted a high-resistance lamp as early as October 1878 and had reached this conclusion by reasoning about his envisaged system of electric lighting. Jehl also states that Edison reasoned to the essentials of his system by applying Joule's and Ohm's laws.

Edison's reasoning can be illustrated with a simple example using approximate, rounded-off values. By 1880 he obtained a carbonized-paper filament with resistance ranging from 130 ohms cold to about 70–80 ohms heated. (He wanted 100 ohms.)[36] Desiring a lamp with candle power equivalent to gas, he found that this filament required—in present-day units—the equivalent of about 100 watts. This meant that the product of the voltage across the lamp and the current must equal 100 watts. Since the resistance was 100 ohms, the current had to be 1 amp because by Joule's Law the heat energy was equal to the product of the C^2 and the resistance ($100 = C^2 \times 100; C =$

---

[34]T. A. Edison, "Beginnings of the Incandescent Lamp and Lighting System," p. 4 (a typescript in the Edison Archives, West Orange, N.J.). It is dated 1926 and identified as an item sent to Henry Ford at his request. The item should be used cautiously because, by 1926, Edison and his patent lawyers had organized history with priorities in mind.

[35]Ibid., p. 5.

[36]Notebook no. 52 (July 31, 1879), p. 229. The entry is dated December 15, 1879.

1). It then followed that the voltage must be 100 in order to fulfill the energy need ($H = C \times V$; $V = 100/1 = 100$). Therefore the specifications of the lamps in Edison's system in today's terminology: 100 watts; 100 volts; 1 amp; and 100 ohms.[37]

Space does not permit an analysis of the way in which Edison invented other components of his system. The analysis of the Edison method as revealed in the invention of the high-resistance filament, however, cuts close to the core of his creativity. As he declared, there was an abundance of patient hunt and try, and even his most superficial biographers grasp this. Furthermore, those who want to discount the Edisonian method, as compared to the so-called scientific method of the laboratory scientists who followed him, choose to stress the empirical approach. This superficiality and distortion are regrettable as one more of many instances of the obfuscation of the nature of creativity. What should be stressed are the flashes of insight within a context of ordered desiderata. By ordering priorities, Edison defined the problem and insisted, as many other inventors, engineers, and scientists have, that to define the problem is to take the major step toward its solution. The prime desideratum was an incandescent light economically competitive with gas; the major flash of insight was realizing that Ohm's and Joule's laws defined the relationship between the technical variables in his system and allowed their manipulation to achieve the desired economy. Having accepted this, we can better understand why laboratory assistant Jehl in his memoirs repeatedly refers to Edison's familiarity with Ohm's Law as the explanation for his creative success and better understand why so many interpretations of Edison have been flawed by the simple assumption that he did not use science because he spoke testily of scientists, especially mathematicians.

The invention and development of the incandescent light seem to have been the leading edge of Edison's systematic approach. After the characteristics of the lamp were established, then the problem of generator design was generally defined. The generator, for instance, had to supply 100 volts for the parallel-wired incandescent lights and

[37]In January 1881 Edison conducted an economy test of his electric light system installed at Menlo Park to demonstrate its practicability. This was a prelude to his installation of a full-size system at Pearl Street, New York City, in 1882. The test showed him using two sizes of lamps, 16 and 8 c.p. The 16 c.p. depended upon an electromotive force of 104.25 volts and a resistance of 114 ohms. C. L. Clarke, "An Economy Test of the Edison Electric Light at Menlo Park, 1881," dated February 7, 1881, and published for the first time in Committee on St. Louis Exposition of Association of Edison Illuminating Companies, *Edisonia: A Brief History of the Early Edison Electric Lighting System* (New York, 1904), pp. 166–78. Therefore the current in the 16 c.p. lamp was .9 amps.

206   Thomas P. Hughes

an amperage equal to the number of lamps times approximately 1 amp. The relationship between generator and lamps was determined by the decision to wire the lamps in parallel which in turn resulted from the need to keep the system voltage at a safe level and to make possible operation of the lamps independent of one another. The Edison system was evolving like a drama with a cast of developing, interacting ideas.

In October 1879, the same month in which he found the first practical filament, Edison announced the generator for his system. Other components followed. In September 1882, the Pearl Street system began to supply light for the Wall Street district. With the opening of the Pearl Street Station of the Edison Electric Illuminating Company, the age of central-station incandescent lighting had begun; the modern age of public electric supply had opened. Edison however, gradually withdrew from the field of electric lighting. In 1892 this became apparent when a new company was formed from a merger of Edison General Electric (his manufacturing companies) and the Thomson-Houston Company, another electrical manufacturer, without his name being used (the General Electric Company). But there were a number of Edison utilities bearing his name and supplying light in the large cities of America. One of these in Chicago was taken over in 1892 by a young and close associate of Edison who proceeded to assume a position of leadership in presiding over the further growth of electric light and power systems.

Samuel Insull: Manager-Entrepreneur

When Samuel Insull (1859–1938) left the newly formed General Electric Company in 1892 to become president of the Chicago Edison Company, he gave up the vice-presidency of a $50 million corporation to become the head of an $885,000 one. He also left the electrical manufacturing business to enter the utility field. Insull recommended himself for the new position, after being asked by the Chicago concern for suggestions. He realized that the new company formed from the merger of Edison General Electric, in which he had played such a prominent role as head of sales and manufacturing, and Thomson-Houston Company, led by the impressive Charles Coffin, would be dominated by Thomson-Houston men.[38] Insull, thirty-two, accustomed to authority, and inspired by Edison's drive to create and construct, was ready to build his own system.

Chicago Edison's acceptance of Insull's nomination is easily understood, for Insull, despite his youth, had an impressive reputation in

[38]Forrest McDonald, *Insull* (Chicago, 1962), pp. 50–52.

the electric lighting field. In 1880 he had emigrated from England to America to become Edison's personal secretary. Insull's initial impression on the inventor was not positive—Insull was young, weighed 117 pounds, and had been seasick for eight successive days. Soon, however, Edison understood why Edward H. Johnson had recommended the young man who had worked as private secretary for the manager of Edison's London office: Insull was knowledgeable, intelligent, ambitious, and bold. Only hours after meeting Edison, Insull showed him how to raise an additional $150,000 needed to establish companies to manufacture the components for the electric light system being built at Pearl Street and to be erected elsewhere in America.[39]

Insull was of great value to Edison, but Edison's influence upon Insull was of greater consequence. From 1880 to 1892, Insull was the inventor's secretary and personal representative and then became the manager of the Edison General Electric plant at Schenectady. These were formative years for the electric light and power industry and for Insull. He was in at the start when Edison presided over the construction and early operation of New York's Pearl Street district; he took part in the establishment of the manufacturing facilities that ultimately became General Electric; he participated in numerous conferences involving inventors, engineers, entrepreneurs, mechanics, financiers, managers, electricians, and others who made the early history of the electrical industry, both utility and manufacturing. In short, he studied in the Edison school and absorbed its creative, problem-solving, inclusive, systematic, innovating, and expansionist approach. He knew its graduates who spread far and wide to take part in the growth of the industry; furthermore, Insull was a leading member of the school—very close to the master. Insull later said that Edison "grounded me in the fundamentals . . . no one could have had a more considerate and fascinating teacher."[40] Insull never tired of characterizing Edison as the greatest man whom he had known and the one who most shaped his character.

Years later, in 1934, when Samuel Insull stood federal trial accused of using the mails to defraud in connection with his bankrupt holding company, his defense was in essence that he, like Edison, was a creative man and one greatly interested in managing productive

[39]Burton Y. Berry, "Mr. Samuel Insull," pp. 33–34, a memoir (typescript) in the Samuel Insull Papers at the Loyola University, Chicago. These papers were used by kind permission of Samuel Insull, Jr., and are hereafter cited as Insull Papers. I appreciate the assistance of Janet Halder in using the Insull Papers.

[40]Samuel Insull, "Memoirs of Samuel Insull," a typescript written 1934–35, Insull Papers, p. 37.

technology; he denied that he was a predatory holding-company tycoon. His son, Samuel Insull, Jr., recalls that after the government prosecutor heard the defense he said privately to the son, "Say, you fellows were legitimate businessmen."[41] Insull was acquitted, but he was exhausted; the immense system of utilities built up by him lay fractured, never to be in his hands again. History has not dealt generously with Insull: The disastrous climax burned itself into the public memory. For newspapers, politicians, and former competitors, Insull was a Depression scapegoat; the decades of complex system building were easier to ignore or forget—they involved difficult concepts, esoteric technology, uncommon economics, and sophisticated management.

The creation of an all-embracing Chicago system of electric light and power was Insull's prime objective for almost two decades. Finally he reached out beyond the city and interconnected the Chicago system with suburban companies and then these with neighboring municipalities. The scope became regional. The histories of the Public Service Corporation of Illinois (a regional system) and of Insull's Middle West Utilities Company (a holding company on a national scale) are important chapters in the history of Insull—and of electric light and power.[42] An account of the Chicago system alone, however, reveals the powerful and effective conceptual synthesis of technology and economics that guided Insull and his associates' actions as they knit together and managed a Chicago system—a utility named Commonwealth Edison and considered by many as early as 1910 the world's greatest.

When Insull arrived, the Chicago Edison Company was but one of more than twenty small electric-light utilities. Within two decades, Insull and his associates had created a single, mass-producing, monopolistic, technologically efficient, and economically operated company for all of Chicago.[43] Insull became a spokesman for the utility industry, and his company was a pacesetter both in technological and business policy. The system and the way in which it was

[41]McDonald, p. 331.

[42]Historical sketches of the Middle West Utilities Company can be found in the annual reports in Insull Papers (see esp. "Report . . . for the Fiscal Year Ending December 31, 1926"). See also, "Presentation of Public Service Company of Northern Illinois in Competition for the Charles A. Coffin Prize Award April 1st 1924," Insull Papers.

[43]H. A. Seymour, "History of Commonwealth Edison Company," unpublished typescript completed in 1935 and in archives of Commonwealth Edison Company, Chicago, p. 200. I am grateful to Helen P. Thompson, librarian, for help in using the archives and library.

created became models for other urban utilities and, for the historian, the leading edge and at the same time a representative case of the waves of development that followed.

Within two years of his arrival in Chicago, Insull demonstrated his bold intentions by raising funds to build the Harrison Street Station during a year of financial panic. This generating station became known for its advanced engineering and rapid growth during its first decade.[44] With his thorough exposure to the technology of electric light and power while he was with Edison and his companies, Insull took far more than an abstract managerial interest in the way in which Harrison Street was equipped. Opening in 1894 with 400-kw direct-current generators totaling 2,400 kw and driven in pairs by highly efficient 1,250 h.p. condensing, reciprocating steam engines, within a decade Harrison Street had swollen to 16,200-kw capacity and 3,500-kw units. Sited on navigable water, the station had coal and cooling water available. The increasing capacity dictated a larger and more diverse market. So, on the business level, Chicago Edison acquired twenty Chicago utilities and their market areas. The culmination of the expansion was the formal merger with Commonwealth Electric to form Commonwealth Edison in 1907. (The two companies had been jointly managed for some years.)[45]

In presiding over the growth of the Chicago system, Insull demonstrated how new technology could be coupled to the old to make an embracing system. Technological change was unceasing but, unlike so much political and social change, the new did not destroy the old—it related to, modified, and sustained the old until age and amortization dictated retirement. As the small and inefficient companies were absorbed, their generating plants were used for a time. After 1896 and the introduction of alternating current, obsolete generating stations were transformed into substations. Insull believed that, after S. Z. de Ferranti who created the Deptford Station system in London in 1890, he was the first to transform inefficient stations into substations. (The very prefix indicates the emerging hierarchical system.) Ferranti, the forward-looking British engineer and entrepreneur, had converted the small Grosvenor Gallery Station into a transformer substation for the mammoth Deptford generating plant across the Thames.[46] Like Ferranti, Insull and his associates not only utilized alternating current, transformers, and high voltage distribution lines, but also employed the newly invented rotary converter.

[44] Ibid., pp. 202–10.
[45] "Commonwealth Edison Company," a typescript prepared ca. 1934 for Samuel Insull in preparation for his federal trial, Insull Papers, pp. 2–3.
[46] Arthur Ridding, *S. Z. de Ferranti: Pioneer of Electric Power* (London, 1964).

The importance of the rotary converter as a coupler of subsystems in system building needs stressing. Essentially a motor-generator combination, the device played a major role in ending the much-publicized "battle of the currents" or "battle of the systems" and, in the late nineties, proved an effective tool in the hands of electrical engineers and managers intent upon growth.[47] A frustration for them had been the heavy investment in downtown direct-current stations of the Edison type, for these, despite the introduction of multiwire distribution, had limited supply areas because of the high cost of low-voltage direct current distribution. The alternating current system with transformers, increasingly common after 1890, permitted high voltages for distribution and low for consumption, but the established utilities could not abandon their heavy investment in urban direct current. So alternating-current stations sprang up in the suburbs, sometimes owned by the older direct-current utilities like Chicago Edison, but of necessity operating independently. This situation goaded into action the system builders like Insull who knew that economies would result from scale whether measured in generator capacity or area of district served. The rotary converter was an answer because it coupled ac and dc systems at their interface, thus integrating the old and the new.

In 1893 the Westinghouse Corporation exhibited a rotary converter at the Chicago Exposition; General Electric had rotary converters as well. Insull and his chief engineer, Louis Ferguson, believed that they were the first to order them (May 1896), but acknowledged that Boston Edison used them first. In the case of Chicago Edison, Ferguson and Insull shut down an inefficient direct-current station on Wabash Avenue, installed a reversed rotary converter at Harrison Street (direct current from the generators was changed to alternating), and sent the high-voltage alternating current over the transmission line three miles to the former generating station, now converted into a substation, where a rotary converter changed alternating current to direct for the consumers in the district.[48]

Another invention, the frequency changer, facilitated Insull's system building and his policy of change by accommodating, inclusion, and amortization. Like the rotary converter, its role in the history of expanding electric-light and power systems has not been appreciated. In essence—again, like the rotary converter—the frequency changer was a motor generator, the motor driven by electricity of one frequency and the generator producing electricity of another frequency.

[47]On development of the converter, see Benjamin G. Lamme, *Benjamin Garver Lamme: Electrical Engineer* (New York, 1926), pp. 59, 64, 81, 102, 104, 109–16, and 133.
[48]Seymour, pp. 308–9.

Just as alternating-current and direct-current utilities had grown up side by side, systems of different frequencies coexisted unconnected until the advent of the frequency changer. The primary reason for the various frequencies was different applications—low frequencies, especially 25 cycles per second, suited stationary motors while high frequencies, especially 60 cycles, kept incandescent lights from flickering.

Insull, who usually depended upon General Electric for electrical equipment, by 1900 had the rotary converter, the frequency changer, and, of course, the transformer. Schooled by Edison, Insull never lost his sensitivity to technological developments; his economic and business acumen allowed him to grasp quickly how to exploit them. At his side were Louis Ferguson, his chief engineer, recognized as a leader by his peers, and Frank Sargent, a consulting engineer whose reputation as a power plant designer rapidly spread as he became primary consultant to Chicago Edison.[49]

Insull not only spread his distribution system, but he increased the scale of the generating units within the system. As in the case of coupling of the old and new, he was motivated by a highly developed economic sense and an aptitude for identifying the technology of the future. A historic technological achievement of which he was immensely proud was the Fisk Street Station completed in 1903; his controversial decision to build the station followed from the fact that large generating units were more efficient than small ones. Also influencing him was the realization that greater capacity reciprocating steam engines than the 5,000 h.p. unit recently installed in the Harrison Street Station were impractical because of the size.[50] Adhering to his objective of massing production in large units, he turned to the steam turbine, which was beginning to win a reputation in Europe for its relatively small size, simplicity of mechanism, and moderate first cost. Insull demonstrated in this decision—as in many others—his awareness of European technological developments and his inclination to transfer and adapt technology.[51]

Having seen steam turbines in a central station in Germany, Insull sent his chief engineer, Ferguson, and his consultant, Sargent, to Europe early in the summer of 1901 to inspect turbine installations.

[49]On Sargent, see Sargent and Lundy, *The Sargent & Lundy Story: 70 Years of Engineering Service* (Chicago, 1961).

[50]Seymour, p. 319.

[51]Samuel Insull, Jr., when traveling in Europe as a young man, investigated technological developments there at his father's request. He confirmed his father's close attention to European technological developments (interview with Samuel Insull, Jr., Chicago, August 14, 1975).

Charles Parsons had introduced the steam turbine for generator drive in 1884. There were small installations in England and in 1900 there was a major central station installation in Elberfeld, Germany.[52] Development brought a marked decrease in weight per horsepower as compared to the reciprocating engine. If Insull had been content to try a 1,000-kw unit, there would have been little controversy about the decision, for the turbines as Elberfeld were 1,000-kw units and the Hartford Electric Company in America had ordered a 1,500-kw turbogenerator from Westinghouse in 1900.[53] Committed, however, to size and efficiency, Insull wanted 5,000-kw units for the Fisk Street Station.

In Europe, Ferguson's and Sargent's enthusiasm for turbines increased. Successful installations of water turbines also encouraged them. They saw the waterpower station at Paderno of the Milan Edison Company with seven 2,000 h.p. turbines; the chief engineer informed them that the turbines had been so successful that he had ordered two steam turbines rated at 4,000 h.p. each. Greatly stimulated, the two engineers then visited the works of Brown, Boveri and Company in Baden, Switzerland, where the Milan Edison turbines would be made. There they also learned of an order from Frankfurt, Germany, for a 4,000 h.p. turbogenerator; they were convinced that Europe "was tending very definitely to the turbogenerator unit."[54]

Although Charles Coffin, president of General Electric, was interested in supplying Consolidated Edison a small turbogenerator of 1,000 kw, he would offer none of the customary guarantees for larger units, certainly not the 5,000-kw unit Insull had in mind. Insull persisted, having decided to go ahead and "bore with a big auger" rather than "living from hand to mouth" by adding a small turbine to a reciprocating-engine station.[55] The compromise reached was that General Electric would take the manufacturing risk and Chicago Edison would assume the expense—success or failure—of the installation. (The building would have to be remodeled extensively, for example, if the turbine proved unsatisfactory.) Insull placed the order December 28, 1901, and the first 5,000-kw unit went into service on

[52]Arnold Th. Gross, "Zeittafel zur Entwicklung der Elektrizitätsversorgung," *Technikgeschichte* 25 (1936): 131; R. H. Parsons, *The Early Days of the Power Station Industry* (Cambridge, 1940), pp. 170 ff.

[53]The Hartford Electric Light Company installed a single 1,500-kw steam turbine in its existing Pearl Street plant in 1901. This Westinghouse Parsons turbine was the first built in America for a utility (Glenn Weaver, *The Hartford Electric Light Company* [Hartford, Conn., 1969], p. 87).

[54]Commonwealth Edison Company, *Edison Round Table* (November 1928), p. 2.

[55]Ibid., p. 11.

October 2, 1903. Three more 5,000-kw units followed by 1905, but then these were removed to be replaced by 12,000-kw units. By 1911 there were ten 12,000-kw units in Fisk Street.[56]

Insull and other Edison officials never hesitated to celebrate the pioneering decision, but Insull later admitted that the first turbines were not economical to operate, perhaps not as economical as the reciprocating engines. "I think it was the fourth turbine," he recalled, "that was a really efficient turbine."[57] However, low first cost and labor savings offset operating problems. The installation also helped introduce the turbine to America; within a year after Fisk Street began operation, General Electric and Westinghouse received orders for 540,000 kw of turbogenerator capacity. Ferguson considered the introduction of the large turbine by his company daring, despite European precedent. "European conditions," he remarked, "individualized design of each machine, the careful coddling of new devices and the infinite pains and complication which is of the nature of European engineering gave an utterly different environment from American conditions."[58]

A look at Commonwealth Edison about 1910 shows how Harrison Street, Fisk Street, and other components had been made into a system. Besides central generating stations, there were sixty-seven substations in the Chicago system, some of them converted obsolete generating stations, others especially built as substations. Fisk Street in 1910 remained the largest of the generating stations, but just across a slip of the Chicago River stood the Quarry Street Station which began operation in 1908 and had six 14,000-kw turbogenerators in 1910. Frequencies and voltages were manipulated with ingenuity. All of the Fisk Street units generated three-phase alternating-current at 9,000 volts and 25 cycles, but three 5,000-kw-capacity transformers raised the output from three of the generators to 20,000 volts. The higher voltage current was for transmission to the more distant substations. In Quarry Street four of the generators worked at 9,000 volts and 25 cycles, which made possible interchange of current with Fisk, but two others generated at 12,000 volts and 60 cycles. Quarry Street had also used frequency changers to take 25-cycle current and change it for the 60-cycle load. Substations received current by direct line from Fisk, Quarry, and the older Harrison Street Station.

The combination of technology made possible the supply of incandescent light load, stationary motor load, and streetcar-elevated

[56]Commonwealth Edison Company, *Principal Generating Stations and Transmission System* (Chicago, 1911), pp. 3, 16.

[57]*Edison Round Table* (November 1928), p. 12.

[58]Ibid., p. 3.

railroad load. Transformer substations reduced the 25-cycle current to usable voltages for the large motors of industrial enterprises; rotary converter substations received very high-voltage, alternating, polyphase current direct from Fisk or Quarry and transformed it into direct current at 600 volts, needed by the streetcars and the elevated; and the 12,000 volt, 60-cycle current was transformed in substations into low voltages for incandescent light consumers. Since large capacity tie lines connected Fisk Street and Quarry Street, they assisted one another according to the varying peak and low loads. The single Chicago system was impressive. The total generating capacity was 219,600 kw; the high-voltage transmission lines extended for 525 miles; and the low-voltage distribution system (mostly underground) spread to all parts of a growing Chicago.[59]

No description of system technology around 1910 is complete without consideration of the load dispatcher, for he functioned as the control center of the system. After his role, or function, was defined, the system concept was consciously and operationally articulated. The load dispatcher's office at Commonwealth Edison dated back to 1903. A company publication said then that "system had assumed sufficiently definite proportion to call for the organization of an operating office prepared to handle at all times the steadily increasing intricacy of detail connected with the operation of a centralized system of electrical supply."[60] The system, in other words, could be operated as a "coherent whole."

A primary responsibility of the load dispatcher was to run Fisk and Quarry Streets in tandem, or parallel, so that each station carried a reasonable share of the varying system load. Also, when generators in one of the stations failed or had to be taken off load for repair, then the load dispatcher found the combination of generators and substations to carry the load in the interim. (There were storage batteries in some of the substations and these facilitated the meeting of peaks under usual and unusual conditions.) The load dispatcher was not only a troubleshooter, but something of a historian. At his disposal were load curves, or graphic records of the varying output from various parts of the system for each hour and each day of past years. By analysis of these the dispatcher anticipated loads resulting from the social customs and industrial routines of the Chicago population.

[59]Commonwealth Edison Company, *Principal Generating Stations*, pp. 21–22.

[60]National Electric Light Association, Commonwealth Edison Section, *How Commonwealth Edison Works* (Chicago, 1914), p. 62. The following section on the load dispatcher is based on the essay by R. B. Kennedy, "Operating Department—Load Dispatching of Old, and of Today," pp. 62–66 in Commonwealth Edison Company, *How Commonwealth Edison Works*.

(The load dispatcher even knew when people left work early on Christmas Eve, for the industrial motor load dropped and that of the streetcar and elevated rose.) He also knew that Chicago observed the Sabbath, and he planned system operations accordingly.

The load dispatcher's operating board provided a graphic model of the system for information and control. It showed all generating units, the frequency changers, the transformers, and the layout of the high-voltage transmission lines and the low-voltage distribution system. Indicators showed the condition of all switches. Measuring instruments indicated continuously output and load in various parts of the system. Of critical importance to the load dispatcher was the direct-connected telephone system providing control over all generating units and switches. Because of the complex control problems involving a number of simultaneous variables, it is not surprising that early analog computers—network and differential analyzers—were designed several decades later to take over some of the load dispatcher's responsibilities.

The technology and organization of the Chicago system were a synthesis of the ideas and activities of innumerable inventors, engineers, entrepreneurs, manufacturers, and managers from all parts of the world. Insull did not invent the Chicago system to the extent that Edison invented the Pearl Street system. Edison acquired patents on the essential components of his system and the organizing concept of the system was clearly his. Insull was not a professional inventor or an engineer. He was, however, a systems conceptualizer comparable to Edison, but on a high level of abstraction. Edison, though deeply aware of the seamless fabric of economics and technology, was relatively naïve about the long-term economic and social factors making up the environment within which his system functioned. Others like Grosvenor Lowrey told him of the expansion of consumer and money markets. Furthermore, Edison did not articulate his technological and economic concepts so that a large organization could make decisions and carry out policy without his immediate supervision.

Insull, by contrast, analyzed and articulated concepts that guided policy not only in Chicago but in other utilities as well. His conceptual syntheses involved social and market needs, financial trends, political (especially regulatory) policies, economic principles, technological innovations, engineering design, and managerial techniques. Insull discussed his concepts, policies, and experiences in addresses to utility groups and to the public.[61] Addresses around 1910 are especially

[61]Samuel Insull, *Central-Station Electric Service: Its Commercial Development and Economic Significance as Set Forth in the Public Addresses (1897–1914) of Samuel Insull,* edited by

informative and important because they were given after he had time to reflect upon the formative years in Chicago and while he was presiding over the development of a large and complex enterprise. Before considering some of his guiding principles in detail, a summary of all the important ones, the actions they dictated, and the results that followed can be given. From the start Insull was committed to large units—generators, steam engines, transformers, and others. In order to use large, efficient units, he massed production and distribution. The results were the large generating stations at Harrison, Fisk, and Quarry Streets. He also adhered to the principle of low rates to increase consumption and of a tariff structure differentiated according to the customers' demand pattern. Insull also consistently sought to improve diversity of load and load factor by carefully planned sales and load management. In order to reduce the cost of borrowing money and thereby the unit cost of output because of interest on capital investment, he sought a regular return on investment and, in this way, hoped to establish a highly favorable market for his company's stocks, bonds, and debentures. To the suprise of some of his peers in other utilities, he also sought regulation by government, for he wanted a monopoly from government. His concepts were interrelated—in sum, he wanted a mass-producing monopoly supplying a diverse load, broadly based on low, differentiated rates, and obtaining a fair return resulting from efficient production and maximum utilization of capital borrowed at reasonable interest. Insull's pursuit of his policies was facilitated by a superb statistics department that kept him informed of output, performance, costs, and other variables in his business and technological system.

The way Insull's utility managed the load deserves special attention. It is comparable to the historic managerial contributions made by railway men in the 19th century and as interesting as the widely publicized managerial concepts and policies of John D. Rockefeller and Henry Ford. He explained his method on many occasions, but never more clearly and graphically than in a 1914 address.[62] He began by explaining diversity of demand. His figure showed a block of apartment buildings in Chicago. The critical items were the "customers' separate maxima" (92 kw) and the "maximum at transformers" (29 kw). The first amount is simply the sum of individual annual maximum demands of the 193 customers in the apartment block.

William E. Keily (Chicago, 1915); Samuel Insull, *Public Utilities in Modern Life: Selected Speeches (1914–1923)*, edited by William E. Keily (Chicago, 1924).

[62]Insull, "Centralization of Energy Supply," pp. 445–75 in *Central-Station Electric Service* (hereafter cited as Insull, "Centralization").

Because of the varied habits and activities of the customers, each made his maximum demand at a different time during the year. Therefore, the maximum simultaneous demand measured at the transformers supplying the apartment block was only 29 kw. The diversity of the various consumers was measured by dividing the separate maxima by the maximum at the transformers, resulting in a diversity factor of 3.2.

Insull also explained the advantages of a diversified load by reference to eleven different classes of consumers. If each of these had made the maximum demand simultaneously, Commonwealth Edison's peak load from these customers would have been 26,640 kw. As a matter of fact, on the day that the utility carried its maximum load in 1914 (January 6), the eleven customers made only a simultaneous demand of 9,770 kw. Insull graphically showed a column indicating for each class of customer the maximum demand during the winter of 1913–14; the shaded portion of the column was the demand made on January 6, 1914, showing thereby the responsibility—or lack of it—of each class of customer for the utility *bête noire*, the annual peak load. Insull enjoyed expanding upon his theme, describing the varying characteristics of each customer or class of customer, and explaining how these could be analyzed and combined to achieve economies. For instance, Insull liked ice manufacturers because of their low level of activity during cold January days when Commonwealth Edison's capacity was strained by the heaviest loads of the year. The brickyards and quarries also fit in well with Insull's scheme of things, for "frost interferes with their business."[63] If it had not been for the complex metering and information-gathering techniques of Commonwealth Edison and the analytical services of Insull's statistics department, he would not have had the indicators that allowed him to exploit diversity and manage the load.

Closely related to the concept of diversity was load factor. Early in his career as a central station manager, Insull recognized load factor as the most important operating principle in central-station economy.[64] The concept was not his invention, but he relentlessly pursued policy following from it. Load factor is the ratio of the average to the maximum load of a customer, group of customers, or the entire system during a specified period. To raise the load factor, Insull sought customers with load curves without high peaks and low valleys. He also wanted customers whose combined load curves were diverse and combined well. To have had customers with good individual load

[63]Insull, "Centralization," p. 452.

[64]Insull, "The Development of the Central Station," in *Central-Station Electric Service*, p. 27.

curves but who made their peak demands simultaneously would have resulted in a poor system load factor despite the good characteristics of the individual loads. By load management and planned sales, Commonwealth Edison improved the load factor from year to year. As a result, the cost per kilowatt resulting from fixed, or investment, charges fell, which greatly pleased Insull—he identified interest on investment as the "most important factor in cost in any public service business."[65]

Insull recalled that he became aware of the importance and complexity of load factor when he first met Arthur Wright, manager of a small municipal station in Brighton, England. Wright also informed him about the critical relationship of metering and load factor. "We had to go to Europe," Insull continued, "to learn something about the principles underlying the sale of the product."[66] Vacationing in England during the Christmas season of 1894, Insull first encountered the results of the Wright system of metering. In Brighton Insull saw small shops burning electric light as though indifferent to the amount consumed. Upon inquiry he found that Wright had developed a metering system measuring not only use but also the extent to which each customer used his installed capacity. So metered, the customer paid a charge that was fair, ingenious, and economical. His bill reflected his share of the central station's fixed costs resulting from the generating capacity the central station had installed in order to carry his peak load; the bill also mirrored the operating cost dependent upon the electricity he consumed. A customer with many, little-used electric lamps was a more costly customer than one who had the same number installed and used them often—unless the rate reflected the installed load as well as use. The Wright metering system in essence took into account the customer's load factor. Insull acknowledged that Boston Edison Company may have considered the Wright system before Commonwealth Edison, but it was his company, he asserted, that developed and promoted it in the United States until it became a widely used basis of charging for energy.[67]

The massing of production and distribution with large high-efficiency units, the raising of diversity and load factor, and the adoption of a metering system that reflected the principle of load factor were among the technological and economic principles Insull clarified, synthesized, and articulated during his first two decades at

[65] Ibid., p. 29.

[66] Insull, "Stepping Stones of Central-Station Development through Three Decades," in *Central-Station Electric Service,* p. 351.

[67] On adoption of Wright meter system, see Insull, "Memoirs," pp. 88–89 (see n. 40 above), Insull Papers; Seymour, pp. 278–79.

Commonwealth Edison. By 1914, after twenty years, his concepts brought order embracing even more complex and abstract variables. For example, he explicated the relationship between kilowatt hours sold per capita, the utility's income per kilowatt hour sold, and the annual load factor. For comparison, he chose municipal utilities in large American cities, Berlin, and London. (Insull, with close contacts abroad, often compared American conditions to those in England and Germany.) His analysis showed that as the income per kilowatt hour fell, the output per capita increased; correspondingly, as the income per kilowatt hour decreased and the output per capita increased, the load factor increased. The Niagara system had the highest output, the lowest income per capita, and the best load factor on the graph; London had the poorest. Cheap hydroelectric power and wide area transmission explained Niagara; London had dispersed, small generating stations and a relatively small industrial, or motor, load among—from Insull's point of view—other problems. Insull took satisfaction in pointing out that Chicago, despite dependence upon thermal plants, compared favorably with Niagara and San Francisco, which utilized hydroelectric power.[68]

Insull usually had at his fingertips an impressive array of statistical data. Besides what has been noted, he could cite for Commonwealth Edison—and often for other utilities—total capitalization year by year (steady growth until 1930); the number of customers and the Chicago population (the former increasing more rapidly than the latter, again until 1930); the gross operating revenue; total operating expenses; total payroll; the connected load by class of consumer (especially a breakdown into light, power, and railway); taxes; and interest charges.

Insull's blend of technology, finance, management, and entrepreneurial drive was by no means exhausted by mass producing cheap electricity for Chicago. As early as 1910 Insull applied his concepts, technology, and managerial techniques to small town and rural electrification in Lake County, Illinois. His biographer, Forrest McDonald, believes that "the Lake County experiment was the first demonstration anywhere that systematized electric service was economically and technically possible in large areas, rural as well as urban. The news of it exploded upon the industry."[69] To fulfill his concept of an integrated centrally supplied system, Insull acquired isolated utilities serving small towns and their vicinities, shut down their inefficient generating plants, erected substations, and supplied

[68]Insull, "Centralization," pp. 461–63.
[69]McDonald, *Insull,* p. 139.

electricity from a few large generating plants by means of transmission lines. In 1911, he had organized the entire Lake County enterprise as the Public Service Company of Northern Illinois. By 1923 the company supplied a territory of 6,000 square miles and 195 communities. In the process, fifty-five municipal or privately owned electrical plants were shut down or dismantled and replaced by substations and 875 miles of high-tension transmission lines from four large, efficient central stations.[70]

Having interconnected the central stations of a Chicago utility, having interconnected Lake County utilities physically by high-voltage transmission lines and then organizationally by merger or acquisition, Insull continued to expand by means of the holding company structure. By the mid-twenties, when the holding companies were flourishing in the American utilities field, Insull had one of the largest, the Middle West Utilities Company, with subsidiaries operating in nineteen states. It supplied 8 percent of the commercial total of kilowatt-hours for the United States.[71] There was, however, a larger, older, more complex holding company with a greater influence on the electrification of America—the Electric Bond & Share Company.

S. Z. Mitchell: Financier-Entrepreneur

If Edison was the entrepreneur-inventor of the earliest period in the history of electric light and power systems, if Insull the entrepreneur-manager of a later period, then S. Z. Mitchell was the financial entrepreneur of a still later stage. Edison fathered the urban district station; Insull presided over the evolution of the urban utility; and Mitchell facilitated development of regional systems of electric power. Edison invented a technological system, which he covered with patents; Insull developed management techniques, which he publicly articulated and widely disseminated; and Mitchell introduced financial and organizational means by which the growth of utility systems might continue on a regional—even a national—level.

Mitchell's organizational innovations were the Electric Bond & Share Company and the family of utility holding companies directly supervised by it. The importance of these organizations is attested by the control they exercised over the industry by the mid-1920s. In 1924, when measured by energy generated, Bond & Share and its

[70]*Presentation of Public Service Company of Northern Illinois* . . . (Chicago, 1924), p. 3, Insull Papers.

[71]U.S. Congress, Senate, *Electric-Power Industry: Control of Power Companies*, 69th Cong., 2d sess., 1927, Document no. 213, p. xxxvii (hereafter cited as Senate Document no. 213 [1927]).

family of holding companies controlled at least 10 percent of the electric utility industry in the United States.[72]

The control followed from direct management, engineering advice and supervision, and financial services. Furthermore, Mitchell's organizations spurred others to employ the holding company form, so that by 1924, when measured by energy generated, American holding companies controlled about two-thirds of the privately owned electric utilities.[73]

Electric Bond & Share Company was established in 1905, but its traceable origins extend back to 1890. In that year, the Thomson-Houston Electric Company, the manufacturer that merged in 1892 with Edison General Electric to form the General Electric Company, established a subsidiary company called the United Electric Securities Company. The function of the new company was to help finance the small, independent electric light and power companies, buying generators and other equipment manufactured by Thomson-Houston. Local sources of capital considered neither the bonds nor the stocks of the small, often indifferently run electric light companies a good risk.[74] As a result, the companies, many of them Thomson-Houston customers, could not respond to a growing market, first for light, later for power, which Mitchell calculated as expanding about 10 percent annually. The problem of the companies was more serious than for many small enterprises because the capital to be invested in relation to income was unusually high. Mitchell estimated that, as a rule, it took \$4–\$6 capital investment to produce \$1 per year gross revenue.[75] The problem was intensified as government regulation hindered growth through reinvestment of earnings.

The United Electric Securities Company financed Thomson-Houston's small customers by taking nonmarketable bonds and using them as collateral for bond issues of United Electric Securities, marketable because of its size, management, and relationship to Thomson-Houston.[76] Despite the considerable market that developed over the years for the bonds, the problem of financing local utilities was not finally solved, however. The panic of 1893 saw many small utilities sink into bankruptcy and surviving ones facing an even less

[72] Senate Document no. 213 (1927), p. 41.

[73] U.S. Congress, Senate, *Utility Corporations: Summary Report of the Federal Trade Commission,* 70th Cong., 1st sess., 1935, Document no. 92, pt. 72A, p. 36 (hereafter cited as Senate Document no. 92 [1935]).

[74] Sidney Alexander Mitchell, *S. Z. Mitchell and the Electrical Industry* (New York, 1960), p. 60 (hereafter cited as *S. Z. Mitchell*).

[75] Mitchell in an address to Association of Edison Electric Illuminating Companies in 1920, quoted in *S. Z. Mitchell,* p. 118.

[76] Senate Document no. 213 (1927), p. 70.

responsive investment market. Investors who had bought the bonds of the small town utilities insisted after the panic that the utilities must raise more of their capital by common stock. Bond buyers wanted a larger margin of safety for their utility bonds in the form of heavier investment in common stock. In short, the bond buyers demanded more owners and fewer creditors. The capital-hungry utilities were advised to cover no more than one-half to two-thirds of the cash cost of new power plants and other facilities by mortgage securities.[77]

General Electric, successor to Thomson-Houston by 1893, inherited the United Electric Securities Company of Boston and subsequently formed the Electrical Securities Company of New York to fulfill basically the same financial function. General Electric advised the smaller utilities to sell more junior securities or stocks, but "merely pointing the way was not effective."[78] The problem for General Electric was further complicated because over the years it had acquired many shares of unmarketable small utility stock in payment for equipment. The electrical manufacturing firm, under the leadership of Charles Coffin, had to find more effective ways of financing its customers.

The inventive solution came in 1905 out of the deliberations of Coffin and Mitchell.[79] Mitchell was involved because he had been dealing with small utilities in the West, companies whose electric service was poor and whose financial structures were crude—companies that were, in the vernacular, "cats and dogs."[80] Mitchell, born in 1862, was prepared for problem solving by engineering training at the U.S. Naval Academy. Finding bleak the prospects in the Navy of 1885, Mitchell, newly commissioned, resigned and found employment and exciting prospects with the Edison Company in New York. (Mitchell became extremely interested in electric lighting while installing incandescent lights on the *U.S.S. Trenton* in 1883.) Mitchell worked at the Edison Machine Works, where dynamos were made, and also did wiring on the distribution system of the Edison Electric Illuminating Company in New York City. Within a year, he seized the opportunity to take over the exclusive agency for Edison products in the northwestern states after Henry Villard, friend and financial adviser of Edison, gave it up. In the Northwest from 1885 to 1905, Mitchell gained varied experience finding financial backers, obtaining franchises, installing equipment, setting up operating and management procedures, and cultivating the growth—or at least insuring the

[77]Senate Document no. 213 (1927), p. 71.
[78]Senate Document no. 213 (1927), p. 93.
[79]*S. Z. Mitchell,* p. 61.
[80]Ibid., pp. 61, 64.

survival—of small electric light and power utilities. He, like Insull, learned in the Edison school; also like Insull, he was in at the start of the electric light and power industry.[81] The call from Coffin to deliberate about cats and dogs was therefore no surprise.

The solution they worked out in 1905 was the Electric Bond & Share Company. The new company exchanged, or sold, its stock for stocks of power and light companies held by General Electric with par value of $2,782,150; for bonds with a par value of $1,476,000; and for $1,300,576.90 in cash. Electric Bond & Share then converted the utility securities acquired into marketable assets by resuscitating and invigorating the utilities that had issued them. This was done by providing management and engineering expertise and by financing the companies. Bond & Share did not, however, hold the majority of stocks in the companies that it advised, managed, and helped finance. Not a holding company in the sense of controlling utilities by voting stock, its control of the companies came through the services provided and was a contractual one that could be cancelled on short notice.[82]

The contractual, fee-based relationship between Bond & Share and the companies with which it had "service contracts" varied and evolved over time. By 1929, the contract provided for a number of general services, but principally management, engineering, and financing. To facilitate the arrangement, Bond & Share recommended that the directors of the client company elect a Bond & Share executive as a nonsalaried officer. This officer—like Bond & Share executives serving other companies in a similar capacity—remained in its New York offices to facilitate coordination, communication, and the transaction of business. From its large staff of men experienced in managing and advising utilities, Bond & Share also designated one to act as "sponsor" manager of the client. The sponsor kept informed of the company by visits, observation, inspection, and correspondence. He applied Bond & Share's general knowledge of the utility business to the problems and opportunities of the utility company. Further, he had at his disposal the large Bond & Share staff of specialists in insurance, taxes, rates, public relations, statistics, and other management functions. The contract also called for a "sponsor engineer." A senior engineer drawing upon his own and Bond & Share's long and generalized experience, he advised the client utility about engineering practices. By visits and correspondence, he too kept informed about the client. Besides the management and engineering specialists, there was a "sponsor accountant" from the New York office. To illustrate

[81]The biographical sketch of Mitchell is based on ibid., pp. 19–42.
[82]*S. Z. Mitchell,* pp. 64–65; Senate Document no. 213 (1927), p. 73.

224   *Thomas P. Hughes*

the kind of problem to which the sponsors directed their attention, the contract singled out expansion of the business, physical interconnection of the utility with systems of other companies, "important contracts for the sale of your service, matters pertaining to your capital structure and financial program, and other questions and problems of importance in the operation and management of your business."[83]

Electric Bond & Share soon after its founding also began to organize holding companies that controlled operating utilities by stock ownership. Bond & Share then established the contractural relationship with the holding company rather than the operating utilities. Bond & Share seems to have been the first of the management and engineering companies to resort to the holding company structure.[84] The primary objective was the persistent one of creating a means by which the utilities could dispose of their nonmarketable junior securities. Their need for capital was constantly stimulated by the ever-expanding market for light and power. Mitchell, seeing every person as customer, told utilities that they needed $50 capital for each baby born in the United States.[85]

The essential financial structure evolved by Electric Bond & Share for the operating utility companies which it organized into holding companies called for 60 percent capitalization by the sale of bonds to the public, 20–25 percent by preferred stock sold to the public, and the remaining 20–25 percent in common stock to be taken by the holding company.[86] The advantage to the holding company of keeping the amount of the common—the voting or controlling—stock low was obvious: Control was less costly. Furthermore—as will be explained—the arrangement could mean high returns on investment for the holding company.

Following this plan, Bond & Share participated in the formation of five holding companies: American Gas & Electric Company (formed in 1906); American Power & Light Company (1909); National Power & Light Company (1921); American & Foreign Power Company, Inc. (1923); and Electric Power & Light Corporation (1925).[87] The first originated before Bond & Share had built up its staff of experts, so American Gas & Electric supplied its own managers, engineers, and accountants to the operating companies whose common stock it held. However, American Gas & Electric Company relied upon Bond &

[83]Senate Document no. 92 (1935), p. 672.
[84]Senate Document no. 213 (1927), pp. xvii–xxxvii.
[85]*S. Z. Mitchell*, p. 59.
[86]Senate Document no. 92 (1935), p. 87.
[87]Senate Document no. 213 (1927), p. xxxvii.

Share for financial services, and, in 1924, the latter company owned about 7 percent of the voting stock of the former.[88]

Because American Power & Light Company was controlled by Bond & Share, as were the holding companies formed subsequently by Bond & Share, its history serves as a representative example of the concepts and actions of S. Z. Mitchell and his associates. They established American Power & Light as a holding company to finance two small utility companies in Kansas. The companies were too small to interest investors in their bonds and preferred stock, and there was no market for the common. The owner of the two asked Mitchell to form a holding company to take their common shares and thereby supply capital for expansion. The American Power & Light Company raised the money for the stock by selling its own securities. Subsequently, the holding company acquired other companies in the immediate vicinity, and Mitchell merged them into Kansas Gas & Electric, a company large enough to interest investors in its preferred stocks and bonds. American Power & Light controlled the Kansas Gas & Electric Company through common stock, and the holding company entered into the customary contractual relationship with Electric Bond & Share. In 1910 American Power & Light, directed by Bond & Share, acquired properties in Washington state and Oregon. In subsequent years, it obtained control of other groups of companies in Texas, Minnesota, Florida, and Nebraska. By the year 1924, the operating companies held by American Power & Light (Bond & Share) generated about 2.4 percent of the commercial total output of electricity for the entire country.[89]

When the companies acquired were in the same geographical area, as was the case in Kansas, they were united by transmission lines into a continuous system and often merged into larger operating companies. As a result, small inefficient plants could be shut down, load factor could be improved, and diversity exploited—as we have seen Insull do in Chicago. But Mitchell went beyond Insull by introducing the diversity principle on a higher level of abstraction—a principle he called financial diversity. The industry called the holding companies of Electric Bond & Share the "diversified investment type."[90]

The principle of financial diversity as fulfilled by the American Power & Light Company and in other Bond & Share holding companies was explained by Mitchell as follows:

> If one owns a light and power plant in a single community, his investment and his earning power is subject to the risk of that

[88]Ibid., p. 7.
[89]Ibid., pp. 93–108, xxix.
[90]Senate Document no. 92 (1935), p. 84.

community. Floods may come and wipe it out; cyclones may hurl it down; crops may fail; business depressions there may be acute. The capital invested in that plant, if owned by a single man, is subject to those contingencies. But if men combined their investments in a large number of plants, widely diversified geographically, the floods will never come to all at once; the failure of crops will never come to all at once; a depression in business is unlikely to come to all at once, if the diversity is widely made. Therefore, a given investment in a group of plants is much safer than an investment in a single plant of similar amount. Not only is the principal safer but the continuity of return is better insured.[91]

The inventive concept is as systematic and embracing as those of Edison's and Insull's on other levels.

Not only might the investor's principal be safer and the continuity of return better insured in a diversified holding company, but his return might well be quite large. Mitchell explained in a simplified example. He said that for each $100 a holding company invested in the common stock of a utility it should earn $9.00. Assuming the holding company had raised the $100 by marketing $60 of its own debentures (nonmortage or collateral bonds), $20 of its preferred stock, and $20 of common stock of its own, then on the bonds it would have to pay 6 percent ($3.60) and on the preferred stock 7 percent ($1.40), which would leave $4.00 in return on the common stock (a rate of 20 percent). As Mitchell observed, "for that possible 20 percent return the holding company common stock should be salable."[92] He might have added that the organizers of holding companies were often ready to purchase the common stock themselves.

The Federal Trade Commission in a report in 1927, at the height of the holding company era, observed that promoters often retained ownership of the common shares of their holding companies and obtained rates of return ranging from 19 to 55 percent.[93] High returns could be achieved by increasing the proportion of low-return bonds and preferred stocks among the holding company's securities. For instance, if the holding company in the example above had increased the bonds to 65 percent and the preferred stock to 25 percent, the rate of return on the $10 of common stock remaining would have been 33.5 percent. S. Z. Mitchell recognized the possibility of high—even excessive—return on investments, but he characterized

[91]*S. Z. Mitchell*, pp. 84–85.
[92]Ibid., p. 80.
[93]Senate Document no. 213 (1927), p. xxiv.

his role as primarily resourceful financier of utility expansion rather than reaper of large profits.

Conclusion

The question arises, in conclusion, why Mitchell with a background in engineering and managing technology concentrated far more intently than Edison and to a greater extent than Insull earlier upon financial affairs. A major reason is that during the twenties when holding companies and men like Mitchell, who organized and directed them, flourished, technological problems were not critical—they were not tenacious obstacles to growth. The record is mostly further elaboration and extension of technology well known—even perfected—before 1920. Chronological lists of inventions, development, and innovations in electric light and power between 1920 and 1930 feature references to the increased size of generating units, increased transmission voltages, better frequency control, higher steam pressures and temperatures in turbines, improved insulation for transmission, power factor correction, fuller understanding of circuitry, lighting protection, and the like. The only indications of technological breakthroughs are the references to automatic control of equipment such as substations. More indicative of the technological developments of the era of the holding companies are the graphs showing steadily rising boiler pressures and temperatures, steadily increasing kilowatt output of turbogenerators, and steadily rising transmission voltages.[94]

During the holding company era, the dominant entrepreneurs used available technology both to extend technological systems that had already gathered momentum and to implement managerial concepts already articulated. For Mitchell, intent upon the growth of electric light and power systems, the essential technology was high-voltage transmission of power. The transmission lines provided physical structure for the corporate ones. The country, by 1932, had a longer network of high-voltage transmission lines than it had main-line railroads.[95] Holding companies were responsible for many of the lines erected after 1920. Yet high-voltage transmission lines had been used decades earlier by Insull and others to form large utilities for

[94]National Electrical Manufacturers Association, *A Chronological History of Electrical Development from 600 B.C.* (New York, 1946), pp. 89–98; *Electrical Engineering: Fiftieth Anniversary AIEE, 1884–1934* 53 (May 1934): 840–41; Charles E. Neil, "Entering the Seventh Decade of Electric Power," *Edison Electric Institute Bulletin* (September 1942).

[95]Mileage of transmission systems in excess of 6,600 volts exceeded 251,000 miles in 1932; single track mileage of steam railroads was 247,595 miles (Senate Document no. 92 [1935], p. 41).

mass production and distribution. The regional systems of the holding companies were technologically but an extension. Mitchell also applied the managerial concepts articulated by Insull and others earlier for massing production and distribution, load factor improvement, load diversity exploitation, and other managerial techniques applicable to regional utilities and holding company operations. The new concept of diversity expounded by Mitchell was a financial one.

Edison was not a simple inventor: He presided over change involving technology, management, finance, and other factors; Insull similarly drew upon his knowledge of technology and proved himself an ingenious financier, but his profession was management. Like the other two, Mitchell was also an entrepreneur presiding over many factors involved in change, including technology and management, but his focus was such that he should be called a financial entrepreneur. Each man flourished during that state of electric light and power development when his drive to "relate everything to a single central vision, one system less or more coherent or articulate" was particularly appropriate.

MIRROR-IMAGE TWINS: THE COMMUNITIES OF SCIENCE AND TECHNOLOGY IN 19TH-CENTURY AMERICA

EDWIN LAYTON

American technology went through a scientific revolution in the 19th century. Technological knowledge was uprooted from its matrix in centuries-old craft traditions and grafted onto science. The technological community, which in 1800 had been a craft affair but little changed since the middle ages, was reconstructed as a mirror-image twin of the scientific community. The artisan was replaced in the vanguard of technological progress by a new breed of scientific practitioner. For the oral traditions passed from master to apprentice, the new technologist substituted a college education, a professional organization, and a technical literature patterned on those of science. Equivalents were created in technology for the experimental and theoretical branches of science. As a result, by the end of the 19th century, technological problems could be treated as scientific ones; traditional methods and cut-and-try empiricism could be supplemented by powerful tools borrowed from science. This change was most marked in the physical sciences and civil, mechanical, and electrical engineering, the subject of this paper. But similar changes were taking place at the same time in the relations of chemistry, biology, geology, and other sciences to their corresponding technologies. The result might be termed "the scientific revolution in technology."

The significance, indeed the very existence, of the scientific revolution in technology has been obscured by a commonly accepted model of the relationships between science and technology. In essence, this holds that science creates new knowledge which tech-

EDWIN LAYTON, of the Graduate Program for the History of Science and Technology at the University of Minnesota, is the author of *The Revolt of the Engineers, Social Responsibility, and the American Engineering Profession,* and over fifty other publications mostly dealing with the history of engineering.

This essay originally appeared in *Technology and Culture,* vol. 12, no. 4, October 1971.

nologists then apply. Jacob Bigelow articulated a common faith in 1831 when he asserted:

> Our arts have been the arts of science, built up from an acquaintance with principles, and with the relations of cause and effect . . . we have acquired a dominion over the physical and moral world, which nothing but the aid of philosophy could have enabled us to establish. . . . The labor of a hundred artificers is now performed by the operations of a single machine. We traverse the ocean in security, because the arts have furnished us a more unfailing guide than the stars. We accomplish what the ancients only dreamt of in their fables; we ascend above the clouds, and penetrate into the abysses of the ocean.

And he concluded that "the application of philosophy to the arts may be said to have made the world what it is at the present day."[1] But when Bigelow came to enumerate the specific instances in which "philosophy" or science had transformed technology, he noted such inventions as the magnetic compass, the printing press, gunpowder, the clock, glass, the cotton gin, the steam engine, and textile machinery. Yet in none of these cases is the influence of science on technology obvious; certainly in none is their relationship explained satisfactorily by the common model.

That this view of science-technology relations has continued into the 20th century was demonstrated by Vannevar Bush and other architects of America's recent science policies. Bush held that basic research, though undertaken without thought of practical ends, generates knowledge of nature's laws which provides the means of technological progress. He maintained that "basic research leads to new knowledge. It provides scientific capital. It creates the fund from which the practical applications of knowledge must be drawn. New products and new processes do not appear full-grown. They are founded on new principles and new conceptions, which in turn are painstakingly developed by research in the purest realms of science." Bush concluded that, "today, it is truer than ever that basic research is the pacemaker of technological progress."[2]

Inspired by Bush's model of the relation of science to technology, the Department of Defense, from 1945 to 1966, invested about $10 billion in scientific research, of which approximately one-quarter went for basic or undirected research. A growing skepticism con-

---

[1]Jacob Bigelow, *Elements of Technology* (Boston, 1831), p. 4.
[2]Vannevar Bush, *Endless Horizons* (Washington, D.C., 1946), pp. 52–53. See also John R. Steelman, *Science and Public Policy* (Washington, D.C., 1947), pp. 4–5.

cerning the technological value of this enormous expenditure caused the department to undertake an investigation, Project Hindsight. This study took eight years and consumed some forty man-years of time on the part of thirteen teams of scientists and engineers who analyzed the key contributions which had made possible the development of the twenty weapons systems that constituted, in large part, the core of the nation's defense arsenal. Some 700 key contributions or "events" were isolated. They were classified as being either technological or scientific. If the latter, they were further subdivided into basic and applied-science "events."[3]

The preliminary results of Project Hindsight, which were released in November 1966, came as something of a bombshell to the scientific community. Of all "events," 91 percent were technological, only 9 percent were classed as science. Within the latter category 8.7 percent were applied science; only 0.3 percent, or two "events," were due to basic or undirected science.[4] Predictably, the publication of these results produced a spate of indignant letters to the editors of *Science.*[5] Many of these missed the point. The investigators had not sought to show that science has no influence on technology. What they did demonstrate was that the immediate, direct influence has been small; they showed that the traditional model of science-technology relations is in need of revision. To correct the misunderstanding of Project Hindsight, a subsequent study, TRACES, demonstrated the dependence of five recent innovations on prior scientific work. The question, therefore, is not whether science has influenced technology, but rather the precise nature of the interaction.[6]

[3]Chalmers W. Sherwin and Raymond S. Isenson, "Project Hindsight," *Science* 156 (June 23, 1967): 1571–77.

[4]Ibid.; D. S. Greenberg, "'Hindsight': DOD Study Examines Return on Investment in Research," *Science* 154 (November 18, 1966): 872–73; Philip H. Abelson, "Project Hindsight," *Science* 154 (December 2, 1966): 1123.

[5]See the collection of letters in "How Perceptive Is Hindsight?" *Science* 155 (January 27, 1967): 397–98. See also Helen L. Hayes, "Project Hindsight: Basic Research," *Science* 154 (December 23, 1966): 1504; Allen M. Lenchek, "Project Foresight," *Science* 155 (January 13, 1967): 150; Lee Leiserson, "Project Hindsight," *Science* 157 (September 29, 1967): 1512; and Robert M. Lukes, "Masquerade of Undirected Research," *Science* 159 (January 5, 1968): 34.

[6]Illinois Institute of Technology Research Institute, *Technology in Retrospect and Critical Events in Science,* 2 vols. (n. p. [Chicago], 1968). While generally adhering to the traditional model of science-technology relations, the authors of this study noted that "a better understanding needs to be achieved concerning the two-way influence between science and technology. The tracings revealed cases in which mission-oriented research or development effort elicited later nonmission research, which often was found to be crucial to the ultimate innovation" (1:22).

The results of Project Hindsight are surprising only if one assumes the validity of the received model of science-technology relationships. This model is not so much false as misleading. It assumes that science and technology represent different functions performed by the same community. But a fundamental fact is that they constitute different communities, each with its own goals and systems of values. They are, of course, similar in that both deal with matter and energy. But these similarities should not be overstated. Each community has its own social controls—such as its reward system—which tend to focus the work of each on its own needs. These needs determine not only the objects of concern, but the "language" in which they are discussed. These needs may overlap; but it would be surprising if this were a very frequent occurrence. One would expect that in the normal case science would beget more science, and technology would lead to further technology. This is precisely the finding of Project Hindsight.

* * *

The difficulties of the traditional model may be illustrated by the relationship, or lack of one, between Newtonian mechanics and the "golden age" of mechanical invention in America in the 19th century. An enthusiastic group of scientists, technologists, and reformers in America, as in Europe, were attempting to foster the application of science to technology. Among them was James Renwick, professor of natural experimental philosophy and chemistry at Columbia College. He wrote two books that were intended to bridge the gap between art and science. The first, *The Elements of Mechanics,* published in 1832, was a conventional exposition of the science of mechanics. In it Renwick followed a well-trodden path in treating systems in equilibrium by the principle of virtual velocities.[7] The second book, his *Applications of the Science of Mechanics to Practical Purposes,* published in 1842, surveyed the field of mechanical technology, including prime movers, clocks, and various types of machinery.[8] But despite Renwick's earnest efforts, the principles of the first book did not carry over to the second to any significant degree.

A mechanic interested in designing a water wheel would have found the methods and principles of the first book of little value,

[7]James Renwick, *The Elements of Mechanics* (Philadelphia, 1832), p. viii.

[8]James Renwick, *Applications of the Science of Mechanics to Practical Purposes* (New York, 1842). Five of the ten chapters deal with what we would now class as mechanical engineering, two with civil engineering (railroads and canals), one with both civil and mechanical technology (wheels and roads), one with marine engineering, and one with mining.

even if he could have understood them. But the same mechanic would have found valuable assistance in Renwick's second book. John Smeaton, the 18th-century British engineer, had used the experimental methods of science to derive a set of "maxims" or design principles for this type of prime mover. Two of these may be quoted as examples:[9]

> In a given undershot wheel, if the quantity of water expended be given, the useful effect is as the square of the velocity,

and

> In a given undershot wheel, if the aperture whence the water flows be given, the effect is as the cube of the velocity.

Neither could be classed as laws of nature; they were lawlike statements about man-made devices. They were not logical deductions from the science of mechanics; they constituted the germ of a new technological science.

Far from constituting a unity, Renwick's two books pointed to two quite different lines of technological development. Technology might, as suggested by his first book, build directly on the foundations of science. The science of mechanics could be extended to create new, technologically oriented sciences such as the strength of materials and hydraulics. Or, following Smeaton, technologists might borrow the methods of science to found new sciences built on existing craft practices.

To some extent inventors helped to develop technological sciences. Oliver Evans attempted to apply scientific methods to technology in his *The Young Mill-Wright and Miller's Guide* published in 1795. Evans began with a survey of the principles of mechanics, and he was able to derive useful design principles directly from them. But his chief reliance was on the application of scientific methods, rather than deductions from existing laws. He derived a set of "rules" for designing mills, including a critical examination of Smeaton's "maxims." But Evans attempted to go further and he devised a set of rules for making inventions in any field. These amounted to applying scientific methods to technology. Included were the discovery of fundamental principles, making deductions from these principles, and testing the results by experiment.[10]

[9]Renwick, *Applications of the Science of Mechanics*, pp. 49–50; John Smeaton, "An Experimental Enquiry concerning the Natural Powers of Water and Wind to Turn Mills, and Other Machines, Depending on a Circular Motion," *Philosophical Transactions* 51 (1759–60): 118–20. I have retained Renwick's phrasing.

[10]Oliver Evans, *The Young Mill-Wright and Miller's Guide* (Philadelphia, 1795), pp. 1–70, appendixes 1–2.

It is, of course, very difficult to discover which works were read by specific inventors; it is even harder to establish a correlation between particular inventions and prior published information. But it is easy to show that there was a vast increase in the volume of written, more or less systematic technical information available to American inventors in the course of the 19th century.[11] This was part of a worldwide movement that had its origins in the great encyclopedias of the 18th century. Oliver Evans, much of whose inventive career came before 1800, recalled that the chief impediment for the inventor was the lack of reliable published information.[12] By the middle of the 19th century, through the efforts of men like Evans and Renwick, this barrier to invention had been largely removed.

Inventors might apply scientific methods; but despite the work of a few like Evans, the inventor was ill-adapted to the task of building up technological sciences. Scientists, on the other hand, had the necessary skills, and they played a vital role in stimulating the development of engineering sciences. But scientists lacked the lasting commitment and the intimate knowledge of technology and its needs that was required. The bulk of the effort to build technological sciences, therefore, fell on the engineering profession itself.

* * *

The engineering sciences, by 1900, constituted a complex system of knowledge, ranging from highly systematic sciences to collections of "how to do it" rules in engineering handbooks.[13] Some, like the strength of materials and hydraulics, built directly on science; they were often classed as branches of physics. Others, such as the kinematics of mechanisms, evolved from engineering practice. In either case, their development involved the adoption by engineers of the theoretical and experimental methods of science, along with many of

[11]There are two excellent guides to this literature: Eugene S. Ferguson, *Bibliography of the History of Technology* (Cambridge, Mass., 1968); and Brooke Hindle, *Technology in Early America* (Chapel Hill, N.C., 1966).

[12]Eugene S. Ferguson, ed., *Early Engineering Reminiscences of George Escol Sellers* (Washington, D.C., 1965), p. 38.

[13]Hunter Rouse and Simon Ince, *History of Hydraulics* (Iowa City, Ia., 1957); Eugene S. Ferguson, *Kinematics of Mechanism from the Time of Watt* (Washington, D.C., 1962); Stephen P. Timoshenko, *History of Strength of Materials* (New York, 1953); Isaac Todhunter, *A History of the Theory of Elasticity*, 3 vols. (New York, 1960); James H. Potter, ed., *Handbook of the Engineering Sciences*, 2 vols. (Princeton, N.J., 1967) is a useful modern survey. For a view of the engineering sciences similar to the one taken in this paper, see James Kip Finch, "Engineering and Science: A Historical Review and Appraisal." *Technology and Culture* 2 (Fall 1961): 318-32.

the values and institutions associated with their use. By 1900 the point of origin made little difference; the engineering sciences constituted a unity. Those derived from practice took on the qualities of a science in their systematic organization, their reliance on experiment, and in the development of mathematical theory. At the same time, sciences like the strength of materials gradually diverged from physics, assuming the characteristics of an autonomous technological science.

The separation of the engineering sciences from physics may be illustrated by the strength of materials and its sister disciplines, the theory of elasticity and the theory of structures. They were the first of the engineering sciences to be cultivated extensively, both in Europe and America. The reasons for this were twofold. The intractable nature of materials constituted one of the most important barriers to the development of technology. The 19th century saw many new uses for materials like iron and steel; a scientific study of their properties would enable designers to avoid costly failures. But another reason for the early emphasis on this science was that it represented one of the most promising avenues for the application of science to technology. It could draw upon a sophisticated body of physics accumulated since the time of Newton. Thus, it attracted scientists and others inspired by the vision of a scientific technology.

Both in Europe and America scientists played a key role in fostering the development of the science of the strength of materials. But once it was established, technologists dominated its further development, although scientists continued to make important contributions. Scientists such as Hooke, Euler, Young, and Coulomb did much to lay its foundations; it is worth remembering that the second of Galileo's "two new sciences" was the strength of materials. But once it reached the stage of being technologically useful, its development was undertaken by engineers. A critical institutional innovation was the development of engineering colleges in which technology would be pursued in the manner of science. The Ecole Polytechnique was the pioneer, widely imitated both in Europe and America. A group of polytechnicians, notably Louis Marie Navier (1785-1836) and Barrie de Saint-Venant (1797-1886), reformulated and extended this science.[14]

As the strength of materials moved from the community of sci-

[14]Frederick B. Artz, *The Development of Technical Education in France, 1500-1850* (Cambridge, Mass., 1966), pp. 81-86, 151-66, 230-53; Timoshenko, *History of Strength of Materials,* pp. 67-80, 135-41, 229-42.

ence to that of technology, it went through an important transformation. Its ties with physics were weakened, and it developed in ways uncharacteristic of the basic sciences. At the same time, its range of technological usefulness was gradually expanded. Scientists tended to explain their findings by reference to the most fundamental entities, such as atoms, ether, and forces. But these entities cannot always be observed directly. To be useful to a designer, however, a formulation must deal with measurable entities, particularly those of importance to the practical man. These need not be fundamental in the scientific sense. The scientists who had done so much to found a science of the strength of materials—notably Young, Coulomb, and Poisson—strove to found this study on the same ontological basis as classical mechanics—that is, they sought to explain their results in terms of molecules and the forces between them. Although not without interest, these efforts were not wholly successful. They were also needless complications from the technological point of view. A few of the engineers poineering in this field, including Navier and Saint-Venant, continued this quest, but in the end the attempt was abandoned.[15] Instead, engineers were content with a simple macroscopic model—for example, viewing a beam as a bundle of fibers.

In America as in Europe the foundation of the science of strength of materials owed much to scientists. Led by Alexander D. Bache, the Franklin Institute in 1830 undertook an investigation of the causes of steam boiler explosions for the federal government. This study was itself one of the first significant attempts to use scientific methods to investigate technological problems in America.[16] One aspect of this multifaceted effort was a systematic study by Walter R. Johnson of the strength of the metals used in boiler construction. This involved building the first testing machine in America and conducting a well-conceived and highly fruitful series of experimental tests.[17] Scientists also fostered the use of mathematical

[15]Timoshenko, *History of Strength of Materials*, pp. 104–7, 231–32.

[16]Bruce Sinclair, *Early Research at the Franklin Institute, the Investigation into the Causes of Steam Boiler Explosions, 1830–1837* (Philadelphia, 1966); John G. Burke, "Bursting Boilers and the Federal Power," *Technology and Culture* 7 (Winter 1966): 1–23.

[17]"Report of the Committee of the Franklin Institute . . . on the Explosion of Steam Boilers . . . Part II . . . ," *Journal of the Franklin Institute* 19 (February 1837): 73–109; ibid. (March 1837): 156–93; ibid. (April 1837): 241–77; ibid. (May 1837): 28–31; ibid. (June 1837): 409–51; ibid. 20 (July 1837): 1–31; ibid. (August 1837): 72–113. See also George E. Pettengill, "Walter Rogers Johnson," *Journal of the Franklin Institute* 250 (August 1950): 93–113.

theory for the study of materials. William Barton Rogers, although primarily a geologist, was well grounded in physics and mathematics. His *An Elementary Treatise on the Strength of Materials* published in 1838 was the first American book in this field.[18]

But perhaps the greatest contribution of scientists like Rogers was to foster institutions to encourage the marriage of science and technology. Rogers himself was one of the foremost; he had apparently become concerned with technology when he lectured at the Maryland Institute, a Baltimore mechanic's institute, in 1827. His treatise on the strength of materials was produced as part of an effort to found a school of engineering at the University of Virginia. While this venture did not succeed, Rogers did not give up. In 1846 he drew up a plan for a "polytechnic school" for Boston—a dream finally realized in 1861 with the chartering of the Massachusetts Institute of Technology.[19] Rogers was, of course, not alone in his vision of a scientific technology. Led for the most part by chemists and geologists, the scientific schools attached to Harvard, Yale, and other colleges instituted engineering programs. Benjamin F. Green reorganized Rensselaer into a polytechnic school after 1847. It was the first to concentrate almost exclusively on engineering, and the first to go beyond one-man departments in this area—a vital step in encouraging the specialization required for the development of the engineering sciences.[20]

While scientists like Rogers, Bache, Renwick, and Green did much to found the scientific study of materials in America, its systematic development lay principally with the engineers themselves. An important role in this was played by West Point, the first American engineering school. It was reorganized after 1818 by Sylvanus Thayer on the model of the great French engineering schools. One graduate, Dennis Hart Mahan, was sent to France to complete his engineering education; on his return, he taught civil and military engineering to cadets from 1832 to 1871. Mahan produced in 1837 the first American textbook based on French engineering practice, *An Elementary Course of Civil Engineering.* Over 15,000 copies of this work were sold; it had an important impact on the teaching of

[18]William Barton Rogers, *An Elementary Treatise on the Strength of Materials* (Charlottesville, Va., 1838).

[19]Emma Rogers, *Life and Letters of William Barton Rogers*, 2 vols. (Boston, 1896), 1: 40–54, 259–62, 420–27.

[20]Samuel Rezneck, *Education for a Technological Society* (Troy, N.Y., 1968), pp. 78–110; Palmer C. Ricketts, *History of the Rensselaer Polytechnic Institute* (New York, 1895), pp. 69–112.

engineering in America.[21] It included a brief survey of the strength of materials. Although Mahan limited himself to elementary mathematics, his treatment of this subject was distinctly professional, in striking contrast to the purely qualitative discussion of the strength of materials in Bigelow's *Elements of Technology*. It is perhaps significant that Bigelow sought explanations for the properties of materials in molecules and forces between them but Mahan made no reference to these fundamental entities of physics. Mahan had read deeply in the European literature and he particularly recommended the works of Navier to his students. A few apparently followed his advice. West Point engineers did much to establish a tradition of the scientific study of engineering in America.[22]

Mahan's work, and the technical works which followed it, provided a basis for introducing European methods into ordinary engineering practice in America. But creative contributions, the founding of a science, required money for laboratories and equipment as well as men trained to use them. The federal government played an important role in supporting experimental investigations in the second quarter of the 19th century. Federal funds had made possible the Franklin Institute's studies of boiler explosions. The same testing machine was used for another pioneering investigation, the study of the causes of the disastrous explosion of a cannon on the U.S.S. *Princeton*.[23] But the institute lacked the funds for developing a sustained program of research. Army officers, particularly in the Ordnance Department, to some extent filled in the gap. A series of experiments on the strength of cannons by Maj. William Wade and Capt. Thomas Jackson Rodman were among the first American contributions to attract European attention. Wade's testing machine was apparently the second to be built in America.[24] Other ex-

[21]George W. Cullum, "Dennis H. Mahan," *Biographical Register of the Officers and Graduates of the U.S. Military Academy at West Point*, 7 vols. (Boston, 1891), 1:319-25.

[22]Dennis H. Mahan, *An Elementary Course of Civil Engineering* (New York, 1837), pp. vii, 44-53, 86-104; Bigelow, *Elements of Technology*, pp. 43-53.

[23]Lee M. Pearson, "The 'Princeton' and the 'Peacemaker': A Study in Nineteenth-Century Naval Research and Development Procedures," *Technology and Culture* 7 (Spring 1966): 163-83.

[24]U.S. Ordinance Department, *Reports of Experiments on the Strength and Other Properties of Metals for Cannon, with a Description of the Machines for Testing Metals, . . .* (Philadelphia, 1856). On Wade's testing machine, see Chester H. Gibbons, *Materials Testing Machines* (Pittsburgh, 1935), pp. 27-28. For a critique of the work of Wade and Rodman, see Todhunter, *History of the Theory of Elasticity*, 2 (pt. 1): 688-96. For Rodman's later work, see U.S. Ordnance Department, *Reports of Experiments on the Properties of Metals for Cannon, and the Qualities of Cannon Powder . . . by Captain T. J. Rodman* (Boston, 1861). (Hereafter cited as Rodman, *Reports of Experiments*.)

perimental investigations were carried out by John Dahlgren and Benjamin F. Isherwood of the navy.[25] But the government was unwilling to make a long-range commitment for research not directed to an immediate mission. In 1872 the American Society of Civil Engineers requested that the government undertake tests of the properties of American iron and steel. Congress authorized a study and created a board of seven engineers to supervise the work. But the president placed the control of the program with the Ordnance Department, which differed with the civilian engineers. In 1878 Congress turned the testing machine over to the army and dissolved the board. The army refused to cooperate with the civilian engineers on the grounds that they lacked the necessary funds.[26]

Large business ventures were also in a position to undertake scientific studies. The proprietors of Lowell supported James Francis's hydraulic experiments; but for his studies of the strength of cast iron he had to rely on European data.[27] The building of the Eads bridge necessitated the adoption of systematic testing of materials, and this practice gradually spread through the steel industry. But these tests were usually geared to the needs of particular projects.[28] Thus, while business and government did much to encourage the adoption of experimental methods in technology, they were unwilling to carry out basic research on a sustained basis.

What engineering needed was not just short-term studies directed to specific problems, but a broad and continuous program of basic research in laboratories specifically dedicated to developing the engineering sciences. Robert Thurston, one of the founding fathers of mechanical engineering in America, was perhaps the foremost champion of basic research in the engineering sciences. He wanted

[25]Edward William Sloan, *Benjamin Franklin Isherwood, Naval Engineer* (Annapolis, Md., 1965); Clarence S. Peterson, *Admiral John A. Dahlgren, Father of U.S. Naval Ordnance* (New York, 1945). Both Isherwood and Dahlgren were active in applying experimental methods to derive design principles for engineering. Isherwood was concerned with the designs of marine steam engines, of screw propellers, and of other subjects. Dahlgren, along with Rodman, used experiment to design the bottle-shaped cannon used in the Civil War.

[26]Charles W. Hunt, *Historical Sketch of the American Society of Civil Engineers* (New York, 1897), pp. 82–83; William F. Durand, *Robert Henry Thurston* (New York, 1929), pp. 79–81.

[27]James B. Francis, *Lowell Hydraulic Experiments* (Boston, 1855), p. xi; Francis, *On the Strength of Cast-Iron Pillars* (New York, 1865), pp. 1–17. Francis derived a classic set of design principles for turbines in the former work (pp. 44–52).

[28]Carl W. Condit, *American Building Art: The Nineteenth Century* (New York, 1960), pp. 9, 139–140; Gibbons, pp. 31, 34–40.

engineering laboratories established in connection with engineering schools. The rise of research-oriented universities and technical institutes after the Civil War gave him his opportunity. He founded two of the earliest and best-known engineering laboratories in America, at Stevens Institute of Technology and Cornell University. Thurston devised two new testing machines and made important discoveries of the properties of materials. With his massive, three-volume work, *The Materials of Engineering*, the experimental study of the strength of materials reached maturity in America.[29]

* * *

Although the experimental approach to technology was readily adopted in America, theory tended to lag behind. American technologists generally lacked the advanced mathematical training needed to make contributions to a sophisticated field like the theory of elasticity. American engineers also tended to pride themselves on their practicality, and regarded mathematical theory as of little real value. The theoretical approach had to prove its utility to be adopted. The difficulty lay with the limitations of existing theory. Although the strength of materials had developed into a science by the 1830s, the range of application of its theory was limited. Very elegant solutions for a limited number of problems were available; but most problems were not solvable.[30] Many problems were indeterminate; they could not be solved because the number of unknowns was greater than the number of equations. Unfortunately, the indeterminate cases included some of the ones most frequently met in American engineering practice: the continuous beam and the truss bridge.[31]

From the 1830s to the 1870s there was a major effort, both in Europe and America, to extend the range of applicability of the engineering sciences. This effort met with remarkable success; by 1880 it was possible to attack a wide range of problems by mathematical theory. In America much of the effort went into the analysis of truss bridges. Squire Whipple's *An Elementary and Practical Treatise on Bridge Building*, the first version of which appeared in 1847, was a homespun product developed apparently in complete innocence of

[29]Durand, pp. 65–73, 236–40; Robert H. Thurston, "On the Necessity of a Mechanical Laboratory," *Journal of the Franklin Institute* 70 (December 1875): 409–18; Robert H. Thurston, *The Materials of Engineering*, 3 vols. (New York, 1883).

[30]Timoshenko, p. 231.

[31]Condit, pp. 6–9, gives a particularly clear sketch of the development of the strength of materials with emphasis on the problem of indeterminate structures.

European work. Whipple employed no mathematics other than elementary geometry and algebra; he did not use the calculus or even trigonometry. While he expressed his results algebraically, the argument was basically geometrical, giving his work a quaint, 17th-century flavor at times. But Whipple's work was a remarkable achievement. He derived mathematical and graphical methods by which he was able to analyze correctly truss bridges which were indeterminate by the usual methods.[32]

A second American effort to establish a mathematical theory for bridges was that of Herman Haupt, whose *General Theory of Bridge Construction* appeared in 1853. A West Point graduate, Haupt had some familiarity with European theory. The British scientist Thomas Young, upon whose work Haupt relied heavily, assumed, like many scientists, that stresses were ultimately reducible to forces between particles. On this assumption, Haupt sought to resolve the forces operating on a beam to a single resultant force acting at the center of an equivalent geometric figure. Unfortunately, stresses are not forces and they cannot be combined in this manner. Although Haupt's assumptions were open to question, his approximations were doubtlessly a vast improvement over the rule-of-thumb methods still in general use. A correct theory of stresses, developed at about the same time by European engineers, did much to further the separation between engineering sciences and physics. It was no longer helpful to attempt to base this engineering science on the fundamental assumptions of physics, atoms, and forces.[33]

Whipple's and Haupt's use of graphical methods to extend the range of engineering science was prophetic of one of principal lines of advance in the science of the strength of materials. In 1866 the Swiss engineer Karl Culmann developed an important graphical

[32]Squire Whipple, *An Elementary and Practical Treatise on Bridge Building*, 4th ed. (New York, 1883). See also Squire Whipple, "On Truss Bridge Building," *Transactions of the American Society of Civil Engineers* 1 (1868–71): 239–44.

[33]Herman Haupt, *General Theory of Bridge Construction* (New York, 1853). Haupt's assumption was that "the weight of any body may be supposed concentrated at its center of gravity; and, in general, any number of parallel forces may be replaced by a single force called the resultant. In the present case . . . the sum of all the forces upon the fibres . . . will be the same, as if a single force equal to its area was applied in the direction of a line passing through its center of gravity" (p. 20). Each normal stress is always accompanied by two components of shearing stress acting at right angles. These cannot be combined by a parallelogram of forces to give a single resultant. In modern terms, forces behave like vectors, but stresses behave like tensors. See also Thomas Young, *A Course of Lectures on Natural Philosophy and the Mechanical Arts* (London, 1807), pp. 135–52.

method in which stresses were represented by segments of circles. Culmann's use of circle diagrams was extended by the German engineer Otto Mohr and others, resulting in a great increase in the range of usefulness of the strength of materials.[34] The development of graduate-level work at American universities after the Civil War produced engineers who had the training to develop and apply methods of mathematical theory to materials. Henry Turner Eddy, a graduate of Yale's Sheffield Scientific School who received his Ph.D. from Cornell, was among the first of this new generation of scientific technologists. In 1878 he published an extension of the new graphical methods in his *Researches in Graphical Statics*. It was one of the first American engineering books to be translated into German; Florian Cajori, the historian of American mathematics, called it "the first original work on this subject by an American writer."[35]

The expansion of the range of application of the engineering sciences was accompanied by a tendency away from analytic solutions, a reliance on approximations, and, to some extent, a lessening of mathematical rigor. A given problem in the strength of materials might be solved rigorously by the theory of elasticity or it might be treated by less rigorous graphical methods. American engineers, beginning with Rodman, pioneered still less rigorous empirical methods using strain gauges and models. The selection of technique depended on economic as well as technical factors, since rigorous treatment, when possible, often involved more time and effort. The development of hierarchies of methods of variable rigor, along with the importance of economic factors in determining their use, served to distinguish the engineering sciences from physics where only the most rigorous methods were normally admitted.[36]

By 1900 the American technological community was well on the way to becoming a mirror-image twin of the scientific community.

[34]Timoshenko, pp. 190–97, 283–88. See also Hans Straub, *A History of Civil Engineering* (Cambridge, Mass., 1964), pp. 197–202. Not all of the changes were in the direction of lessening rigor; Saint-Venant for one was opposed and his development of the "semi-inverse" method extended rigorous analytic solutions.

[35]Florian Cajori, *The Teaching and History of Mathematics in the United States* (Washington, D.C., 1890), p. 177; Henry Turner Eddy, *Researches in Graphical Statics* (New York, 1878); idem., *Neue Constructionen aus der Graphischen Statik* (Leipzig, 1880); Arthur E. Haynes, "Henry Turner Eddy," *Minnesota Engineer* 20 (March 1912): 104–7; *Dictionary of American Biography*, s. v. "Henry Turner Eddy."

[36]Eddy's and Mohr's methods rested on rigorous mathematical proofs. But graphical methods usually involved simplifying assumptions about the distribution of stresses. For Rodman's pressure gauge, see Rodman, *Reports of Experiments*, pp. 299–300 (see n. 24 above).

The rise of engineering sciences had played a vital role. They gave technology equivalents to the theoretical and experimental departments of physical science. They were fostered by engineering colleges which, by 1900, had virtually displaced apprenticeship as a means of training engineers. Scientifically inclined engineers like Thurston played an important role in the founding of professional engineering societies after the Civil War, and an even more important role in producing worthwhile technical literature for engineering journals to publish. But despite the structural similarities between science and technology, the two were further apart in some respects. In many important areas engineering and physics had ceased to speak the same language.

* * *

In the case of mirror-image twins there is a subtle but irreconcilable difference which is expressed as a change in parity. Between the communities of science and technology there was a switch in values analogous to a change in parity. One way of putting the matter would be to note that while the two communities shared many of the same values, they reversed their rank order. In the physical sciences the highest prestige went to the most abstract and general – that is to the mathematical theorists from Newton to Einstein. Instrumentation and applications generally ranked lowest. In the technological community the successful designer or builder ranked highest, the "mere" theorist the lowest. These differences are inherent in the ends pursued by the two communities: scientists seek to know, technologists to do. These values influence not only the status of occupational specialists, but the nature of the work done and the "language" in which that work is expressed.

An indication of the gap between science and technology is provided by two discoveries, one by Henry Rowland the American physicist and the other by Francis Hopkinson, a British electrical engineer. Rowland, starting from an idea of Faraday, published a paper on magnetic permeability in 1873. James Clerk Maxwell, to whom Rowland sent the paper, recognized its importance, and arranged to have it published in *Philosophical Magazine*. Hopkinson in 1879 published the results of his investigation of the efficiency of electric dynamos. By graphing his results, he discovered the "characteristic curve" of the direct-current dynamo, a vital key to rational design. Hopkinson could show, for example, how the Edison dynamo could be radically improved by simply changing the dimensions of some of its parts. It was not discovered until several years later that, in a certain sense, Rowland and Hopkinson had made the same discovery.[37]

There was an irony in the fact that Rowland had "discovered" a key to the design of electric dynamos without realizing it. For Rowland's only earned degree was in engineering; and while he had transferred his primary loyalty to physics, his laboratory at Johns Hopkins was an important center for the training of electrical engineers. Rowland missed the significance of his discovery because he was looking for a law of nature, not a design principle. Each man expressed his work in the terms appropriate to his quest; Rowland discovered a relation between the entities of electromagnetic theory; Hopkinson between basic engineering parameters, such as the input and output of a dynamo. The method of approach, the argument, and the form of presentation differed, each according to its purpose and the audience for which it was intended. The two might be considered equivalent because the engineering variables of Hopkinson could be expressed as functions of the electromagnetic entities employed by Rowland.[38]

Perhaps no scientist has had a greater impact on technology than James Clerk Maxwell. But his influence was indirect, since few engineers could understand him. It required a creative effort almost equal to Maxwell's own by the British engineer Oliver Heaviside to translate his electromagnetic equations into a form usable by engineers.[39] Yet Maxwell was one of those scientists who consciously attempted to contribute to technology. Thus, he developed an important method for solving indeterminate problems in the theory of structures. But this work, too, had to be "translated" for technologists. A British engineer, after quoting Maxwell's conclusions, commented that "few engineers would, however, suspect that the two paragraphs quoted put at their disposal a remarkably simple and accurate method of calculating the stresses in a framework."[40]

The cases of Rowland and Maxwell suggest how the interchange between science and technology may take place. For information to pass from one community to the other often involves extensive

[37]James E. Brittain, "B. A. Behrend and the Beginnings of Electrical Engineering, 1870–1920" (Ph.D. diss., Case Western Reserve University, 1969), pp. 6–18.

[38]Ibid., p. 41.

[39]James E. Brittain, "Heaviside and the Telephone: A Case Study of the Interaction of Science and Technology in Nineteenth Century Telephony" (Master's essay, Case Western Reserve University, 1968), pp. 1–3, 7–14. See also Oliver Heaviside, *Electrical Papers*, 2 vols. (Boston, 1925).

[40]Quoted in Timoshenko, p. 203.

reformulation and an act of creative insight. This requires men who are in some sense members of both communities. These inter- mediaries might be called "engineer-scientists" or "scientist- engineers," depending on whether their primary identification is with engineering or science. Such men play a very important role as channels of communication between the communities of science and technology. It is perhaps significant that of American physical scien- tists of the 19th century, Joseph Henry, Alexander D. Bache, Henry Rowland, and J. Willard Gibbs were all trained as engineers. Admin- istrators of scientific agencies of government and those engaged in teaching science to engineers would be more effective if capable of understanding and reconciling the competing demands of science and technology.

* * *

It is worth noting, however, that the relationship between science and technology is a symmetric one. That is, information can be transferred in either direction. The flow of technology into science in the form of instrumentation has long been recognized; but the traditional model does not provide for the possibility that tech- nological theory might influence science. Yet the rise of engineering sciences such as the theory of elasticity and hydrodynamics did have an influence on science. The theory of elasticity provided a means of constructing models of the ether, a favorite occupation of Lord Kelvin, among others. The maturing of hydrodynamics was one cause of the proliferation of vortex theories of matter in the second half of the 19th century. Thermodynamics presents a somewhat more complex interaction. This science began with the French engi- neer Carnot as a design principle. It was not a law of nature but a statement of the limits of the efficiency of the steam engine. Its development relied on a simple macroscopic model – Carnot's ideal heat engine; it did not rely on molecular hypotheses. It was dis- covered in the engineering literature by scientist-engineers like Kelvin, Rankine, and Helmholtz, and translated by them into the language of science. As thermodynamics was absorbed by physics, Carnot's ideal heat engine tended to be replaced by the molecular model of statistical mechanics.

Because of the status differentials, one would expect engineers with the appropriate training to attempt some work directed at the scientific community. Theory ranks high in science but low in engi- neering. Many examples of American engineers contributing to ba- sic science could be cited. De Volson Wood used elasticity considerations in an attempt to determine the density, pressure, and specific heat of the ether. Although one of the weakest of his works, Wood apparently took inordinate pride in it since he published an expanded version as a book. Henry Turner Eddy did some interesting papers in which he concluded from kinetic considerations that the atom must have some form of internal motion and he postulated the existence of a subatomic particle.[41] The only earned degrees of J. Willard Gibbs and Henry Rowland were in engineering. Considering the low status of academic theorists in engineering, not to mention the low status of engineers as a group, their identification with physics is not surprising.

The most important influence of technology on scientific ideas, however, was more indirect. Engineering sciences did not postulate unobservables. Their example was, therefore, a challenge to physics. They contributed to the critical reexamination of the foundations of physics which took place in the latter 19th century. But the engineers themselves contributed little to this movement; it was carried forward by physicists under the banners of positivism and energeticism. The influence of technology on science, like that of science on technology, was an indirect, "second-order" effect.

The coupling of science and technology in the 19th century had at least two important social consequences in the twentieth. It accelerated the pace of technological change, and consequently of social dislocation. It also encouraged engineers to adopt a self-image based on science which served to discourage them from assisting society in meeting the problems they had done so much to create. The scientific self-image caused engineers to portray themselves as logical thinkers, free of all bias and emotion, and it promoted a sort of "above-the-battle" neutrality on the part of the profession. Although engineers gave lip service to the idea of social responsibility, their definition of this responsibility served to prevent effective action. When faced with an actual social problem, engineers have sought objective or "scientific" solutions. In practice, this made the discovery of methods of social engineering a precondition to social action, thus substituting an impossible task for a difficult one. The delusive quest for social engineering led more than one engineer down the blind alley of technocracy.

The reversal of "parity" between science and technology further

[41]David R. Topper, "The Development of the Kinetic Theory of Gases in America: An Analysis of the Ideas of Three Key American Figures prior to Gibbs" (Master's essay, Case Western Reserve University, 1968), pp. 35–62; De Volson Wood, *The Luminiferous Aether* (New York, 1886).

reduced the engineers' ability to respond effectively to social problems. The scientific community has been better able to act on social issues because those with the greatest prestige were in universities where they were relatively free from pressures from corporations and government. The engineers who enjoyed a corresponding independence lacked sufficient prestige to lead their profession. Prestige and power in engineering went to the "doers," not the "theorists." This had the practical effect of giving the control of the engineering profession to men who were linked by ties of self-interest to those who were using, and in some cases, misusing technology. This conflict in interest between the leaders of the profession and rank-and-file engineers did much to frustrate the legitimate professional aspirations of American engineers.[42]

[42]The author's *The Revolt of the Engineers: Social Responsibility and the American Engineering Profession* (Cleveland, 1971) deals with the engineering profession's concern for social responsibility.

Practice

LOCAL HISTORY AND NATIONAL CULTURE: NOTIONS ON ENGINEERING PROFESSIONALISM IN AMERICA

BRUCE SINCLAIR

On the face of it, we know more about the engineering profession than practically any other large topic in the history of American technology. There are eight historical monographs explicitly concerned with the subject, half a dozen relevant biographies, and at least as many books that deal with the profession indirectly.[1] So it might seem fanciful to argue that we lack important details or that major questions remain to be addressed.

Yet those are just the claims I want to make. After a quarter-century of historical attention, we still know very little of the vast majority of American engineers. The actual case is that this extensive literature tells us mostly about the profession's central characters and the organizations those kinds of men established and perpetuated. The rank and file of any large group are always harder to apprehend than its

Bruce Sinclair is the Melvin Kranzberg Professor of History of Technology at Georgia Institute of Technology.
[1]The studies that deal particularly with the engineering profession in America are, in order of publication date: Daniel H. Calhoun, *The American Civil Engineer: Origins and Conflict* (Cambridge, Mass., 1960); Monte A. Calvert, *The Mechanical Engineer in America, 1830–1910: Professional Cultures in Conflict* (Baltimore, 1967); Raymond H. Merritt, *Engineering in American Society, 1850–1875* (Lexington, Ky., 1969); Edwin T. Layton, Jr., *The Revolt of the Engineers: Social Responsibility and the American Engineering Profession* (Cleveland, 1971 [and Baltimore, 1986]); David F. Noble, *America by Design: Science, Technology, and the Rise of Corporate Capitalism* (New York, 1977); Bruce Sinclair, *A Centennial History of the American Society of Mechanical Engineers* (Toronto, 1980); Terry S. Reynolds, *75 Years of Progress—a History of the American Institute of Chemical Engineers* (New York, 1983); and A. Michal McMahon, *The Making of a Profession: A Century of Electrical Engineering in America* (New York, 1984). A complete listing of works concerned to one degree or another with engineers and engineering would be very long indeed. But for an important and different set of arguments about the nature of technical work, see Eugene S. Ferguson, "The Mind's Eye: Nonverbal Thought in Technology," *Science* 197 (August 26, 1977): 827–36, and Brooke Hindle, *Emulation and Invention* (New York, 1981).

This essay originally appeared in *Technology and Culture*, vol. 27, no. 4, October 1986.

leading figures, but in this well-studied profession the discrepancy is especially glaring. Nor do we know much of anything about what other Americans thought of these engineers. We claim that they symbolize technology, or reflect America's commitment to it, or in some other way provide insight into big issues like that. But, in fact, none of us has more than a slight grasp on the way the work and lives of engineers might illuminate the study of American culture.

Think about the paradoxes. More American men follow engineering than any other profession. Yet the vast majority—75–80 percent—of those identified with the principal branches of the field do not belong to their national societies and do not participate in the activities of those organizations. That was true in the past, and it is still so. Furthermore, those organizations—which ostensibly exist to serve their members—not only have no historical records of their memberships, they know surprisingly little about those who currently belong and nothing at all about the great numbers of engineers who might logically join.

If this state of affairs conjures up a mental picture of one of those buildings of the Old West, where there is more in front than there is behind, the difference between the claims engineers make for themselves and the way they appear to others is equally anomalous. Ever since John Alexander Law Waddell started telling engineering students to clean their fingernails and comb their hair, it has been easy to caricature the profession. Herbert Hoover's biographer said of the Great Engineer, for example, that he had all the emotions of "a slide rule."[2] Or, as the editor of *Toronto Life* recently described the grammar and punctuation programs of his word processor, "they seem to have been designed by engineers, not writers; they force your prose into a stuffy and predictable style."[3] Instead of the portrait of a profession, what we have is a grab bag of stereotypical images and they picture a group that seems politically inflexible, socially awkward, culturally limited, and ethically inert.

In the late 1890s, the civil engineer George R. Morison argued that the coming potential for generating unlimited amounts of electrical energy would inaugurate a revolutionary stage in human development, unlike anything that had gone before, and that engineers would be the "priests of the new epoch."[4] Why, then, didn't Theodore Dreiser

[2]A sampling of Waddell's ideas can be found in *Addresses to Engineering Students* (Kansas City, Mo.: Waddell & Harrington, 1912). The remark about Hoover comes from Robert S. McElvaine, "An Uncommon Man: The Triumph of Herbert Hoover," *New York Times, Book Review Section*, September 2, 1984, p. 4.

[3]*Toronto Life*, May 1984, p. 5.

[4]As quoted in Layton, *Revolt of the Engineers*, p. 59.

or Frank Norris or Sinclair Lewis write a novel about these powerful characters in the American drama? How is it that Eugene O'Neill's 1929 play, *Dynamo*, which so caught the spirit of that great transformation Morison had in mind, mentions engineers only to dismiss them as irrelevant? If technology really stands at the center of the American experience, if its history tells us something both novel and essential about the country's past, as Brooke Hindle has said, why are engineers so invisible in American culture?[5] Is it because, after all, they are not synonymous with technology? Or is it that, in any event, elevated literature is the wrong place to look for them? And if existing historical scholarship does not answer these kinds of questions, what sort of an approach would?

Population biologists struggle with the problem that, in groups large enough to be statistically significant, individual complexities are lost, while aggregations small enough to make individuals significant become numerically irrelevant. Our difficulties with the existing literature are analogous. Edwin Layton and David Noble deal with essentially the same group of people, and, by making national engineering societies and corporate industry the principal settings for their stories, each implies a scale of historical action grand enough to describe important truths. However, they characterize this group in two quite different ways. Those engineers who for Layton are riven by ambivalence—pulled in opposite directions by science and business and consequently unable to realize either their professional aspirations or their economic ambitions—are for Noble a powerful and cohesive group, quick to identify their interests and to plot strategies that gain them their objectives. So it is apparently difficult simply to characterize the profession's leadership and the effect of occupational circumstance. But even if we could, we would still be talking about a small fraction of the total population and one in most respects unrepresentative. Nor does either analysis reflect that more individualistic, solitary, creative, and aesthetically satisfying side of engineering that Eugene Ferguson and Brooke Hindle talk about and that is presumably an important element of the engineer's self-image as well as of his work. Thus—and to overstate the case for the sake of argument—the existing literature not only fails the test of statistical validity; it yields an insufficient amount of information about individual variety.

There is another, more promising, avenue of attack open to us. I think the vital core of the profession might best be discovered at the level of local engineering associations. The membership, activities, and

[5]Brooke Hindle, ed., *Technology in Early America: Needs and Opportunities for Study* (Chapel Hill, N.C., 1966).

orientation of these groups are more representative of the profession than are the national societies. And it also strikes me that a study of them will most probably lead to a synthesis of our knowledge of this subject as well as to its integration into main themes of American culture. It may seem unpromising to seek in parochial associations the national dimensions of engineering professionalism, but there is where we will find most of the country's engineers, and there is where we are also more likely to get an enriched sense of their lives.

It is not, of course, that the concerns of national society leaders are irrelevant to the rank and file or that there is not an overlap in their interests, but rather that, besides the congruences, there are differences. For example, it has been clearly shown that the officers of national organizations are more conservative in their economic and social views than the membership. Conversely, city engineering societies in places like Cleveland, St. Louis, and Boston, institutions more likely to enroll those men who were not members of a national organization, were leading elements in the profession's reform movement of the early 20th century. The discourse in these local clubs is less self-conscious, too, and more likely to suggest what people feel as well as what they think.

There is to hand a neat case study to support the proposition. In 1930, when the American Society of Mechanical Engineers wanted to mark its fiftieth anniversary (and also to combat those critics of mechanization who blamed it for causing unemployment and dehumanized working conditions), the New York officers and staff planned an elaborate, week-long celebration carefully designed to publicize the claim that engineering was the basis of modern civilization. The festivities featured a gathering of prominent engineers from all over the world, a special banquet addressed by the president of the United States, and a unique theatrical production entitled *Control* that aimed to dramatize the connection between engineering and human progress.[6] Besides the novel use of light, sound, and motion pictures, that pageant employed a cast of allegorical characters—Curiosity, Intelligence, and Beauty, among others—to illustrate engineering's professional and intellectual development. The proofs of that maturation, according to the pageant, were the readiness of engineers to assume a leading role in the solution of the world's economic and social problems and the ability of engineering to provide consumers with aesthetic satisfaction as well as material abundance.

Now, it happens that in 1930 the Engineers' Club of St. Louis also staged a play about the profession. Topical, funny, irreverent, and

---

[6]That pageant and the other ceremonials of the ASME's fiftieth anniversary celebration are described in my *Centennial History of the American Society of Mechanical Engineers.*

sardonic, it conveys a very different message than the pageant organized by ASME's elite, and thus we have the ingredients for an unusual
comparison. Peter Gay, in his *Education of the Senses*, reminds us of
Freud's argument that "institutions, whether of society or of the mind,
at once control passions and satisfy human needs."[7] In these two
theatrical presentations, then, what can be discovered about the passions and needs of engineers?

The St. Louis engineers titled their production *Every Engineer: An
Immorality Play.*[8] It was, of course, to be a sort of *Pilgrim's Progress*, and it
depicts the career of a naive young engineering graduate as he discovers what a professional life is really like. This play also has its allegorical
characters—Youth, Ambition, and Ingenuity—who are Every Engineer's companions on his journey, as well as a cast of villains, called
"robbers" on the program, which identifies them as St. Louis private
utility corporations.

As the play opens, we learn two things about Every Engineer, that
he is powerfully educated and enormously, indeed brashly, self-
confident. Here is how he describes himself:

> Building a bridge is merely childish play,
> Electric theories are at my finger's ends
> The methods of the laboratory, say:
> I know how every beam of concrete bends!

Ambition echoes the extent of the engineer's learning with the observation, "All the professors passed him in their courses / *He* knows the laws
which govern mass and forces." Nor is this the extent of his knowledge,
as Youth advises us, "And he can juggle chemistry to boot / And as for
handling men, that's his long suit." These lines may sound like one of
Gilbert and Sullivan's patter songs, but it is not difficult to hear in them
the language engineers of that era used when seriously describing
themselves.

In Every Engineer this sense of commanding knowledge generates a
considerable audacity, and he says,

> Before me mighty work is all I see
> Perhaps some trifling task to fill the hour
> Until the ginks with money come to me
> And give me a position of much power.

[7] Peter Gay, *The Bourgeois Experience—Victoria to Freud*, vol. 1, *Education of the Senses*
(New York, 1983), p. 459.

[8] I am grateful to the Engineers' Club of St. Louis for a copy of the original typescript of
the play.

So, full of himself, Engineer breezily approaches Corporation I, a private utility company of the city, for a job. The dialogue makes it plain, however, that these sorts of firms are put off by his independent cockiness, that they want experienced men, and that they want them cheaply, too. There is also, behind these lines, something of the painful knowledge of personal experience, of having learned the difference between school and the world, between mastery of knowledge and control over one's life.

Chastened by this rebuff, Every Engineer next approaches "Municipality" for a job and in that exchange is taught the realities of local politics. He is hired only because he knows someone and then discovers that, besides having to make a contribution to party funds out of his salary, he will be judged on his ability to win votes rather than on his technical skills.

After a brief piece of dialogue that satirizes the laziness and incompetence of city engineers, all the corporations reappear on stage, swaying gently to the "Flower Maiden" music from *Parsifal*. And now we come to the central confrontation. Attracted by Every Engineer's moral pliability, they introduce themselves one after another in a wonderfully scurrilous fashion:

Corporation II	We are bold Corporation
	The terror of this nation.
	In Jersey we incorporate
	But take our compensation
	From every man in this wide land
	Of high or lowly station.
Corporation III	Our stock on its own water
	Floats, though it shouldn't oughter,
	Our rights they are inviolate
	As Pharo's only daughter . . . .
Corporation IV	Precision such as yours, sir,
	Efficiency so sure, sir,
	We yearn to hire and consecrate
	To uses high and pure, sir;
	Come! find your ends in dividends.

As clearly ironic as their blandishments are, Every Engineer is easily persuaded. His acceptance speech reveals his awareness of the potent consequences in the combination of capital and technical skill, just as the corporations know that too, and they rejoice:

> We are together now and will
> Make the dear people foot the bill.

And by alchemic methods surer
Squeeze dividends from Aqua Pura.

Yet even as Every Engineer contemplates the bargain, he feels a sense of responsibility to his new employer, and his words reflect the profession's claim of ethical obligation to the client:

I hold a job, but bet your cash,
They'll get their money's worth.
No more on petty work
Have I a minute's leisure
No time to eat, no time to sleep,
No hour consigned to pleasure.

Youth joins the engineer in this commitment with the pledge, "I willingly will give my finest days / If for their wasting Corporation pays." His loyalty is met with scorn, however, as, *sotto voce*, the corporations mock Youth's poignant declaration with derisive laughter.

Oblivious to these portents, Every Engineer now calls on another allegorical familiar, Ingenuity, who with Youth and Ambition will raise him to success in his new job. Naturally, Ingenuity finds Every Engineer's situation appealing, and in a bit of stage business characteristic of this broad farce, slips an idea under his hat. When he sees it, Every Engineer exclaims:

Now will fat corporation
Be pleased with me. He'll pat me on the back,
And raise my pay: I should worry now!
Here's the stunt and it's a cracker-jack
A scheme to bring a joy to Jonny Hunter's heart,
A plan to use electric currents in plumb tarts!
T'will flatten out the peak, the hollows fill
And we'll get profits from every grocery bill!

But the corporation, in this case the United Electric Light and Power Company of St. Louis, has been looking over the young man's shoulder and snatches the idea from him. "Here, give me that you mutt," the corporation says, "I own the product of your festering nut." Youth and Ambition are pretty badly jolted by the experience, but Ingenuity slips another idea under the engineer's hat, with the advice, "Next time, my friend, make corporation *buy!*"

The second idea is directed toward one of the city's transit companies:

A plan to make the seats so darn unpleasant
That no one, whether Lord or lowly peasant,
Will stand for them, but *on* them, then you see
They'll hold not two unfortunates, but three.

This corporation, too, has sneaked up on the engineer and says, "I'll take that stunt, so come across / You are my hireling and I am your boss." When the young man tries to hold it back, the corporation knocks him to the floor and takes the idea anyhow. In the struggle Youth has been struck down. Indeed, that is the end of him and, as the subsequent dialogue makes plain, of innocence besides.

Every Engineer is momentarily saddened by this turn of events, but Ambition cheers him on and Ingenuity gives him yet another idea, which this time, with a craftiness matching that of his adversaries, he hides for safekeeping. Boldly, then, he goes up to the local gas utility and tells the corporation, "This is a stunt, a peacherino true, / for multiplying all gas bills by two." As the others did, this corporation also grabs the engineer and searches him for the idea. When it isn't found, however, the corporations realize they must revise their tactics, and they invite the engineer to lunch where, after some bargaining, they agree to make him their consulting engineer. As one of them puts it, "My man, you've got a nifty little thinker / With all our properties we'll let you tinker."

The post of consulting engineer is the pinnacle of achievement for Every Engineer, and, in unison, the corporations sing with him a brief but deliciously ironic chorus that makes it seem as if he has acquired the position as a result of an arduous though honorable climb to the top. Then Municipality, politically reformed now, appears back on stage and joins in to say, "But talent such as yours I cannot buy," to which Every Engineer adds the refrain, "Can't buy." Municipality promises, however, in another of the play's topical references, that, if the new city charter is adopted, things will change for the better. Here again, the message—so reminiscent of Morris L. Cooke's anti-utility campaign two decades earlier—is obvious.

At that point the play's focus shifts as the last of the allegorical characters, Success, comes on stage, laurel wreath in hand. Every Engineer turns eagerly toward Success, casually leaving Ambition, his youthful companion, to depart the stage alone. As if it were not already clear, Success then describes the human costs of Every Engineer's achievement:

Few men attain my friendship without sin,
My presence is no mark of purity

No guaranty of firm security,
The gaunt wolf "Want" may yet be heard
Outside your door.

In the 1930s these references to the unpredictable nature of economic life were real enough, and Success continues to mix harsh
imagery with idealism. Indeed, Success's speech is a curious one. It
ends the play and one might expect an upbeat, lighthearted tone.
Instead, Success compares engineering with the other professions, to
its detriment. The format is a familiar one in the contemporaneous
engineering literature and in that context should have produced the
old joke about doctors who bury their mistakes. But Success tells the
engineers that they cannot hope to enjoy the status or financial rewards
of the independent consultant, despite the learning and labor their
profession demands. They have in their hands the "instruments that
lay the *real* truth bare"; they can make "the poets' dreams" come true,
he says. But few will value their achievement, and if they fail—an idea
that consistently appears in these kinds of professional comparisons—
they alone will bear the burden of it.

The play's authors meant by this astringent, Grail-like characteriza-
tion of the engineering profession to close on an elevated, though
somewhat elegiac, note. One could decide, despite those intentions,
that this stark contrast with the funny and pointed material that came
earlier was simply due to a failure of imagination. But there are other,
more interesting ways to look at the play, the most obvious of which is
that to some degree the production mirrors in both its humorous and
serious modes the actual circumstances of St. Louis engineers during
the 1930s. There is, for instance, an inescapably rueful undertone in
the way they kidded themselves about their educations, employers, and
careers. And in a similar fashion, the discontinuity between those jokes
and the play's somber ending also suggests that, beneath the surface,
there are attitudes and ideas worth exploring. But most of all, the play
indicates how important such ephemeral, local sources can be in getting us closer to our subjects. And that possibility points toward
another simple truth, namely, that the specifics of time and place are
still the essential ingredients of the historian's work and that even such
conventional tools are useful in the understanding of engineering
professionalism.

It must be admitted, however, that in selecting this odd and fugitive
document as case study, I also want to argue that the history of engineering professionalism is ripe for new adventures in analysis. One
example of the different kind of interpretive modes that lie waiting for
us is that book of Peter Gay's. The first of a projected multivolume

258   *Bruce Sinclair*

analysis of the bourgeoisie in the 19th century, *Education of the Senses* is full of ideas and approaches that seem valuable. Gay alerts us to the fact that documents like *Every Engineer* carry latent meanings, that people orient themselves by cultural signals, and that out of "varieties of experience," the historian can construct "a recognizable family of desires and anxieties."[9] This approach encourages us to see the full nature of people; it helps correct the tendency to typecast engineers in the flat, one-dimensional terms we so often resort to; and it indicates how we might more successfully deal with the contradictions that currently hamper our efforts to describe the profession and its relation to American life.

Gay's use of Freudian psychoanalytic concepts, particularly aggression, suddenly made me realize how much an engineer's ordinary experience is dominated by adversarial relations of all sorts. And they are an accepted part of life; at one point in his speech, Success tells Every Engineer, "So I the men of all professions seek, / Saving the strong and grinding down the weak." Aggression also encompasses the notion of mastery—I am reminded of Sally Hacker's study of the function of the calculus in engineering education—and it includes domination over the environment.[10] George Babcock, the founder of Babcock and Wilcox, provided a telling example of that kind of attitude when he claimed that engineering's principal mission was to bring about the day "when every force in nature and every created thing shall be subject to the control of man."[11]

What Gay makes us realize, however, is that in these respects engineers do not stand apart from the rest of American culture. To the contrary, engineering professionalism is a cultural artifact, just as fashion, family life, or the language of the marketplace is, too. And if materialism and a certain difficulty with ideals are hallmarks of bourgeois culture, as Gay claims, then we can begin to recognize characteristics of the engineering population in terms that connect them directly to American history. The sense of impotence that *Every Engineer* expresses is not then a simple function of the terms of employment of engineers but a result besides of the pressures most Americans felt to get ahead and their fear of the consequences if they failed. Or, to put it somewhat differently, it was not simply corporate power or professional status that disturbed engineers but also the rapidly evolv-

[9]Gay, p. 5.
[10]Sally Hacker, "Mathematization of Engineering: Limits on Women and the Field," in *Machina ex Dea: Feminist Perspectives on Technology*, ed. Joan Rothschild (Elmsford, N.Y., 1983), pp. 38–58.
[11]American Society of Mechanical Engineers, *Transactions* 9 (1888): 37.

ing nature of their work, and they felt themselves ground between the millstones of past and present—between an old mechanic arts tradition that spoke to enduring American values and the engineering science of the 20th century that promised insulation both from corporate cupidity and the condescension of aesthetes. So it is out of the processes of bourgeois culture that we get professionalism and specialization, but also, as the St. Louis play suggests, conflicting feelings of helplessness and confidence, of loyalty and isolation.

Thus, to insist on the complexity of human experience and on broad definitions of culture as our points of departure means, for the historian, access to a stock of emotional responses as well as political reactions or economic concerns. And that fuller kind of information yields, I think, better insight into the ways engineers and people like them tried to manage their lives during periods of great change.

THE INTRODUCTION OF THE LOADING COIL: GEORGE A. CAMPBELL AND MICHAEL I. PUPIN

JAMES E. BRITTAIN

The introduction of loading coils into telephone circuits was probably the most important single technical innovation in telephony during the forty-year period between the time of Bell's original inventions and the introduction of electronic amplifiers. By 1900 long-distance telephony had reached what seemed to be a practical limit in a 1,200-mile circuit between Boston and Chicago. The loading innovation, in effect, doubled the practical distance and also led to substantial reductions in the construction costs of underground cable circuits used in urban areas. Exploitation of the loading coil saved the American Telephone and Telegraph Company an estimated one hundred million dollars during the first quarter of the 20th century.[1] Despite its great economic importance and the fact that it has been frequently cited to illustrate the impact of science on technology, the early history of the loading coil has been subject to considerable misunderstanding. This is due not only to the somewhat esoteric nature of the innovation but also to a persistent myth regarding the role of Michael I. Pupin in its introduction.

The coil-loaded telephone circuit may seem deceptively simple, since it consists of inductance coils connected in series with the conductors of telephone wires at periodic intervals. However, an examination of the historical details reveals that this was one of the most sophisticated electrical innovations of the 19th century. It is significant that each of the claimants to priority of discovery was proficient in mathematical physics. Unlike earlier communications advances, even including the telephone itself, loading was based on an understanding of the Maxwell-Heaviside electromagnetic theory and required an "inventor" having a considerable facility in applying

JAMES E. BRITTAIN is a historian in the Department of History, Technology, and Society at Georgia Tech. He is among six coauthors of *Engineering the New South*. A specialist in the history of electrotechnology, he edited *Turning Points in American Electrical History*, and his biography of E. F. Alexanderson is in press.

[1] Thomas Shaw and William Fondiller, "Developments and Applications of Loading for Telephone Circuits," *Transactions of the American Institute of Electrical Engineers* 45 (1926):291–92 (hereafter cited as *Trans. AIEE*).

This essay originally appeared in *Technology and Culture*, vol. 11, no. 1, January 1970.

and extending that theory. Consequently, it is not surprising to find that many engineers and patent specialists experienced difficulty in grasping some of the subtleties of the analysis and explanation of loading. The successful practice of loading required that strict limits be placed on the design and spacing of the loading coils. If these limits were not satisfied, transmitted currents were subjected to more distortion than if loading had not been used at all. Also, inductance coils had often been used in other applications to impede rather than enhance the transfer of alternating-current energy. For this reason, many engineers were not convinced that loading would really improve telephonic transmission until they had witnessed demonstrations on actual circuits.

Many important technical inventions or discoveries have been controversial, and loading was no exception. Michael I. Pupin, a Columbia University professor, won both fame and a considerable fortune for his role in the introduction of loading. Pupin's version of the history of the loading coil was included in his autobiography, *From Immigrant to Inventor*, which won him a Pulitzer prize in 1924. This version seems to have been generally accepted by recent historians of electrical technology.[2] A second claimant was the brilliant and eccentric British electrical theorist, Oliver Heaviside. A few of Heaviside's advocates argued that the Bell company had conspired to deny him the credit for this and other discoveries.[3] A third and probably the least-known

[2] See, for example, Harold I. Sharlin, *The Making of the Electrical Age* (London, 1963), pp. 65–68, and Percy Dunsheath, *A History of Electrical Engineering* (London, 1963), p. 245.

[3] For example, B. A. Behrend wrote to G. F. C. Searle, a British scientist and friend of Heaviside, that he had succeeded in persuading the directors of the American Institute of Electrical Engineers to award an honorary membership to Heaviside only by agreeing to give support to "the recognition of certain parties by the award of an Edison Medal." Presumably, Behrend was alluding to John J. Carty, chief engineer of the American Telephone and Telegraph Company, who was awarded the Edison Medal in 1918 while Behrend was a member of the Edison Committee. Behrend also informed Searle that "certain interests" had tried to persuade him to act as intermediary in offering Heaviside a few thousand dollars in belated recognition for the importance of his contribution to communication technology. Searle had suggested that a Cambridge chair be endowed by contributions from industry in honor of Heaviside, but Behrend opposed this being done "by the very people who destroyed his life and prevented his recognition." Norbert Wiener, another Heaviside advocate, wrote Behrend in 1931 to thank him for a guide to his collection of "Heavisidiana" and added that it was being photostated for "antipupinian" friends. See Behrend to Searle [?] 1928, and Wiener to Behrend [?] 1931, Behrend Papers, Clemson University Archives, Clemson, S.C. Wiener later wrote what appears to be

claimant was George A. Campbell, an engineer with the American Bell Telephone Company. Despite his comparative obscurity, the evidence indicates that Campbell was the first to load actual telephone circuits successfully and that his theoretical analysis of loading formed the basis for subsequent loading practice by the Bell company.

* * *

The three claimants mentioned had very dissimilar personalities and professional affiliations. As a young man, Heaviside had been employed briefly as a telegrapher but never again accepted employment after retiring in 1874 at the age of twenty-four. He came to consider himself as an interpreter and extender of the energy transmission theory of Maxwell for the benefit of telephone and telegraph engineers, and he published frequently in engineering and scientific journals. Heaviside never attended meetings of professional societies, although he corresponded with many leading scientists and engineers. He was eager to receive credit for what he considered to be his intellectual property, although he apparently made no effort to patent any of his many suggested innovations in communications technology.

Pupin was a flamboyant and gregarious individual. He was active in the affairs of the American Institute of Electrical Engineers and published frequently in its *Transactions*. Pupin became his own best publicist and, unlike Heaviside, proceeded to take full advantage of the patent system. Most of his experimental work was carried out with the aid of his students in a small basement laboratory at Columbia. From the sale of the rights to a small number of electrical patents, Pupin achieved a financial success which has seldom, if ever, been equaled by a full-time engineering professor.

Campbell was a member of the engineering staff in the Boston

<hr />

a fictional version of the "conspiracy" against Heaviside. See Norbert Wiener, *The Tempter* (New York, 1959).

Heaviside himself claimed credit for the invention in his essay on telegraphy included in the tenth edition of the *Encyclopaedia Britannica*. See Oliver Heaviside, *Electromagnetic Theory* (New York, 1950), p. 343 (this is an unabridged reprint of the edition originally published in three volumes in 1893, 1899, and 1912, respectively). Heaviside also discussed the priority question in a letter to Behrend in 1918. Heaviside wrote that no one had questioned his priority until Pupin sold his patents, but that, following that event, a great change had occurred and he (Heaviside) had been "repudiated in a most emphatic manner." Heaviside suggested that commercial considerations had been responsible for the change, since "prior publication . . . if admitted, might have interfered with the flow of dollars in the proper direction and have led to much trouble and expense." See Heaviside to Behrend, February 20, 1918, Heaviside Papers, Institute of Electrical Engineers, London.

laboratory of the American Bell Telephone Company. He was an intellectual introvert with a dislike for administrative duties. Unlike Pupin, he was never active in professional societies and published comparatively little in technical journals.[4] Although he remained almost unknown outside the telephone company, Campbell was the company's leading electrical theorist for many years and was responsible for the introduction of several of the most important innovations in modern communications technology.[5]

Guided by his study of Maxwell's *Treatise on Electricity and Magnetism*, Heaviside by 1881 had arrived at a differential equation for the analysis of uniform transmission lines in terms of four distributed line parameters.[6] In 1887 Heaviside announced his discovery of a criterion for "distortionless transmission." His theory indicated that the product of resistance and capacitance of a circuit should be made equal to the product of leakage and self-induction for minimum transmission distortion. Since the product of resistance and capacitance was much larger on normally constructed circuits, Heaviside pointed out that transmission quality should be improved if either the leakage or self-induction were to be artificially increased. He suggested that

[4] This was at least partly due to the general policy of the company, during Campbell's early years in Boston, of not encouraging the publication of results of current research which might be of use to independent competitors. The first report of Campbell's theoretical and experimental work on loading did not appear in a technical journal until approximately four years after it had been done. This paper and a few of his previously unpublished early reports on loading research were included in a volume of his collected papers published in 1937 on the occasion of his retirement. See *The Collected Papers of George Ashley Campbell* (New York, 1937). The policy of secrecy was evidently somewhat modified by 1904. An editorial in the *Electrical World and Engineer* early in 1904 commented on the first paper by Campbell to be published in that journal. The editor stated that Campbell's paper was important both because of its technical content and because it indicated a policy change on the part of the Bell system. The editor suggested that the company's previous policy had been "short sighted, imprudent, and immoral." He argued that, while the company had the right to withhold trade secrets and manufacturing data, it should not keep secret matters related to general technology or applied science "which, if made known, only redound to the credit of the industry disclosing while adding to the world's general stock of knowledge." He continued that, if important discoveries were kept secret and later discovered by outsiders, "the original work done earlier by the industry has lost credit, and has lost the evidence of progress" (*Electrical World and Engineer* 42 [1904]:633).

[5] Frank B. Jewett, "Dr. George A. Campbell," *Bell System Technical Journal* 14 (1935):553–57.

[6] See Oliver Heaviside, *Electrical Papers*, 2 vols. (Boston, 1925), 1:139–40 (this is an unabridged reprint of the original edition published in 1892). The four parameters were resistance, capacitance, leakage, and self-induction.

an increase in self-induction seemed preferable, since an increase in leakage would result in greater attenuation despite the predicted reduction in signal distortion. One way to increase self-induction was to increase the spacing of wires in a given circuit. Another possibility, which Heaviside suggested as worthy of experimental investigation was to employ a dielectric material containing iron dust for insulation in telephone cables.[7]

In 1893 Heaviside suggested that the effective self-induction of a transmission circuit might be increased by the insertion of discrete coils in series with the circuit conductors at periodic intervals. He admitted, however, that there was as yet no experimental evidence that this would prove successful.[8] To his great disappointment, these suggested methods for improving transmission were largely ignored by British telephone engineers. One writer later attributed the neglect to Heaviside's failure to provide "such detailed instruction as to compel the attention of practical telephone engineers."[9] The instructions needed for the effective use of loading coils were the coil design specifications and the optimum coil spacing in terms of the shortest wavelengths of interest in telephony. There is no evidence that Heaviside undertook the analytic or experimental work which would have been required to provide such instructions, perhaps because he was not interested in obtaining patents and felt that such details could best be worked out by telephone engineers using actual circuits. Despite the omission of engineering details, Heaviside had set the stage for the loading coil innovation by his demonstration of the theoretical advantages of increased self-induction in transmission circuits.

John S. Stone, an engineer employed by the American Bell Telephone Company during the 1890s, was apparently the first engineer consistently to apply Heaviside's transmission theory to the practical problems encountered in telephony.[10] In 1897 Stone was granted a

[7] Ibid., 2:123–24.

[8] Heaviside, *Electromagnetic Theory*, p. 112.

[9] See J. A. Fleming, *The Propagation of Electric Currents in Telephone and Telegraph Conductors* (New York, 1911), pp. 108–9.

[10] Stone was one of the first men with a university-level education in science and mathematics to be employed as an engineer by the American Bell Telephone Company. He enrolled at the School of Mines at Columbia in 1886 but transferred to Johns Hopkins University in 1888. Through the influence of Gardner G. Hubbard, a friend of the Stone family and one of the original financial backers of A. G. Bell, Stone was able to spend the summer of 1889 in Paris as an assistant to W. D. Sargent, who was in charge of the American Bell company's exhibition at the Paris Ex-

patent on a proposed new telephone cable which was to use bimetallic wire of iron and copper to give a higher self-induction than available in standard cable.[11] Although Stone's patent made no explicit mention of beneficial effects on transmission other than through the elimination of reflections, he was fully aware of the potential use of a high-induction cable to achieve low-distortion transmission.[12] At Stone's request, the director of the laboratory, Hammond V. Hayes, authorized further research to determine whether bimetallic circuits could be used to obtain a useful improvement in distance and transmission quality.[13]

Campbell's work on the loading coil innovation was a direct outgrowth of the bimetallic cable investigation. When Stone left the company early in 1899, Campbell was assigned to continue work on the high-induction cable.[14] However, Hayes extended the scope of the original work order and authorized Campbell to proceed in any direction which seemed promising. Since no experimental work had so far been done on a long bimetallic circuit, Campbell decided to undertake some preliminary transmission tests for comparison with results obtained on circuits currently being used. The cost of an actual bimetallic circuit of the length needed for conclusive results would have been large, whereas the funds available for research on

position. Stone was hired after his graduation the following year. See George H. Clark, *The Life of John Stone Stone* (San Diego, Calif., 1946), pp. 14–26.

[11] Stone's patent was no. 578,275, issued March 2, 1897. The claims in the patent were confined to the use of the new cable to eliminate wave reflections at junctions with high-impedance overhead lines. Heaviside had shown that this would be accomplished if the ratio of self-induction to capacitance of a cable were made equal to that of the overhead line.

[12] Stone had attempted to calculate the degree of improvement which might potentially be achieved by means of a high-induction cable and had discussed the problem in personal correspondence with Heaviside in 1894. See Stone to Heaviside, May 12, 1894, Heaviside Papers.

[13] See the *Record for George A. Campbell*, Interference no. 20,699 (Boston, 1902), pp. 9–10. (Hereafter cited as *RGAC*.)

[14] Campbell had received a bachelor's degree in engineering at the Massachusetts Institute of Technology in 1891. He had then spent two years at Harvard, where he received a master's degree in 1893. He was then awarded a fellowship which enabled him to spend a year at Göttingen, where he studied advanced mathematics under Felix Klein, a year at Vienna, where he studied electricity and mechanics under Boltzmann, and a year in Paris, where he studied under Henri Poincaré. Campbell was awarded a doctorate at Harvard in 1901, his dissertation being based on his research on loading coils at the Bell laboratory. See *RGAC*, pp. 4–5.

the project were quite limited.[15] In the interest of maximum economy, Campbell proposed the construction of an artificial transmission line which would serve to simulate an actual circuit for his preliminary tests.[16]

In the process of working out the design of the artificial line, Campbell asked himself what proved to be a crucial question. If one could use discrete inductance coils to simulate distributed inductance in an artificial line, might it not also be possible to use similar discrete inductors periodically spaced along an actual line? If so, this would make possible a much greater increase in the effective line inductance than seemed to be theoretically possible using Stone's bimetallic cable design.[17] Excited by the new idea, Campbell immediately abandoned further work on bimetallic circuits and turned his attention to what soon became known as the loading coil. Campbell's familiarity with the Heaviside transmission theory and an extraordinary mathematical ability enabled him to make rapid progress in working out the engineering details needed for a full-scale experiment with a loaded telephone circuit.

The most important immediate problem which Campbell faced was to determine the maximum allowable interval between coils, since the ultimate practical value of the innovation would largely depend on this factor. His preliminary calculations indicated that approximately ten coils per wave at the maximum telephonic frequency employed would be sufficient. This meant that loading coils could be added to the existing underground cable system, since there were normally several manholes per mile. Hayes was especially impressed by this possibility when he discussed the progress of the investigation with Campbell early in 1899. At the request of Campbell, who now wished to test the effect of loading on a standard commercial cable, Hayes arranged to have three reels of the so-called Pittsburgh cable

[15] In a progress report written in July 1899, Campbell stressed the need for economy, since a single experimental setup would probably exhaust the entire project appropriation. See ibid., p. 58.

[16] Artificial transmission lines normally consist of a large number of discrete inductors and capacitors connected in an iterative circuit. The advantage of this transmission structure, in addition to its economy, is that it requires a comparatively small amount of laboratory space. However, considerable care is required both in the manner of construction and in the test procedure employed if the results are to be correlated with the performance of an actual line being simulated.

[17] Although he was familiar with the advocacy of high line self-induction by Heaviside and Stone, Campbell stated that he was not at the time aware that the use of discrete inductors to achieve the increase had already been discussed by Heaviside. See *RGAC*, pp. 12–13.

sent to the Boston laboratory.[18] After Campbell had worked out the design specifications for the loading coils to be used in the proposed tests, an order for 400 of them was placed with a local manufacturing company.[19]

While waiting for delivery of the loading coils, Campbell continued his theoretical work. He succeeded in deriving a general formula for loaded lines which became known as "Campbell's equation." His equation may be written[20]

$$\cosh \gamma'd = \cosh \gamma d + \frac{z}{2z'} \sinh \gamma d, \tag{1}$$

where γ' is the propagation constant of a loaded line, γ is the propagation constant of the same line without loading, d is the interval between coils, z is the loading coil impedance, and z_0 is the characteristic impedance of the line without loading. This equation was immediately used by Campbell and his assistants to make numerical calculations over a range of assumed loading conditions on standard telephone cables. The results were presented in the form of detailed graphs which clearly showed the optimum coil values and spacing. These graphs were included in a report prepared by Campbell dated July 18, 1899, and entitled "Statement of Progress on Cable Problem."[21] Another report, entitled "On the Introduction of Inductance on Telephone Lines: Description of Loaded Lines for Examination into Patent Claims," was submitted by Campbell the following day.[22] However, Hayes decided that it would be best to wait until

[18] This was called the Pittsburgh cable because it had been used in transmission tests in Pittsburgh. It consisted of standard 19-gauge cable pairs. Connection of the pairs in series provided a circuit with a total length of 35 miles. See *RGAC*, pp. 38–40.

[19] Ibid., p. 42.

[20] For a copy of four pages of Campbell's original notes containing the derivation of his equation, see *RGAC*, pp. 313–16.

[21] Ibid., pp. 57–58, 319–20. In addition to being the first graphs to show the effect of various degrees of loading on attenuation, these were probably the first clearly to depict the cutoff phenomenon which is characteristic of electrical filter networks as well as periodically loaded transmission lines. The loaded lines were found to exhibit a sharp rise in attenuation as the transmitted frequency approached the characteristic cutoff frequency determined by the amount of loading added. Campbell's analysis showed that this critical condition was reached at a frequency of 11,000 cycles per second for the loading used on the Pittsburgh cable. This was reduced to approximately 2,000 cycles per second on some later circuits which were more heavily loaded.

[22] *RGAC*, pp. 58–59.

the tests had been made on the Pittsburgh cable before contacting the company's patent attorney.[23]

The first successful tests ever made on a coil-loaded telephone cable were carried out by Campbell and his assistant, Edwin H. Colpitts, on September 6, 1899.[24] These tests were made using loading coils on the Pittsburgh cable in the Boston laboratory. Campbell and Colpitts were able to carry on a conversation over a 46-mile circuit with a transmission quality equivalent to that of a 23-mile circuit without loading. The tests were repeated the following day for the benefit of Hayes and others in the laboratory. Campbell's theoretical calculations were completely confirmed by these experiments, which demonstrated for the first time that the use of loading coils on standard telephone circuits could achieve an important improvement in distance for a given transmission quality or in quality at a given distance.[25]

Hayes was now convinced that Campbell had made a potentially valuable discovery. He wrote to the company president, John E. Hudson, to inform him of Campbell's results and to request that the use of loading coils be investigated for patentability by the company's patent attorney, William W. Swan.[26] Shortly afterward Campbell was assigned to help Swan in the preparation of a patent application. However, there was a long delay before the application was finally submitted in March 1900. When questioned about the reason for the delay, Campbell later explained it had been necessary for him to spend a considerable amount of time in attempting to explain the loading graphs in his July report to Swan. He stated that they had also undertaken an extensive survey of the relevant literature, including American and foreign patents. Finally, Swan had insisted that the patent specification not contain mathematical formulas but that it be written "in language which would be at once understood by telephone engineers." The subsequent history of the patent application indicates that both the delay in applying and the total deletion of graphs and equations from the application were prejudicial to Campbell's attempts

---

[23] This delay, in conjunction with the long delay in filing after the Pittsburgh cable tests, made his attempt to establish priority over Pupin's work far more difficult, since the first applicant in a disputed case enjoyed certain procedural advantages.

[24] Colpitts had been hired as Campbell's assistant in 1899. He had completed three years of graduate study in physics and mathematics at Harvard and held a master's degree. See *RGAC*, pp. 171–72.

[25] Ibid., p. 50.

[26] For a copy of the letter from Hayes to Hudson dated September 11, 1899, see *RGAC*, p. 253.

to establish the priority of his work. The technical details contained in his reports prepared in July 1899 were of fundamental importance for the successful introduction of loading.[27]

After Campbell's impressive demonstration of loading on the Pittsburgh cable, Hayes and his engineering staff prepared to exploit the discovery even before the patent application was filed. One member of the staff, Chester H. Arnold, was assigned to carry out a study of the potential advantage of loading a circuit from an economic standpoint.[28] Arnold's calculations indicated that the probable saving in construction costs of new installations proposed in New York and New Jersey alone would amount to approximately $700,000 if loading were used.[29] The first loaded circuits open to public use were two circuits connecting Jamaica Plains and West Newton, suburbs of Boston. These circuits were loaded under Campbell's direction and were first opened to commercial traffic on May 18, 1900.[30] Only six days before, Campbell received from the patent examiner a notification of a rival applicant for a loading coil patent. The rival applicant was Michael I. Pupin.[31]

* * *

Pupin had filed an application for a patent on loading in December 1899. Nearly a quarter of a century later, he wrote an account of his discovery and its significance in his autobiography. It is somewhat surprising that this version has been so widely accepted by historians,

[27] See *RGAC*, pp. 71–72. It is also possible that this was an example of "incomplete disclosure" wherein essential information was deliberately withheld. Whether this was a factor in Swan's insistence on deleting mathematics and graphs has not been determined. For a discussion of the practice of incomplete disclosure, see Floyd L. Vaughan, *The United States Patent System* (Norman, Okla., 1956), p. 218.

[28] Arnold had been with the Bell company since 1893. He had been a classmate of Campbell at Harvard and held a master's degree. See *RGAC*, p. 213.

[29] For a copy of Arnold's report dated January 2, 1900, see ibid., pp. 223–24.

[30] Ibid., p. 66.

[31] Ibid., pp. 268–69. Pupin had arrived in America in 1874 as a sixteen-year-old immigrant from Serbia. After working for several years as a farm laborer and in a cracker factory, he had enrolled at Columbia University, where he received a bachelor's degree in 1883. During the following year, Pupin studied mathematics at Cambridge, England, under John E. Routh. He was then awarded the Tyndall fellowship, which enabled him to study at the University of Berlin under Helmholtz and G. R. Kirchhoff. Pupin was awarded a doctorate at Berlin in 1889 and then accepted a teaching position in the newly created department of electrical engineering at Columbia. See *The Dictionary of American Biography* (New York, 1944), 21:611–15.

since personal memoirs written long after the events described have so often proved to be unreliable as historical sources. In this instance, a comparison of the version in Pupin's book with available contemporary sources reveals a number of distortions and omissions in Pupin's account, which contributed to the growth of a persistent myth depicting Pupin as one of the heroic electrical inventors. Pupin's contributions to the electrical engineering profession as an educator and popularizer of the use of mathematical analysis of alternating-current phenomena were not unimportant. However, the value of these contributions has been largely overlooked because of the notoriety surrounding his electrical "inventions."

In his book, Pupin wrote that the idea of the loading coil invention had first occurred to him while climbing a mountain in Switzerland during the summer of 1894. He stated that he had been preparing lectures on the theory of sound, to be given at Columbia the following fall, and that the idea of loading had been stimulated by his reading of the problem of a vibrating string in Lord Rayleigh's treatise, *The Theory of Sound.*[32] According to Pupin, he had conducted some experi-

[32] Michael I. Pupin, *From Immigrant to Inventor* (New York, 1960), pp. 330–31. It is interesting that Pupin was issued two patents in May 1894 in which he proposed to extend greatly the possible range of transmission over telephone circuits by adding capacitors in series with the circuit conductors at periodic intervals. An abstract of his patents was published in *Electrical World*, along with an editorial which suggested that the proposed method might greatly reduce the cost of long-distance telephone circuits. The editor also pointed out that the invention might extend the possible distance of telephony, which seemed to have reached its practical limit in the recently completed circuit from New York to Chicago. The patents included the claim that the proposed method might even permit transatlantic telephony. See *Electrical World* 23 (1894):666, 678–80.

The publication of the claims in Pupin's patents stimulated a critical editorial in the British journal, *Electrical Review*. This editorial stated that the method proposed by Pupin was not new and was based on fallacious reasoning. It predicted that a test of the invention on an artificial cable would be sufficient to demonstrate that it would not achieve the effect claimed by Pupin (see a digest of this editorial in ibid., p. 873). The critical comments of the British editor in turn stimulated an editorial in *Electrical World* entitled "Criticism vs. Abuse." The editor argued that Pupin had at least given a very logical discussion in support of his proposal and that, even if his reasoning were fallacious, his claims should not be dismissed so arbitrarily without some definite evidence to refute his suggested method (see ibid., p. 824). This brought another response from the editor of the *Electrical Review*, who pointed out that experiments which had previously been carried out by engineers of the British Post Office had shown that the invention proposed by Pupin had an effect exactly opposite to that which he predicted (see *Electrical World* 24 [1894]:97). Pupin made no mention of these events in his book, which seems somewhat surprising because his suggested capacitor loading arrangement was ostensibly intended to achieve the same improvement in transmission as coil loading. In any event, nothing more was heard of capacitive loading after 1894.

mental tests related to loading shortly after returning to his laboratory at Columbia but had dropped the project in order to experiment with X-rays soon after hearing reports of Roentgen's discovery. Pupin wrote that he had suffered a nervous breakdown in the spring of 1896 following the death of his wife but had been able to resume work on the loading coil later the same year. His account gave no details of this work other than a statement that he had used apparatus which had enabled him to conduct experiments without using actual telephone circuits. Pupin also referred to a paper which he had read at a meeting of the American Institute of Electrical Engineers early in 1899, a paper which had been devoted to the theory of his apparatus but which had contained no explicit mention of loading. Pupin continued that his friend Cary T. Hutchinson, a consulting engineer, had approached him in October 1899 concerning the technical feasibility of designing an underground cable from New York to Boston for an independent telephone company. In his autobiography, Pupin wrote that he had expressed a willingness to design such a circuit, "guaranteeing to employ a wire not bigger than in ordinary cables capable of operating satisfactorily over a distance of only twenty miles." It was at that time, Pupin stated, that he had decided to apply for a patent on loading.[33]

Pupin's account contained no explicit mention of Campbell, but he alluded to the priority conflict:

> There were others besides Hutchinson, who had recognized that in my communication to the American Institute of Electrical Engineers there was hidden an invention for which telephone engineers had been eagerly waiting ever since the birth of the telephone art. This created an interference in the Patent Office which annoyed me, but not nearly so much as I had been annoyed there before. About a year from the date of my application for patent, the American Telephone and Telegraph Company acquired my American patent rights, treating me most generously. It gave me what I asked.[34]

[33] Pupin, pp. 332–37.

[34] Ibid., pp. 337–38. The earlier interference which Pupin mentions in this passage was a three-way contest between Pupin, John S. Stone of the American Bell Telephone Company, and two French applicants, Hutin and Leblanc. This dispute had involved a proposed method of transmitting several messages simultaneously over a single circuit with separation to be achieved at the receiving end by means of resonant circuits. Although hearings in this interference began in 1896, the final decision, which awarded the priority to Pupin, was not reached until 1902. Somewhat ironically, the method proposed proved to be impractical until after the development of electronic amplifiers and the discovery of a new type of wave filter by G. A. Campbell. The first successful system, which became known as multiplex or carrier telephony, was not placed in operation until 1918. See Julian Blanchard, "A Pioneering

Pupin went on to discuss the great impact which his invention had had on the public, the scientific community, and the telephone industry. He gave some credit to the Bell system for having exploited the loading coil successfully, but he added another somewhat condescending reference to the Boston laboratory:

> Twenty-five years ago the American Telephone and Telegraph Company had a small laboratory in Boston, where it did all its scientific research and development. But, presently, faultfinders like my invention moved into the peaceful and drowsy precincts of that tiny laboratory, and stirred up the engineers and the board of directors. I am very happy whenever I think that, possibly, my inventions have contributed some to this healthful stirring up. What was the result? To-day the American Telephone and Telegraph Company and the affiliated Western Electric Company employ about three thousand persons at an expenditure of some nine million dollars annually in their research and development work.[35]

In addition to an absence of experimental and theoretical details of his work on loading, Pupin's version of the invention and its development gives a distorted estimate of the actual impact of his personal contribution and omits many significant historical details brought out in the interference proceeding. It is therefore necessary to consider other evidence in order to achieve a more objective view regarding Pupin's role in the introduction of loading in telephony.

*   *   *

There were important qualitative differences in both the experimental and theoretical contributions of Campbell and Pupin insofar as their influence on the eventual commercial success of loaded telephone circuits was concerned. For example, Pupin did no experimental work on actual telephone circuits comparable with the dramatic and convincing Pittsburgh cable tests performed by Campbell. It is clear that Pupin and some of his students at Columbia conducted intermittent experiments using artificial transmission lines during the years 1895–1900. However, it seems quite doubtful whether these experiments were recognized as being relevant to the loading problem before late in 1899.[36] The paper

---

Attempt at Multiplex Telephony," *Proceedings of the IEEE* 51 (1963):1706–9; and E. H. Colpitts and O. B. Blackwell, "Carrier Current Telephony and Telegraphy," *Trans. AIEE* 40 (1921):205–300.

[35] Pupin, p. 340.

[36] The problem of evaluating the work before 1899 is complicated by the fact that an artificial line designed to simulate an ordinary telephone circuit or to simu-

which Pupin read early in 1899, in which the loading invention was supposedly "hidden," contained a mathematical analysis of an artificial line having *n* identical sections. In this paper, he sought to determine the conditions under which such an arrangement would "become equivalent with sufficient approximation to a uniform transmission line when n is not infinitely large."[37] The experimental results reported in the paper included only wave length and propagation velocity measurements, not attenuation. Pupin admitted that the leakage factor of the apparatus was much greater than that of a normal circuit. Since this would have made the attenuation much greater as well, it is probable that loading experiments would have been inconclusive even if attempted with this apparatus.[38]

Charles S. Bradley, who was not mentioned in the account of the invention in Pupin's book, was one of Pupin's more important witnesses in the interference proceeding.[39] Pupin testified that he had discussed the loading invention with Bradley shortly after the presentation of the March 1899 paper.[40] In addition to supporting Pupin's testimony on this point, Bradley stated that he had then attempted to interest the General Electric Company in the loading coil but without success. None of the General Electric officials mentioned were questioned during the proceeding.[41] And Pupin's behavior during the months following his con-

---

late a loaded telephone circuit would appear substantially the same insofar as its constructional features are concerned. Pupin was unable to establish that any of his theoretical or experimental work prior to 1899 had been on the problem of coil loading and had not merely been intended to determine the conditions under which normal telephone circuits could be adequately simulated by artificial lines.

[37] M. I. Pupin, "Propagation of Long Electrical Waves," *Trans. AIEE* 16 (1899): 120.

[38] Pupin attempted to demonstrate the effect of coil loading using one of his artificial lines in May 1901 during the interference hearings. The results were apparently erratic and marginal. See *RGAC*, p. 49; see also the patent examiner's decision of December 9, 1903 in the U.S. Patent Office file for Interference no. 20,699, pp. 70–72.

[39] In 1899 Bradley was president of the Ampere-Electric Chemical Company, with Pupin serving as one of the company directors. See the *Record for M. I. Pupin*, 2 vols., Interference no. 20,699 (Washington, D.C., 1902), 1:288 (hereafter cited as *RMIP*). Bradley had also founded the Bradley Electric Power Company to exploit his rotary converter but had sold out to General Electric in 1893. See W. Paul Strassmann, *Risk and Technological Innovation: American Manufacturing Methods during the Nineteenth Century* (Ithaca, N.Y., 1959), p. 167.

[40] *RMIP*, 1:76.

[41] Ibid., p. 291. Why the General Electric Company rather than a telephone company should have been approached was never made clear. The fact that Bradley had been concerned primarily with the technology of electrical power rather than

versation with Bradley does not indicate that he felt the sense of urgency which might be expected of a man in possession of a potentially valuable technological innovation. The available evidence seems to support the view that Pupin did not undertake to work out the necessary theoretical conditions for loading or to obtain experimental evidence that loading would actually improve transmission quality even on artificial lines until after he was approached by Cary T. Hutchinson late in 1899.

Hutchinson, retained as a consultant by the Telegraph, Telephone, and Cable Company of America to provide technical advice on a proposed underground cable between New York and Boston, requested an interview with Pupin to discuss the problem.[42] Pupin later testified that he had offered to design a loaded cable circuit and also to demonstrate its feasibility using an artificial line if Hutchinson's clients were willing to pay for its construction. The company had agreed to pay up to $5,000 to support Pupin's proposed experimental demonstration. Pupin stated that he had begun preparations for the demonstration but had subsequently broken off negotiations and had decided to continue preparations for the experiment at his own expense. He gave as his reason that he had decided that Hutchinson's clients were attempting to gain control of the loading invention for an inadequate sum.[43] The aborted negotiations and the concurrent calculations which Pupin carried out seem to have been decisive stimuli in causing him to realize the potential value of loading in telephony and to undertake to obtain a patent on it.[44] Pupin stated that he had contacted his attorney on the

---

with electrical communications suggests that Pupin and Bradley might have been thinking in terms of a method for improving the efficiency of power transmission. If so, this would in itself be evidence that Pupin had not yet gone beyond the qualitative stage, since the later theoretical work of both Pupin and Campbell clearly demonstrated that loading was of no practical value in power transmission.

[42] The Cable Company was one of a group organized in 1898–99 for the purpose of forming a national system of circuits to compete with the Bell system. The attempt soon collapsed despite substantial capitalization. See J. W. Stehman, *The Financial History of the American Telephone and Telegraph Company* (New York, 1967), pp. 57, 81. Hutchinson was acquainted with Pupin through their activities in the American Institute of Electrical Engineers. Hutchinson was aware of Pupin's interest in telephony and could, of course, expect no assistance from Bell system engineers.

[43] *RMIP*, 1:85.

[44] Although it is possible that Pupin may have learned of Campbell's loading tests on the Pittsburgh cable a month before, there seems to be no convincing evidence of stimulus diffusion in this case.

matter in November and had spent the remaining time before the filing date in December in explaining the invention to the attorney and in helping to write the specifications.[45]

Pupin's patent application contained his so-called sine rule, which Pupin called "the foundation upon which the invention described in this application rests."[46] The rule stated that the degree of equivalence between a coil-loaded line and a uniform line having the same total self-induction and other parameters could be determined by comparing the quantities $\sin \phi/2$ and $\phi/2$ of the loaded circuit. The angle ϕ was defined as the angular or radian interval between coils expressed as a fractional part of a wavelength. For cases which the rule indicated to be sufficiently equivalent, one could employ Heaviside's equations for a uniform line to compute the attenuation and other properties of the loaded circuit. For example, if a line were loaded with ten coils per wavelength at a given frequency, one half of the radian interval would be 0.314 or 18 degrees. Since the sine of 18 degrees is 0.309, the rule would predict that the error involved in using the Heaviside uniform line equations to calculate the performance of the loaded circuit would not exceed 1.6 percent. The patent application did not include a derivation of the sine rule.

Two weeks after filing the patient application, Pupin presented a paper on the mathematical theory of transmission circuits at a meeting of the American Mathematical Society.[47] The paper included an analysis of three different transmission structures. These were a uniform transmission line, a "non-uniform conductor of the first type," and a "non-uniform conductor of the second type." Pupin pointed out that he had previously published an analysis of the first two structures in his March 1899 paper in the *Transactions of the American Institute of Electrical Engineers* but that his analysis of the third structure was new.[48] A nonuniform conductor of the first type was simply an artificial transmission line, but a nonuniform conductor of the second type was an actual telephone circuit with identical loading coils inserted at equal intervals along the circuit. Pupin stated that the principal aim of the paper was to determine the "conditions under which a non-uniform conductor of the second type is approximately equivalent to its corresponding

---

[45] *RMIP*, 1:86. Pupin's attorney, Thomas Ewing, later became the commissioner of patents. See Vaughan, p. 292.

[46] See a file wrapper included in *RMIP*, 1:2.

[47] M. I. Pupin, "Wave Propagation over Non-Uniform Electrical Conductors," *Transactions of the American Mathematical Society* 1 (1900):259–86.

[48] Ibid., p. 261.

uniform conductor."[49] He then proceeded to derive the sine rule which gave the degree of approximate equivalence.[50]

Pupin's loading-coil analysis was first brought to the attention of the electrical engineering profession when he read a paper at a meeting of the American Institute of Electrical Engineers in May 1900. The paper included the theoretical material from his earlier papers before the American Mathematical Society and, in addition, some experimental results obtained on a recently constructed artificial line.[51] He also reported that he had used telephone apparatus provided for him by the New York Telephone Company, a Bell system operating company.[52] By way of contrast, Campbell had carried out similar tests approximately a year earlier and had loaded an actual telephone cable eight months earlier. Campbell had also designed the loaded circuits which were opened for public use the day before Pupin's paper was read on May 19.[53]

Pupin's claims in the patent application were initially rejected by the patent examiner, who cited the earlier patents of Silvanus P. Thompson and John S. Stone.[54] Pupin then filed an amendment and

[49] Ibid., p. 275.

[50] This paper does not seem to have stimulated any comment in engineering journals, so the effective disclosure of his loading theory to the electrical engineering profession was delayed until his paper was read before the American Institute of Electrical Engineers in May 1900.

[51] M. I. Pupin, "Wave Propagation over Non-Uniform Cables and Long Distance Air-Lines," *Trans. AIEE* 17 (1900):445–507. See also the discussion of his paper in ibid., pp. 508–13, and an editorial about it in *Electrical World and Engineer* 35 (1900):773–74.

[52] Pupin, "Wave Propagation over Non-Uniform Cables and Long Distance Air-Lines," p. 472. There was apparently little coordination between work in New York and that carried out by the staff of the Boston laboratory. It is doubtful whether the engineers of the Bell affiliates in New York were yet aware of Campbell's success in loading the Pittsburgh cable or the experiments which were in progress on circuits in the Boston area. If they were aware of the success achieved by the Boston engineers, it is difficult to understand their interest in supporting Pupin except perhaps in the context of the existence of a degree of rivalry between the Bell engineers in New York and Boston.

[53] *RGAC*, pp. 58–59.

[54] See n. 11 above. Thompson had read a paper at the Chicago meeting of the International Electrical Congress in 1893 suggesting that the use of periodic coils in shunt with underwater cables might compensate for cable capacitance and make transatlantic telephony possible. Thompson was granted four patents on "compensated cables" between 1891 and 1894. The American rights to his patents were purchased by the American Bell Telephone Company, but the use of shunt coils for compensation never became a practical success. See Silvanus P. Thompson, "Ocean Telephony," *Electrician* 31 (1893):474. Also, see J. A. Fleming, pp. 133–41.

a written argument in support of patentability. The examiner again rejected his claims and added a reference to Heaviside's papers. Pupin replied with an argument which emphasized the importance of relating the coil spacing to the wavelength of the transmitted signals. In a letter dated May 12, 1900, the examiner, in effect, accepted Pupin's argument but also informed him of a possible interference with the claims of another applicant. In a later letter, the examiner requested that Pupin divide his claims into two groups, since all of them could not be allowed in a single application. Pupin complied with this request and was then notified that the applications were being sent to issue. Pupin was issued two loading patents on June 19, 1900. An announcement of an interference with the still pending applications of Campbell followed in August.[55]

Meanwhile, Campbell, whose application had been filed in March, received notification that his claims were being rejected, based on a reference in Heaviside's papers. He made no immediate reply and received another letter from the examiner on May 12 which informed him of a possible interference if his claims were found patentable. Campbell was given ten days to respond, with the warning that if he failed to do so, the application of the rival claimant would be sent to issue. The deadline was allowed to pass without response.[56] An amendment and an argument in support of patentability were finally filed in Campbell's behalf late in June. The examiner then notified him that loading patents had already been issued to Pupin. In August, Campbell filed another amendment and three affidavits in support of his application. He was thereupon notified that the priority of invention would be determined by an interference proceeding.[57]

* * *

There were a number of interesting developments long before a final decision was reached in the interference proceeding early in 1904. W. W. Swan, the Bell patent attorney, had written to the company president, John E. Hudson, in the summer of 1900 to express

[55] See a file wrapper included as an exhibit in *RMIP*, 1:32–53. The patents issued to Pupin were nos. 652,230 and 652,231.

[56] The responsibility for handling the case was in the hands of the Bell patent attorney, with Campbell acting primarily as a technical consultant. The failure to respond to the examiner before the deadline was never explained. This placed Campbell at a greater disadvantage than if both applications had still been pending. Since Campbell became the junior contestant, Pupin and his attorney were able to read Campbell's deposition before preparing Pupin's brief.

[57] *RGAC*, pp. 273–93.

the belief that patents might be issued to both Pupin and Campbell without an interference being declared. Swan suggested that, in that event, it would be in the company's best interest to control both patents. However, since it was uncertain whether either man's patents could be successfully defended if contested, because of previous patents and publications on loading, Swan recommended that provision be made to give yearly payments to Pupin only so long as his patents should remain undefeated. Hudson thereupon decided to obtain an option on the patents, and Pupin was given an initial sum of $15,000 in October 1900.[58] The option was exercised the following January with Pupin receiving an additional payment of $185,000 despite the fact that the first testimony had not yet been taken in the interference proceeding.[59]

Following extensive testimony on behalf of each claimant, the examiner's decision favoring Pupin was handed down in December 1903.[60] The Bell company president, Frederick P. Fish, then contacted Campbell to request that he read over the examiner's decision and give advice on what further action the company should take. Fish

---

[58] A copy of the letter from Swan to Hudson and a number of other letters and documents related to the activities of Hammond V. Hayes and other members of the staff of the Boston laboratory are contained in the files of Professor Edward L. Bowles of Wellesley Hills, Mass. (hereafter cited as Bowles file). I am indebted to Professor Bowles for permitting me to examine this material. John Hudson died suddenly on October 1, 1900. Alexander Cochrane served as president pro tem until Frederick P. Fish, an eminent patent lawyer, was selected as the new president on July 1, 1901.

[59] According to extracts from Pupin's contract in the Bowles file, Pupin continued to receive annual payments until 1917 for a total of $455,000. Pupin also agreed to assign to the Bell system any other inventions which he might make related to loaded circuits. Reports in the technical press at the time varied widely on the amount received by Pupin for the rights to his patents. See, for example, *Electrician* 46 (1900):454, 532. The editor pointed out that Pupin was only following the suggestions made by Heaviside and expressed surprise that it had taken so long to submit them to a practical test. The editor wrote that the great financial reward received by Pupin might finally lead engineers to pay more attention to the results of mathematical analysis even though the details of Heaviside's theoretical work had been beyond the comprehension of most of them. Pupin was granted a loading patent in Germany in 1904 after extended tests and hearings. He sold the right to this patent to the Siemens and Halske Company. See *Electrical World and Engineer* 43 (1904):324. For an account of Pupin's negotiations with Siemens and Halske, see George Siemens, *History of the House of Siemens,* trans. A. F. Rodger, 2 vols. (Freiburg, 1957), 1:178–81.

[60] The decision gave Pupin a disclosure date of March 22, 1899 and Campbell a disclosure date of March 24, 1899. See the examiner's decision of December 9, 1903, pp. 11, 21.

stated that it was his wish that "there should be no cloud upon the title of the successful contestant."[61] Apparently, Campbell recommended that the matter be pursued, since the examiner's decision was appealed to the board of examiners-in-chief in January. The board's decision was handed down in April and also awarded priority to Pupin. The board decision might then have been appealed to the patent commissioner and, if necessary, to a federal appeals court. However, Fish decided against further action in the case, leaving Pupin the legal winner.[62]

In examining the basis for the decision favoring Pupin's priority, it appears that the strategy of Pupin and his attorney was to concentrate attention on Pupin's early work on "non-uniform conductors of the first kind," or artificial lines. They were able to take full advantage of the superficial similarity of this structure to the loaded circuits which Pupin astutely called "non-uniform conductors of the second kind." There is no convincing evidence that Pupin began either theoretical or experimental work related to the latter structure until after his interview with Hutchinson in October 1899. It is perhaps not surprising that the patent examiner failed to grasp the crucial difference between the two structures in view of the conflicting testimony of the expert witnesses in the case. The notoriety achieved by Pupin following the sale of his patent rights for such an impressive sum may also have helped to cloud the issue and increased the tendency to give Pupin the benefit of any remaining doubts.[63]

61 Fish to Campbell, December 18, 1903, Bowles file.

62 Campbell wrote a memorandum on the matter in 1911 in which he stated that Fish had informed him that the decision not to appeal to the commissioner was based on Fish's belief that the commissioner was not only likely to render a decision against Campbell but might even declare the invention to be unpatentable. The reason given for the commissioner's probable prejudice was that Fish had previously declined to recommend the commissioner for his position. A copy of Campbell's memorandum is in the Bowles file.

63 Pupin was even given public credit for the invention by John J. Carty, a leading Bell engineer, at a meeting of the American Institute of Electrical Engineers in 1902 several months before the interference was settled. See *Trans. AIEE* 19 (1902): 411–12. Carty had also supported Pupin in 1900 in an abortive attempt by a dissident group in the American Institute of Electrical Engineers to elect Pupin as president of the institute instead of the official nominee, Carl Hering. See a letter from Carty in *Electrical World and Engineer* 35 (1900):709. The campaign to elect Pupin took the form of a letter, circulated among the members, which repudiated the official slate chosen by the institute council and solicited votes for Pupin. See ibid., pp. 591–92. The dramatic coincidence which might have occurred if Pupin had been elected as the new president of the institute at the annual meeting in May, where he also read his paper disclosing his work on loading, was averted when he was de-

The favorable decision in the interference proceeding further enhanced Pupin's image as a great inventor and exemplar of American opportunity for immigrants. Despite this, the evidence strongly suggests that Campbell's contributions were actually much more influential insofar as their direct impact on communications technology is concerned. Campbell's theoretical analysis of loading was more exact than Pupin's and also proved more applicable in practice. The loading graphs which Campbell prepared for his July 1899 report were in the form needed by engineers in making practical decisions on loading, while Pupin's work produced nothing comparable. Pupin's famous "sine rule" was an approximation giving results which were least accurate for the maximum allowable coil interval, the case where maximum accuracy was needed.[64] Because of his great concern with the problem of establishing the degree of equivalence between loaded circuits and a fictitious uniform line having the same electrical parameters, Pupin also overlooked the important discovery made by Campbell of the cutoff phenomenon.[65] This became the basis of Campbell's later work on the electric wave filter, a key innovation in 20th-century electrical technology.[66] Finally, there seems no question that Campbell was the first to demonstrate loading on an actual telephone circuit and that he and his assistants were already in the process of exploiting the discovery commercially before Pupin's work was disclosed.

It does seem probable that loading could have been introduced on the basis of Pupin's theory if Campbell's work had not already been available. However, in that case the introduction of loading would have been considerably delayed and would have led to somewhat different procedures than were actually used. In particular, more coils per wavelength would have been indicated, with a considerable increase in loading costs, if Pupin's sine rule rather than Campbell's

---

feated by a vote of 381 to 179. See ibid., p. 791. Apparently, the faction supporting Pupin's candidacy was motivated at least in part by opposition to proposed changes in the institute constitution, changes which would, among other things, limit the rights of associate members to hold office. See ibid., p. 709. Presumably as a consequence of the "independent" campaign of 1900, the "gentleman's agreement" that Pupin would be the official candidate the following year was not carried out, although he was finally elected president of the institute in 1925. See ibid., p. 592.

[64] For a comparison of results predicted by Campbell's equation and by Pupin's sine rule, see Fleming, pp. 130–31.

[65] See n. 21 above.

[66] See J. D. Tebo, "George A. Campbell, 1870–1954," *Bell System Technical Journal* 34 (1955):1–4. Also, see Lloyd Espenschied, "The Electric Wave Filter," *IEEE Spectrum* (August 1966), p. 162.

equation had been used as a guideline for engineers. Although it does not appear that Pupin's theoretical work had any significant influence on loading as practiced by the Bell engineers, the possession of the rights to his patents was of value in that it enabled the company to maintain an effective monopoly on loading until 1917. During their lifetime, the Pupin loading patents were probably the most important which the company controlled.

* * *

The analysis of the events in the early history of loading in telephony suggests some more general conclusions of significance in the history of technology. This technical advance was symbolic in marking the appearance of a new type of "invention" and a new type of "inventor." Both contestants for legal priority were men with graduate-level training in mathematical physics. The theoretical aspects of loading were such that the necessary details for its successful introduction could not have been worked out by anyone not having a similar degree of analytical sophistication.

The rather procrustean nature of the patent system, which required that a single inventor be determined without the exercise of a value judgment on either the comparative quality or the influence of each man's work, has sometimes been, as in this instance, more conducive to legends than to accurate technological history. Furthermore, the general policy of secrecy which seems to have been followed in the Bell system at that time also tended to exercise a distorting effect on the professional reputation of its own research engineers. The apparent decision of the company management to give Pupin all of the public credit for the invention of the loading coil, while failing to publicize the many important contributions of the staff of the Boston research laboratory, may have seemed quite rational in the competitive situation in which the company then found itself. However, this policy tended to prolong the "heroic age of invention" and to postpone public realization of the rapidly growing importance of industrial research laboratories, staffed by graduate scientists and engineers. The introduction of loading might have become an early symbol of achievement of organized industrial research and development, but instead it became the basis of a persistent myth associated with the independent "immigrant inventor."

CHARLES F. KETTERING AND THE COPPER-COOLED ENGINE

STUART W. LESLIE

Visitors to the 1923 New York Auto Show were greeted by the following statement in the Chevrolet exhibit area: "Chevrolet Motor Company announces an important development in economical transportation, consisting of a motor embodying new application of established principles governing the efficient control of motor temperatures under all weather conditions."[1] This advertisement announced General Motors' latest strategic innovation in the fight against Ford: the copper-cooled engine. When development began on an air-cooled engine for Chevrolet in 1919, General Motors sought a light-weight, inexpensive alternative to the Model T. The new air-cooled engine promised numerous advantages over its water-cooled competitors: It weighed less per horsepower, neither froze in winter nor overheated in the summer, dispensed with the radiator and other cooling accessories, and achieved better fuel economy. Prospective owners no longer needed to garage their vehicles to insure easy cold-weather starts. All in all, the new automobile seemed the ideal antidote for GM's slumping sales position.[2] But Chevrolet's secret weapon was a failure. Less than six months later, the copper-cooled car was abandoned. A dejected Charles Kettering, the man who had directed the project as head of the General Motors Research Corporation, offered his resignation. Air-cooling of General Motors automobiles was cast aside until the appearance of the Corvair in the 1960s.

After four costly years of development toward a promising goal, the

---

STUART LESLIE teaches the history of science and technology at Johns Hopkins University. Among his publications are *Boss Kettering: Wizard of General Motors* and *The Cold War and American Science.*

[1] Chevrolet Division Publicity Release, January 1923, Kettering Archives, Warren, Mich., p. 1.

[2] Alfred Chandler and Stephen Salsbury, *Pierre S. Du Pont and the Making of the Modern Corporation* (New York, 1971), p. 513. These authors report that GM sales declined from 20.8 percent of the market in 1919 to 12.8 percent in 1921. This work is also an excellent source on the copper-cooled project as seen from the management perspective.

This essay originally appeared in *Technology and Culture*, vol. 20, no. 4, October 1979.

failure of the project invites our attention and may offer some surprising lessons. Both the maturing of consumer tastes beyond inexpensive and strictly utilitarian automobiles and, more important, the various technical difficulties help account for a lack of success. But the latter reason may be, in fact, only symptomatic of deeper problems relating to the role of, and the tensions between, competing engineering groups within General Motors. During the project, poor coordination and lack of effective dialogue between design and production engineers resulted in crippling disputes and delays. Each group felt that it alone was capable of successfully developing the new engine. Production engineers remained blind to the problems of perfecting a relatively new and radical technology, while design men failed to appreciate the difficulties involved in moving from a carefully tuned prototype to a trouble-free, assembly line automobile ready for mass production. While this internal dissension alone effectively sabotaged the project, the combination of this factor with declining profit possibilities was lethal. For as Alfred Sloan, president of General Motors, once remarked, "the principal object of the corporation was not only to make cars, but to make money."

State of the Art in Engine Cooling

Before proceeding with the story of the copper-cooled engine, it is necessary to briefly outline the general state of the art of engine cooling. A major problem with an automobile's internal combustion engine was that it produced excess heat which needed to be drawn off in some fashion. On small versions the cooling effect of the surrounding air was sufficient. In larger engines the heat flow from cylinder gases into piston heads, valve heads, and cylinder walls was slightly less per unit of exposed area than in small engines, but the inside temperatures of larger engines were higher due to longer heat flow paths to the cooler outside parts of the engine. Larger and more powerful engines, therefore, required a more sophisticated cooling system. Two methods were generally employed on automobiles. The first was relatively simple. Metal fins were cast onto the cylinders. Heat was conducted from the cylinders into the fins, and as the car moved forward air rushed over the fins and cooled the engine. On larger models the surrounding air could not absorb quickly enough all of the heat that was generated, and soon the attendant problems of overheating appeared—"burned" valves, sticking pistons, loss of power. If an engineer wanted to increase the power of an automobile, he had to employ a more efficient cooling system. One answer was to use a cooling medium with a higher specific heat than air and the ability to "wet" the surface for better heat transfer. Many substances might do, but water was the cheapest and most readily available. Water thus conducted the heat away from the cylinders and was then circulated through a radiator where air provided the final cooling. Although water cooling effectively prevented both local and general overheating and allowed the construction of more powerful engines, unfortunately it required a more expensive and complex design.

If some way could have been found to cool an engine by air as efficiently as by water, a major breakthrough in automotive design would have resulted. An air-cooled engine would have eliminated the radiator, hoses, and ducts needed to circulate the water throughout the engine and would have greatly simplified the design. Further, it would not have frozen in cold weather or overheated in hot and would have offered greater power for a given weight.

The potential of air cooling for automobile engines was realized long before 1923. Numerous European makers used this system on their smaller cars. John W. Wilkinson developed the Franklin air-cooled engine in the United States by 1901. While this particular model was not completely successful, subsequent efforts provided a marketable automobile.[3] Aircraft-engine builders during the First World War also had experience with air-cooled designs. Since the power-to-weight ratio of an airplane engine is obviously quite crucial, air cooling offered considerable promise. Several designs were successfully developed in Britain, but the high velocity of an airplane gave it considerable cooling advantages over an automobile and made the problem entirely different.

Charles F. Kettering had also known of air cooling for several years. A successful inventor with National Cash Register, Kettering had founded the Dayton Engineering Laboratories Company (Delco) in 1909 to exploit his invention of a battery ignition system for automobiles. A year later he began work on an electrical self-starter which appeared on a 1912 Cadillac and brought the inventor widespread recognition and a great deal of money. Kettering hired a number of promising young engineers and put them to work on various projects of commercial importance.

In 1919 Kettering gave his research staff a 1915 publication on the use of copper in the air cooling of engines.[4] As copper was thought to conduct heat nine times more readily than iron, the chances for success appeared favorable. It was hoped that copper's superior thermal properties would result in an engine with a higher compression ratio

[3]C. P. Grimes, "Air-cooled Automotive Engines," *SAE Journal of Automotive Engineering* (August 1923), p. 125.

[4]Thomas A. Boyd, interview with Fred W. Davis and R. V. Hutchinson, March 10, 1947, Kettering Archives Orgal History Project, Kettering Archives, p. 363.

than other air-cooled engines and thus greater economy and more power for a given displacement. Kettering's team was also working on another approach to engine efficiency at the time, one involving the search for a chemical additive to prevent knock—a destructive ping which appeared at full engine load, prevented the use of high compression ratios, and had no known cause or cure that the best engineers of the day could ascertain. Unfortunately, the knock project was stalled at the time, and the copper-cooled program looked like the better bet.

Copper is too soft a metal for cylinder construction, so some method was needed to join the iron cylinder to the copper fins. R. V. Hutchinson, in charge of the air-cooled project at Delco, attempted to cast the copper fins directly onto the iron cylinders. Unfortunately, the copper fins were "too greatly embrittled" to be used. Thomas Midgley, another young engineer at Delco (who would later discover tetraethyllead [TEL] and solve the mystery of knock) hit upon the idea of machining the cast-iron cylinder on the outside, then attaching the fins with molten zinc.[5] This effort, too, was unsuccessful. Next came the idea of brazing on the fins with a blowtorch, but the difference in the coefficients of expansion of the copper fins and the iron prevented bonding with heat alone.

The solution to the problem eluded the research team until Charlie Lee designed and built a special electrical furnace to do the brazing in January 1920. They constructed a wooden model of an eccentric finning machine for mass producing the fins, and Midgley believed the team was now "in [a] position to complete [the] motor design very rapidly with [a] high degree of intelligence."[6] The model proved successful, and by the end of February, Lee had drawn up specifications for "a machine to fold copper into the shape of a fin of any height automatically and any number of fins to the set and cut them off. . . ."[7] Lee thought the machine could eventually produce sixty fins a minute but held early runs to only thirty.

Attaching the fins to the cylinder remained a problem. In February Kettering suggested piercing a number of holes in the fins. Wire could then be fed through the holes during brazing.[8] However, developing a furnace to do the actual brazing took the rest of the year. By April Lee had designed a horizontal furnace, but various difficulties involving danger and discomfort to the operator cased the

[5] Ibid., p. 363.

[6] Telegram, Midgley to Kettering, January 23, 1920, Kettering Archives.

[7] Charles Lee, "Automatic Finning Machine," Charles Lee's Laboratory Notebook, February 25, 1920, Kettering Archives.

[8] Ibid., February 16, 1920.

design to be scrapped. Only in January 1921 did Lee succeed in perfecting the device. The operator placed the cylinder to be brazed into the furnace in a vertical position. He then rolled the furnace into a horizontal position, pushing it over the burner at the rear of the machine, where it engaged a rotating and timing device to complete the brazing. The fins were pressed against the thin brass bonding metal during the heating operation to create a union of the metals. A special electrical furnace later replaced the original gas-fired version. Success cost $40–$50 thousand dollars but demonstrated the commercial practicality of the device. The copper-cooled cylinders had been produced for about $2.00 each.[9]

In final form each cylinder resembled a daisy—the center being the cylinder, the outlines of the petals representing the elliptical copper fins with air spaces between them. Kettering decided to test the new cylinders in a Chevrolet chassis, and constructed three four-cylinder engines. To provide additional cooling, fans were positioned on either side of the block. While test reports are no longer available on these cars, they were apparently considered only marginally successful. Work on the project continued, for as yet Kettering's team was unable to advance the existing state of the art.

Development at General Motors

While the Delco staff continued to search for a solution to the cooling problem, a number of organizational realignments had placed the copper-cooled project in a new environment. William Durant, president of General Motors, had purchased the Delco laboratories from Kettering in 1916 for some $9 million in cash and stock.[10] In keeping with his general policy of a loose-reined organization, Durant pretty much left Kettering to his own devices, without regard for other corporate affairs. John Raskob and Pierre S. Du Pont, anxious to obtain the new air-cooled engine, asked John T. Smith, head of General Motors' legal department, to bring Kettering and his various companies into the corporation. Kettering met with the financial committee on August 26, 1919, to explain his work on the air-cooled engine.[11] They suggested he continue work on the project, but made no definite plans for production. Development proceeded, the only major change being a shift to a "square-cylinder" design of 3½ × 3½ inches.

Financial difficulty forced an organizational realignment at General

[9]Kettering to Alfred P. Sloan, July 12, 1923, Kettering Archives.

[10]Thomas A. Boyd, *Professional Amateur: The Biography of Charles Franklin Kettering* (New York, 1957), p. 85.

[11]Alfred P. Sloan, *My Years with General Motors* (New York, 1963), p. 81.

Motors during the next year. Pierre S. Du Pont replaced Durant as president at the end of December 1920, and introduced a new corporate policy the following June. Basically, GM's line of cars would range from the high-volume but low-priced Chevrolet to the low-volume but high-priced Cadillac. Models would not overlap—a car for each purse and taste. But common parts and technology were used by all divisions.[12] Management had new plans for Kettering's laboratory, too. They planned to centralize research at Detroit, with Kettering as head. Unfortunately, Kettering protested. He wanted to remain at Dayton, and his sizable holdings of General Motors' stock made him a force to be reckoned with in management circles. Research remained in Dayton, and Kettering became both a vice-president in charge of research and a member of the board of directors.[13]

Kettering's project seemed ideally suited for General Motors' new marketing strategy. Promising good performance and little maintenance at low price, the copper-cooled engine could provide the basis for the new Chevrolet models. The final go-ahead was given in January 1921, when the executive committee resolved "that it is our intention that the air-cooled engine be developed first for a low-priced car and that it be made in the Chevrolet division."[14] Perhaps it might ultimately prove useful for higher-priced models as well, they guessed. Sloan, Du Pont, and George Haskell, another top executive, agreed on December 24, 1920, to develop a six-cylinder, copper-cooled engine for the Oakland division for production in late summer or early fall of the next year.[15] Fred Warner, general manager for Oakland, was to send Kettering six motors of the new type as soon as they became available. Kettering was to evaluate the engines and discuss any problems which arose with Warner.[16] No one anticipated any difficulties, for, as Sloan observed, "any differences that may exist I feel sure will, of course, be ironed out in conferences."[17]

Since the new Chevrolet was to compete directly with Ford's Model T, low weight and low price were essential. A special committee from GM drew up a list of standards for the new engine. Fred W. Davis, a key man in the application and testing of the copper-cooled engine, recalled the strict guidelines imposed on the project.[18] Size of the engine was limited to 135 lbs. At least 14 h.p. was required at 1,000 rpm. Additionally, the cost could be no higher than $75.00. Since the

[12]Chandler and Salsbury, p. 517.
[13]Boyd, *Professional Amateur*, p. 118.
[14]Sloan, p. 83.
[15]Sloan to Fred Warner, December 27, 1920, Kettering Archives.
[16]Sloan to Kettering, December 27, 1920, Kettering Archives.
[17]Sloan to Warner, December 27, 1920.
[18]Boyd, interview with David and Hutchinson, p. 366.

entire car was to weigh only 1,450 lbs., pounds were pared from the frame, axles, and steering gears as well. A six-cylinder model was prepared for the Oakland division to assist their lagging sales effort.

Development progressed smoothly throughout the first half of 1921. By summer, Kettering and his staff felt sufficiently confident to begin road testing the vehicles. At this time GM lacked a testing facility. New products were either taken to the Indianapolis Speedway for trials or tested on public roads. Kettering preferred the latter method and assembled a fleet of copper-cooled cars, standard Chevrolets, competitors' cars, and parts truck for an assault on the back roads of Ohio and Kentucky (fig. 1). The expedition set off in late July for what the participants would later recall as one of the most grueling tours of their lives. Most of the driving was over bad roads, which "showed up the deficiencies in the industry's cars as well as the copper-cooled."[19] O. T. Kreusser, who went along on various test runs and who later headed GM's testing grounds, declared: "The cars took it a lot better than the drivers, because after you came off one of

FIG. 1.—The testing caravan on the back roads of Ohio in August 1921. (Courtesy of General Motors.)

[19] Interview with O. T. Kreusser, June 3, 1947, Kettering Archives Oral History Project, Kettering Archives, p. 463.

those road test trips to Southern Ohio and Kentucky, it would take at least three baths to get all the dust out of the inside and outside of you. . . ."[20] Axles and driveshafts broke, teeth pulled out of pinion gears, and none of the cars survived this severe test without damage. After several weeks of testing, however, Kettering was at least satisfied with the performance of the new engine.

Design specifications were completed in the fall of 1921, and the copper-cooled engine emerged in its final form. A continuous strip of thin sheet copper surrounded each cylinder, designed so that the bases of each crimp made contact with one another (fig. 2). Thirteen fins per linear inch appeared the most satisfactory arrangement, and

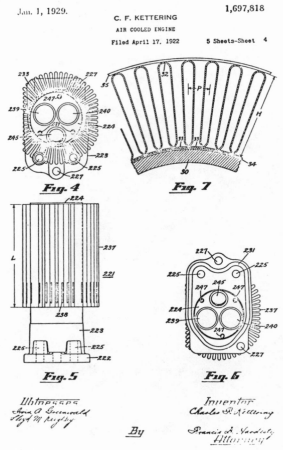

Jan. 1, 1929.   1,697,818

C. F. KETTERING

AIR COOLED ENGINE

Filed April 17. 1922   5 Sheets-Sheet 4

FIG. 2.—The final design of the copper-cooled cylinder from Kettering's patent of 1922. (Courtesy of General Motors.)

[20]Second interview with Kreusser, 1962, p. 10.

the optimum ratio of fin length to the distance between the fins was established at 70:1. With a lesser gap, the air turbulence was suppressed, which reduced velocity at the fin surface, lowered the heat transfer from the fins to the air, and caused the fin temperature to rise.[21] A front-mounted fan, driven by a special V-belt at faster than engine speed, pushed air through the copper fins and out through louvers in the hood to provide more effective cooling (fig. 3).[22] Additional improvements to the car included a tubular frame with one-

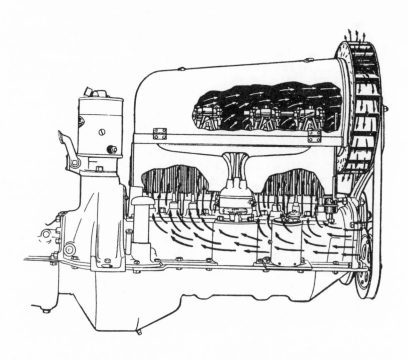

Sectional Perspective View of the
Copper-Cooled Motor

This drawing indicates the circulation of air up through the copper fins, then through the draft tube to the suction fan, which discharges it into the hood, from which it exits through the hood louvres.

Fig. 3.—A sectional perspective of the copper-cooled engine showing the flow of cooling air through the engine. (Courtesy of General Motors.)

[21]U.S. Patent No. 1,697,818, Charles F. Kettering, *Air Cooled Engine*, April 17, 1922, Kettering Archives, p. 6.
[22]Chevrolet Division Publicity Release, p. 3.

quarter elliptical springs, and a new oiling system consisting of a groove in the flywheel which picked up the oil and delivered it through a series of tubes to the rest of the engine.[23] The oiling system was an interesting idea but probably an unnecessary addition to an already overcomplicated engine.

The divisional engineers who were to receive the new engine were in full cooperation with Kettering. George Hannum of Oakland and Karl Zimmerschied of Chevrolet agreed completely with Kettering's plans for a final production schedule.[24] They did not wish to interfere with design. Zimmerschied recognized that "the responsibility for design being Mr. Kettering's, the manufacturing Divisions are in no position to do more than make suggestions on this subject, and in this respect the normal relations are reversed, the line organization becoming an advisory body."[25] But he also argued that "while the advice of practical men may be of negligible value on those questions of design which affect the functioning of a part, it cannot be so considered with relation to factors affecting ease and cost of production."[26] The door was open to future tensions, but for now everyone was on the same side.

Since the Oakland division was in financial trouble, it was to be given the new model before Chevrolet. George Hannum, head of the Oakland division, received the new copper-cooled, six-cylinder auto for final testing before full production in November 1921. Plans were already pending to end the production of water-cooled cars at Oakland within six weeks and replace them with the new design. Assuming that the initial trials were successful, Chevrolet would follow suit and introduce a four-cylinder, copper-cooled version in the early months of 1922. General Motors thus prepared to launch the most ambitious innovation of its short history.

But disaster struck on November 8, 1921. Hannum believed the new Oakland was not yet ready for production. He wrote to P. S. Du Pont and informed him that six months would be needed in order to make the necessary changes. To soften the blow to Kettering, the executive committee wrote him, offering encouragement and a vote of confidence to the discouraged engineer: "In the development and introduction of anything as radically different from standard practice as the air-cooled car is from a regular water-cooled job, it is quite natural that there should be a lot of 'wiseacres' and 'know-it-alls' standing around knocking your development."[27] They expressed

[23] Boyd, interview with Davis and Hutchinson, p. 365.
[24] Zimmerschied to Du Pont, July 11, 1921, Kettering Archives.
[25] Ibid.
[26] Ibid.
[27] Sloan, *My Years with General Motors*, p. 88.

confidence that Kettering would eliminate the problems and bring the engine onto the market by 1923.

The complaints at Oakland were vague—overheating, loss of power, and the like. Controversy filled the office memos. Louis Ruthenberg, an engineer under Kettering who eventually became the liaison between the research laboratory and the operating divisions, sided with Hannum. Ruthenberg believed that the Oakland test only confirmed the troubles already discovered at Dayton. The research engineers, he argued, were fooling themselves. When they read their own reports, they naturally concluded that the troubles could be corrected with slight adjustments. Hannum, on the other hand, merely reported the defects of the engine without comment.[28] Kettering, of course, disagreed completely, but he resolved to examine the engine and correct any difficulties. He feared, however, a certain resistance to his innovation. "If we can ever get the right receptive attitude into the minds of the General Motors people, so that they will really find out what the air-cooled motor means to the future of our industry," he confided to Du Pont, "we will not spend time and money laying out garden variety motors of the ordinary type."[29]

Kettering called a special conference with his top engineers on December 21 to consider any necessary modifications. The conference covered the most minute details, including the engine, drive train, and chassis. Were the oil pan bolts spaced too far apart to prevent leakage? Was the gas tank properly supported? Might redesigning the fan reduce costs? Should alloy bolts be used to hold the flywheel onto the crankshaft?[30] They discussed design changes for two days, then began the slow process of implementing them.

An interesting aspect of the design program was the great attention paid to frame, axle, and transmission details and the very little attention devoted to the engine itself.[31] Perhaps this could be expected, for the Oakland test disclosed as many problems with these details as with engine performance. Chevrolet received the new chassis in March, offered a few minor suggestions, but was generally very pleased.[32] By the middle of the month, the research laboratory turned over detailed prints of the chassis but retained design information of the motor.[33]

---

[28]Louis Ruthenberg, "Ten Great Years with 'Boss' Kettering," *Ward's Auto World* (April–July 1969), p. 50.

[29]Kettering to Du Pont, January 5, 1922, Kettering Archives.

[30]"Minutes of Conference at Mr. Kettering's," December 19, 20, 1921, Kettering Archives.

[31]Report of conference held January 27, 1922, Kettering Archives.

[32]William S. Knudsen to Kettering, April 1, 1922, Kettering Archives.

[33]"Conference on 'C/C 4-cylinder'," April 10, 1922, Kettering Archives.

Everyone seemed pleased with the progress of the project. William S. Knudsen, vice-president in charge of operations at Chevrolet, outlined a tentative production schedule, calling for over 1,000 cars a month by the next January.[34]

The copper-cooled auto seemed ready for another trial. On May 24, 1922, Kettering gave P. S. Du Pont the results of a 20,000 mile test. They were impressive. The valve tappets had been adjusted twice, but no other maintenance had been necessary. Kettering claimed that "the valves and all of the bearings were in perfect condition. There was no carbon deposit in the motor and the oil mileage had been in the neighborhood of a thousand to twelve hundred miles per gallon. In all of our road tests here, we have never seen anything like this; yet I am not surprised, because this simply proves out the fundamental principle."[35]

Under the influence of this optimistic report, Du Pont wrote to his friend Harry McGowan, a young executive with the Nobel Company in England. "All goes well in the copper-cooled car development," he said. "We still expect to start manufacture in September and hope that developments will warrant production of 500 per day in March."[36]

Kettering pressed for immediate production of the new car. Some divisional executives were less sanguine. The major problem, in Kettering's eyes, was the resistance of some corporate executives who were "looking upon this work as being just a slight advance over the present thing instead of being a tremendous asset."[37] Like all research men, Kettering felt his work was not properly appreciated. He was also perceptive enough to realize that Oakland and Chevrolet felt the project had been shoved off on them. Many divisional engineers probably felt Kettering had used his influence in high places to circumvent normal engineering channels and had forced a favorite project of his on the divisions. To remedy this situation, Kettering suggested that several Oakland engineers come to Dayton and offer their design suggestions.[38] He already recognized the growing gulf between the design engineers in Dayton and the production engineers at Oakland, Chevrolet, and Oldsmobile. He attributed this rift to an organized resistance within the divisions—a notion which, in Kettering's mind, quickly grew into the appearance of a full-fledged conspiracy.

Meanwhile, however, sales of water-cooled cars, especially Chev-

[34]Knudsen to Du Pont, April 5, 1922, Kettering Archives.
[35]Kettering to Du Pont, May 24, 1922, Papers of P. S. Du Pont, Eleutherian Mills Historical Library.
[36]Du Pont to McGowan, July 21, 1922, Papers of P. S. Du Pont.
[37]Kettering to Du Pont, May 24, 1922.
[38]Ibid.

rolets, had increased rapidly during 1922.[39] Division executives now openly questioned the wisdom of adopting the new engine at such a time. The original strategy was retained, though, and in September 1922, a copper-cooled "4" was placed in a Chevrolet Superior and sent to O. E. Hunt for testing. Hunt was the chief engineer for Chevrolet at the time and was also responsible for developing a chassis for the copper-cooled automobile. After the engine was tested, however, low power was reported. Kettering quickly dispatched Hutchinson and Walter Geise to the Chevrolet plant in Flint. In an effort to systematize testing procedures, GM had given each division two dynamometers which measured engine horsepower on a test stand. The engineers found Chevrolet's instruments still resting in their shipping crates.[40] After unpacking them, Hutchinson and Geise discovered that the problem had nothing to do with the engine. The wrong axle had been assembled in the car! The incident marked the beginning of a long series of mishaps, errors, and misunderstandings which plagued the subsequent history of the copper-cooled project. It was already evident that divisional engineers were not going out of their way to assist with a project in which they felt they played only a minor role.

While dealers continued to sell standard Chevrolet products in near record numbers, development of the copper-cooled car proceeded. On October 24, 1922, William Knudsen drew up a plan for the introduction of the new automobile. One of two options would be followed. One plan called for the manufacture of copper-cooled cars to begin on July 1 of the following year at the rate of 1,000 per day, increasing to 2,000 per day by October 1. The production of water-cooled cars was scheduled to end in June. Option number two envisioned the manufacture of only 700 copper-cooled cars per day beginning July 1. The production of water-cooled cars would continue at the rate of 1,000 per day.[41] Since water-cooled Chevrolets were selling well, the second option seemed wiser and was adopted. A new model called the "Chevrolet Superior" would be sold as an option, not as a replacement for the standard Chevrolet, until the design was perfected. As additional insurance, the executive committee requested that development continue on the water-cooled engine. Since the research lab was preoccupied with the copper-cooled work, the divisions were assigned the task of improving the water-cooled design. The Dayton team, under Kettering, and the divisional engineers at Chevrolet and Oakland, each now had a vested interest in a different

[39]Chandler and Salsbury, p. 530.
[40]Boyd, interview with Davis and Hutchinson, p. 366.
[41]Chevrolet Division Memorandum, October 24, 1922, Papers of P. S. Du Pont.

design, a situation which could only lead to further tension between them.

In some respects this resembles a normal competitive situation within a large corporation. Each team develops a design, with the best to become standard. Yet this case is different in two important respects. The decision already had been made to market the copper-cooled car—it was no longer supposed to be in the development stage. To begin work on a competitive engine at such a time showed a lack of faith in Kettering's team. Second, the very divisional engineers engaged in developing the water-cooled design were responsible for testing Kettering's product—certainly not a situation guaranteed to result in objective evaluations.

Conflict reached a head in November 1922. On November 8 the executive committee—including Sloan, Kettering, and Du Pont—met to consider the future of the copper-cooled project. Kettering confided to Du Pont before the meeting that he believed the divisions were out to scuttle his innovation.[42] Sloan argued that caution should be exercised before adopting an "untried" innovation during a peak sales period, while Du Pont countered by reminding the participants that adoption was no longer the issue, only when and how.[43] Du Pont prevailed, and the committee decided to continue the project. An unveiling would occur at the New York Auto Show in January. Two days after the meeting, the first orders for copper-cooled cars arrived in Detroit.[44] Irénée Du Pont requested six of the new coupes—a purchase which provided endless headaches during the following year.

Chevrolet's Copper-Cooled Superior proved the smash hit of the New York show (fig. 4). Information on the new car "leaked" to the trade press in late December 1922. Reviewers expressed greatest surprise at the successful brazing of copper to iron. As late as April 1922, the *SAE Journal of Automotive Engineering* had noted that "copper has been used extensively for air-cooled cylinders and was employed on some of the earliest air-cooled cars. The high conductivity is of considerable advantage, but there are many practical objections to its use." Principal among these was the difficulty of attaching sheet copper fins to the cylinder head,[45] but the Chevrolet had overcome this difficulty. In addition to praising the technical merit of the new engine, reviewers stressed the low price of the automobile—only $200 more than the standard Chevrolet Superior. For the first time the

[42]Chandler and Salsbury, p. 532.

[43]Ibid.

[44]Du Pont to Knudsen, November 10, 1922, Papers of P. S. Du Pont.

[45]S. D. Heron, "Some Aspects of Air-cooled Cylinder Design and Development," *SAE Journal of Automotive Engineering* (April 1922), p. 246.

$710
F. O. B.
Flint, Mich.

1923 Superior Chevrolet 2-Passenger Roadster, Copper-Cooled

$725
F. O. B.
Flint, Mich.

1923 Superior Chevrolet 5-Passenger Touring, Copper-Cooled

Fig. 4.—Two of the most popular copper-cooled models displayed for visitors at the New York Automobile Show in January 1923. (Courtesy of General Motors.)

American public was offered a reasonably priced air-cooled car.[46] At first this price differential appears surprising. After all, the point of the copper-cooled engine was to cut initial as well as operating costs. But a new product which offers advantages not available from competitors was usually sold at value rather than cost. The price differential merely reflected the novelty of the new engine. The copper-cooled car made as great an impression on the public as it had on the journalists. Success seemed assured.

Innovation and Collapse

Unfortunately, Chevrolet's effort to introduce the copper-cooled automobile to the public soon encountered problems of technical malfunction, intracorporate communication breakdowns, and changing market conditions. In all, only 759 copper-cooled Chevrolets were manufactured. Some 239 were scrapped by the production men. The other 520 were delivered to the sales organization. Of these, 100 were sold to retail customers while the rest were either driven by factory representatives or remained in inventory.[47] A couple eventually found their way to museums.

After initial sales, complaints came in concerning excess noise, clutch problems, wear on cylinders, carburetor malfunctions, axle breakdowns, and fanbelt trouble. The failure of the car had three primary causes. First, because the standards committee demanded light weight, numerous components were redesigned. As Kreusser remarked, "When you tried to make automobile frames, drive-shafts, rear axles, and steering gears you got into so many unconventional methods in order to get weight and price out that the car wasn't too successful."[48]

The second cause of failure was more crucial. It involved a lack of communication between Kettering's team and the production engineers at Flint and Pontiac. Without adequate coordination of the two groups, minor difficulties grew into major problems, and small rivalries escalated into crippling disputes and delays. Even careless assembly procedures may well have been caused by this tension.

The third cause is more elusive but perhaps the most important of all. Product development in this case was only of value to General Motors if it promised reasonable return within a relatively short period. The copper-cooled car was designed specifically to compete with the Model T, yet Chevrolet was holding its own without the new

[46] J. Edward Schipper, "Chevrolet Copper-cooled Car Ready for Market," *Automotive Industries* (December 28, 1922), p. 1259.

[47] Sloan (n. 11 above), p. 98.

[48] Second interview with O. T. Kreusser, p. 8.

engine. American consumers showed less enthusiasm for the copper-cooled car on the showroom floor than they had at automotive shows. Tastes were changing. Buyers no longer seemed to want an extremely light weight and inexpensive automobile, but looked instead to the style and comfort which water-cooled Chevrolets provided quite adequately.

Some of the difficulties related to engineering, but not necessarily to the air cooling of the engine. Several customers and division testers complained that oil worked its way through the cylinder walls and disabled the vehicle. Detractors of the copper-cooled engine proposed the following theory. Brazing of the copper fins sometimes changed the structure of the cast iron, and allowed it to become porous enough to permit the passage of oil.[49] Little evidence was presented in support of this theory except the fouled cylinders themselves. Other explanations seem much more plausible. More likely the oil was pumping past the piston rings and entering the combustion chamber, where it fouled the plugs and formed carbon deposits. Oil pumping was not limited to the copper-cooled Chevrolets. Alexander Laird wrote to Frank McHugh, an insurance agent for the corporation, on July 19, 1922. He informed McHugh that his new Chevrolet touring car was pumping oil—a defect caused by scored cylinders, badly fitted rings, and cylinders out of true, in other words, a poorly assembled engine. "Mr. Taylor of the Wilmington Auto Company informed me it is pumping oil, and added that the last three carloads of Chevrolets received by them have been giving similar trouble, a very serious defect for the reputation of a car."[50] Kettering never reported oil pumping in his test runs, nor do others associated with the trials mention it. The problem appeared to be related to quality control in assembly and not design. Certainly production engineers, and not Kettering's staff, had responsibility for this aspect of the project.

The fanbelt also created considerable anguish among owners. Wrote Du Pont to Knudsen in March 1923, "It has been reported to me that copper-cooled car No. IC-1187, one of the four shipped to my brother, developed trouble with the fan belt at less than 100 miles."[51] William de Krafft reported a similar defect in his car.[52] At first these complaints appeared slightly mysterious, especially since none of the testers mentioned such a problem. The explanation was really quite simple, however. Two months later de Krafft explained

[49]Interview with Joseph Butz, December 18, 1946, Kettering Archives Oral History Project, Kettering Archives, p. 7.
[50]Laird to McHugh, July 19, 1922, Papers of P. S. Du Pont.
[51]Du Pont to Knudsen, March 30, 1923, Papers of P. S. Du Pont.
[52]Du Pont to de Krafft, March 30, 1923, Papers of P. S. Du Pont.

the situation to Knudsen. The belt had failed because it was placed on the car in the reverse direction. "The replaced belt has been correctly installed on the car," he added, "and so far . . . has not broken."[53]

Among other small defects which surfaced was the clutch. Irénée Du Pont's car went in for service once again in April 1923. (His driving time in this car was severely limited!) Replacement on the clutch collar ended the trouble.[54] Apparently, clutch problems were quite common, but the design came out of Chevrolet and not the research lab. Another incident involved rapid wear of the copper-cooled engine's cylinders, and an investigation revealed loosely adherent core sand left in the intake manifold![55]

Another design problem was never adequately solved: excess engine noise, a common problem with air-cooled engines. Joseph Butz, responsible for road testing the car, discovered that the problem lay in the camshaft. He removed the offending part, gave it slightly different contours and a high polish, returned it to the engine, and was astonished to find it ran almost silently.[56] Kettering asked Butz to duplicate the design, but he had little success. Eventually the original modified camshaft was misplaced, but since the project terminated shortly thereafter, further efforts in this direction ceased.

Chassis design improvements also proceeded slowly. Hunt, of Chevrolet, had primary responsibility for this development but had made only very limited progress by the summer of 1923. Even Pierre S. Du Pont began to wonder about the delays. He wrote to Sloan, expressing his disapproval: "I am inclined to believe that part of the trouble lies in the lack of support of the copper-cooled car from the top," that is, from the top of the divisions.[57]

Yet, many complaints about the new car were centered on items only marginally related to the copper-cooled design. Irénée Du Pont's case seems most typical: "Trivial items were not in order, indicating careless work; for instance, fasteners on carpets were placed at wrong point, steering wheel out of adjustment, and similar items not having anything to do with the design of the car or the copper-cooling."[58] Lack of attention to assembly details, the responsibility of the divisions, again appeared to be the villain.

Kettering believed that organized resistance on the part of the divisions constituted the major problem. He was only partially correct.

[53]De Krafft to Knudsen, June 1, 1923, Papers of P. S. Du Pont.
[54]J. B. Tyler to P. S. Du Pont, April 6, 1923, Papers of P. S. Du Pont.
[55]Interview with Butz, p. 7.
[56]Ibid., p. 5.
[57]Du Pont to Sloan, July 7, 1923, Papers of P. S. Du Pont.
[58]P. S. Du Pont to Knudsen, May 12, 1923, Papers of P. S. Du Pont.

Evidence indicates a poorly coordinated engineering effort rather than a grand conspiracy. Take, for example, the troubles with the rear axle and the carburetor. A three-month feud took place over the proper design of the rear axle. "When the thing [the copper-cooled car] went to Detroit, immediately the axle which we [the research lab] [had] developed and proven on road test was thrown away and an axle which Mr. Barbon had been promoting was put in."[59] A similar misunderstanding surrounded the carburetor. Division engineers objected to a lack of power as the engine warmed up. "We told you," Kettering wrote to Sloan, "that their real trouble was in the carburetor system and valve setting. Trips to the Indianapolis Speedway have proven these points which have been supplemented on the dynamometers."[60] It must be mentioned that at least one design engineer disagreed with Kettering's assessment: Fred Davis believed a design flaw in the ducting system, not the carburetor, was responsible for the trouble.

A general pattern emerges from these examples. Design and production engineers seemed unwilling to cooperate on these problems, and each was ready to place the blame on the other. Knudsen put it best when he stated, "Unhappily, the engine troubles when reported divulged a wide range of differences in opinion between the creators and the producers, both as to their causes and their remedies."[61] Kettering was sure that he was trapped by an "organized resistance" within the corporation and openly expressed his belief that "because of the way in which the Research Corporation is regarded by some of the General Motors Divisions, we can never expect to carry out the constructive programs which we have planned."[62]

Divisional engineers were equally disappointed. One wrote to Pierre Du Pont, explaining that his report on the project was very negative because Chevrolet found it impossible to "produce the car within the limits set by Dayton, and rightly or wrongly demonstrated by their sample car."[63] The letter's tone was more reticent than Kettering's, for no one had his corporate clout. But the Chevrolet engineers obviously felt somewhat betrayed by their counterparts in research, for as Knudsen complained, "we ourselves were at sea as to the existence of troubles which had previously been lauded [by Kettering's group] as non-existent, which put the burden of proof on

[59] Kettering to Sloan, June 30, 1923, Correspondence of Charles F. Kettering, Kettering Archives.

[60] Ibid.

[61] Knudsen to Du Pont, September 1, 1923, Papers of P. S. Du Pont.

[62] Kettering to Sloan, June 29, 1923, Kettering Archives.

[63] Knudsen to Du Pont, September 1, 1923.

us."[64] Division engineers feared Kettering had fallen prey to an over-inflated ego and was trying to push a pet project, regardless of the best interests of the corporation. Perhaps they had a point. Louis Ruthenberg, who served as the liaison between Dayton and the divisions, found Kettering quite dogmatic on the subject of the copper-cooled engine. "Kettering was too emotionally involved in the controversy to appreciate my position," he claimed. Ruthenberg eventually left General Motors, for the tasks of compromise proved too demanding.[65]

Before pursuing the third and final cause of failure, it is worthwhile to review the chronology of the preceding events. These problems arose during the spring of 1923. At the end of May a final test for the copper-cooled six-cylinder engine was ordered. DeWaters from Buick and Hunt from Chevrolet, along with A. L. Cash, general manager of the Northway engine-producing division, were in charge of the trials. They submitted a negative report to the executive committee.[66] Sloan had replaced Du Pont as chief executive at the latter's retirement the month before. Where Du Pont had been receptive to the copper-cooled project, Sloan was more skeptical. Sloan, the cold analyst, had less personal involvement with the new engine. Du Pont, on the other hand, was a close personal friend of Kettering, and was himself keenly interested in the project's success.

This managerial shift cut off much of the executive support for the copper-cooled engine. After considering the test results and the general position of the corporation, Sloan decided to terminate the project. During the summer the executive committee agreed to increase production facilities for water-cooled Chevrolets and end the development of the six-cylinder copper-cooled engine for the Oldsmobile division.[67] By August the copper-cooled engine had been relegated to the position of a research project.

As a personal concession to Kettering, the research lab was permitted to study copper-cooling on a part-time basis. The complaints remained the same (loss of power, etc.), and no cooperation was in sight. All copper-cooled automobiles already sold were recalled. Kettering, deeply hurt, offered his resignation. Once again, encouragement by various executives and colleagues enticed him back into the fold. Although Kettering was obviously bitter about the decision, it is rather doubtful that he actually would have left GM. Not only did he have a large financial interest in the company, but his staff had by this time

[64]Ibid.
[65]Ruthenberg, p. 51.
[66]Sloan, p. 87.
[67]Chandler and Salsbury, p. 545.

made considerable progress on the antiknock problem, and Kettering was anxious to pursue this work. On other occasions Kettering had spoken out of anger but had not acted on his hasty words.[68]

Du Pont also disagreed with the decision but abided by it. His primary concern was the negative influence the recall might have on future prospective customers: "We should stand by our product even though we have offered an exchange for other car when our customers have expressed their preference for copper-cooled. . . . I advise strongly that those still owning copper-cooled cars be written a letter stating that the Chevrolet Division will continue to service these cars and assume any other responsibility usual with their product."[69] Nonetheless, the decision stood. Copper-cooling ceased to be an active concern of GM.

The project was not a total loss. As in any technical failure, numerous lessons were learned and ideas for future projects salvaged. The first successful steel-backed aluminum bearings came out of the research effort. Alfred Boegehold, one of the research engineers, noted that Durex oilless bearings were developed for the overhead camshaft on the engine,[70] and new cylinder metals emerged from the project. These efforts contributed to the eventual use of titanium and various alloys in this capacity. The metallurgical experience gained in producing the brazed-on fins later proved invaluable for making brake drums in permanent molds. The research project also developed a system of torsionally balancing highly stressed rotating parts like the crankshaft.[71]

Perhaps a greater lesson was learned concerning the value of coordinating the efforts of research and production teams. In response to the tension between design and research raised by the controversy, Sloan suggested the creation of a General Technical Committee to initiate contact between the operating divisions and research efforts at the laboratories. In a letter circulated to a number of executives in September 1923, Sloan proposed: "1. That cooperation shall be established between the Car Divisions and the Engineering

---

[68] An active philanthropist, Kettering gave considerable support to Antioch College. Disturbed by reports of radical activity on campus in the thirties, Kettering once announced his intention to withdraw support unless "you people quit playing around with your pink teas"—that is, stopped supporting Communist ideologies. Not satisfied with reassurances, Kettering wrote President Henderson and declared that all support to the college had ended. But Kettering's checks continued to arrive, along with further complaints! (Correspondence of Charles F. Kettering, Kettering Archives.)

[69] Du Pont to Colin Campbell, August 18, 1923, Papers of P. S. Du Pont.

[70] Interview with Arthur Underwood, Kettering Archives Oral History Project, Kettering Archives, p. 466.

[71] Interview with O. T. Kreusser, June 3, 1947 (n. 19 above), p. 466.

Departments within the Corporation, including the engineering and research activities of the General Motors Research Corporation and that cooperation shall take the form of a Committee to be established to be termed the General Technical Committee. 2. The Committee will consist as to principle, of the Chief Engineers of each Car Division and certain additional members."[72] The chief divisional engineers, Kettering's top staff members, plus selected general officers of the corporation met (under Sloan's chairmanship) as the General Technical Committee on September 14, 1923. At first the committee served primarily as a low-key engineering seminar. Topics of current interest were discussed: brakes, steering mechanisms, lubrication, metallurgical improvements. Longer-ranged plans included a proving grounds (a welcome change after the back roads of Ohio) and research on new fuels.[73] The administrative experiment proved quite successful and bore fruit in subsequent research on TEL, freon, and the two-stroke cycle diesel for locomotives. The kinds of difficulties demonstrated in the copper-cooling project were thus minimized if not eliminated; cooperation between designers and producers, so vital in the success of a new technical innovation, was now achieved.

A cause of failure in the copper-cooled project mentioned earlier was as important as engineering tensions or technical "bugs": It did not offer potential profit in the foreseeable future. Mission-oriented industrial research has the one primary objective of producing marketable innovations with valuable sales potential. When the copper-cooled engine appeared to offer this capacity, it was supported. As the market for water-cooled automobiles expanded, a secret weapon in the sales war with Ford became unnecessary, and support was withdrawn.

Kettering and Du Pont maintained that the engine was a technical success long after the project was abandoned. But there is no accurate gauge for judgment. We are as handicapped in our assessment as the engineers who were directly involved. The research staff claimed the engine fulfilled its technical requirements well. They provided data which showed less fuel consumption, more power, and a simpler design than its competitors. The divisional engineers provided evidence that the new engine was a failure: too little power, oil pumping, overheating, and preignition indicated that the engine was less than successful. Neither side was capable of providing an objective evaluation. This disagreement alone may have effectively sabotaged the project. It became obvious to executives like Sloan that the project could not

[72]Sloan, p. 123.
[73]Ibid. Chandler and Salsbury discuss the organizational strategy of the technical committee (pp. 547–48).

succeed whatever its potential technical merit and would best be phased out. Kettering's team continued to work on the project, but it no longer received top priority.

The copper-cooled automobile lingered until 1925, but Kettering increasingly turned his talents to the TEL project and to a host of incremental improvements for the automobile. The fruits of these investigations finally gave General Motors a product to defeat the Model-T in the marketplace, the "secret weapon" the copper-cooled car had never become. In the spring of 1925 Sloan terminated the project. Kettering did not object and the copper-cooled car received a a quiet burial with a private ceremony.

Conclusion

We have already seen how resistance on the part of divisional engineers goes a long way toward explaining the failure of the copper-cooled project. Kettering's team was no less guilty. By insisting that only research engineers were capable of solving technical problems, Kettering alienated the divisional men who were essential if the project was to be carried to successful completion. Nowhere is Kettering's attitude better shown than in a letter sent to Du Pont in late September of 1923 in which he severely criticized the new water-cooled Oakland engine:

> These designs have been influenced entirely by production and a great many fundamentals of engineering have been entirely neglected. The reason for this is that some of the men who were instrumental in this designing have no conception whatever of these principles ... on account of organizational technicalities our new products are designed by men who have never had any broad experience or technical knowledge along the line of motors and that men who have had the opportunity of making a life-long study of these things are kept at arms length and not allowed to contact of the jobs.[74]

Kettering's group showed a common myopia of design engineers. They grossly underestimated the difficulty of converting a prototype into a reliable, mass-produced article ready for sale to the public.

The ultimate effect of all this controversy was to eliminate the copper-cooled car as a potential product. No executive, with the possible exception of Du Pont, seriously considered the copper-cooled automobile after the original disaster because it was obvious that there was no money to be made with it. The controversy not only points out certain aspects of engineering behavior but also helps illuminate the

[74]Kettering to Du Pont, September 28, 1923, Papers of P. S. Du Pont.

role of the engineer in a modern corporate structure. The business of engineering is the creation of profits. Despite his enthusiasm for any particular project, an engineer often confronts a stone wall of economic calculation. The conflict between the research laboratory and the divisions created a situation in which the technical success, and thus the profitability, of the copper-cooled car was endangered. Consequently the project was dropped, despite Kettering's considerable enthusiasm. As J. Brooks Jackson, one of Kettering's close associates at the Delco lab, remembers, "I recall his enthusiasm and interest in an air-cooled engine. . . . He really wanted to drive that along and he did."[75] Yet conflict and timing defeated him. With so much controversy among engineers themselves, it is not surprising that GM executives moved on to more promising technical ventures.

The copper-cooled automobile was a failure, but one which provided General Motors with a number of important lessons. The development process was considerably more difficult than GM initially expected. When Kettering first conceived of the new air-cooled engine in 1919, he hardly expected that incremental improvements would require so much time or be so painful. Neither did he suspect that the project would end in failure. It seemed only a short step from the initial inspiration to a successful engine. Not that Kettering was always so sanguine about overcoming the obstacles of industrial research, but at least in this case he felt that the fundamental principles were so sound that the process from research laboratory to market would follow almost automatically. This is not an uncommon view. Many students of technology often overlook the incredible difficulty involved with bringing even the best of ideas from the laboratory to the marketplace.[76] Obviously, in this case, good research and design were not enough, for the obstacles of development and production proved to be the project's undoing.

The project also suggests a characteristic of professional engineering behavior which is easily overlooked. It seemed that pride, jealousy, and simple petulance played as large a role in determining the success of the project as did more measurable criteria like thermal efficiency or rated horsepower. Although the motives of all of the project's participants can never be fully known, it does seem that objective evaluation of hard data did not completely determine the outcome. Ironically, these motives and subjective evaluations usually

[75] Interview with J. Brooks Jackson, Kettering Archives Oral History Project, Kettering Archives p. 2.

[76] Elting Morison, *From Know-How to Nowhere* (New York, 1974), for instance, tends to discuss the innovations of the General Electric laboratory as if they flowed smoothly and automatically from conception to commercial operation (pp. 114–46).

remained concealed behind technical jargon and well-ordered equations. Perhaps these tensions were and are an inherent problem in corporate research. General Electric faced similar problems with its scientific and engineering staffs.[77] The basic problem may lie in group organization. Engineers may develop group loyalties which are stronger than corporate ties. Certainly in the case of Kettering's staff, the loyalty to the research lab was at least as strong as the ties to General Motors. More than one ex–research man spent his retirement years extolling the blessings of industrial research under "Boss Ket" without a great deal of praise for the production departments so necessary in completing research developments.

General Motors also derived a greater awareness of how to manage research and development. The conflict between design and production engineers paralyzed much of the corporation's technical staff. The new technical committee helped coordinate future engineering projects and opened better, and more formal, channels of communication between the technical staffs. While General Motors was not the first corporation to institutionalize research, it was one of the first to realize the necessity of integrating research with production and marketing. The technical committee improved the management of technological innovation by integrating design and production throughout the corporation. That, in turn, increased the prospects of success for future innovations as they traversed the treacherous path from laboratory to market.

---

[77]James Brittain, "C. P. Steinmetz and E. F. W. Alexanderson: Creative Engineering in a Corporate Setting," *Proceedings, Institute of Electrical and Electronics Engineers* (September 1976), pp. 1413–17. Brittain points out that Steinmetz created a Consulting Engineering Department so engineers who shared his philosophy would not feel constricted by the specialized requirements of the research laboratories.

THE SCIENTIFIC MYSTIQUE IN ENGINEERING: HIGHWAY RESEARCH AT THE BUREAU OF PUBLIC ROADS, 1918–1940

BRUCE E. SEELY

Historians of technology have shown a continuing fascination with the relationship between science and technology, an interest derived in large part from the manner in which technological change over the last century has been perceived to result from scientific advances. Their analyses, however, have frequently produced divergent views regarding the interaction of these two fields, and, as one observer commented at the 1983 meeting of the Society for the History of Technology in Washington, D.C., one formula capable of explaining all situations is unlikely to emerge.[1] Moreover, no significant effort has been made to understand the meaning of this interaction for 20th-century engineering; as Nathan Reingold and Arthur Molella have pointed out, most historians implicitly assume that engineering has benefited auto-

BRUCE SEELY is associate professor in the Department of Social Sciences at Michigan Technological University and secretary of the Society for the History of Technology. The material in this article is based on work supported by the National Science Foundation under grant SES 80-07899. The author would like to express his deep appreciation for helpful comments and valuable suggestions to James Rosenheim, John Smith, Roger Beaumont, Eugene Ferguson, David Channell, and Stuart W. Leslie—all helped to improve the article.

[1]See in particular two special issues of this journal: "Science and Engineering," *Technology and Culture* 2 (Fall 1961): 305–99, and "The Interaction of Science and Technology in the Industrial Age," *Technology and Culture* 17 (October 1976): 621–742. Also, Edwin T. Layton, Jr., "Mirror-Image Twins: The Communities of Science and Technology in 19th-Century America," *Technology and Culture* 12 (October 1971): 562–80. Other examples include Michael Fores, "Scientists on Technology: Magic and English-Language Industrispeak," in *Research, Development, and Technological Innovation: Recent Perspectives on Management*, ed. Devendra Sahal (Lexington, Mass., 1980), pp. 239–250; *The Dynamics of Science and Technology; Social Values, Theoretical Norms, and Scientific Criteria in the Development of Knowledge*, ed. Wolfgang Krohn, Edwin T. Layton, Jr., and Peter Weingart (Boston, 1978); Walter Vincenti, "Control-Volume Analysis: A Difference in Thinking between Engineering and Physics," *Technology and Culture* 23 (April 1982): 145–74; and A. R. Hall, "Of Knowing How to . . ," in *History of Technology*, Third Annual Volume, 1978, ed. A. R. Hall and Norman Smith (London, 1978), pp. 91–103.

This essay originally appeared in *Technology and Culture*, vol. 25, no. 4, October 1984.

matically from the adoption of scientific approaches and methods.[2] A more cautious conclusion concerning the value of science for certain aspects of engineering is suggested by the post–World War I highway research program conducted by the federal government's Bureau of Public Roads (BPR).

Road building early in the century resembled many fields in engineering that depended on empirically derived understandings of nature; these fields were "low technology" when compared with the science-based "high-tech" areas of chemistry and electricity. Before the First World War, civil engineers at the BPR had developed relatively simple tests for determining the physical properties of construction materials by relying on traditional engineering practices. Yet in the period 1918–40, federal highway engineers embraced scientific approaches to research and radically altered the research methods they had used successfully since the agency's establishment in 1893. In the wake of these changes, BPR investigators encountered serious difficulties in producing information of direct utility to engineers in the field. The evidence indicates that the BPR's infatuation with attitudes and experimental methods usually considered typical of science hindered the development of practical answers to engineering questions while failing to enhance theoretical understandings of the problems under investigation.

This finding runs counter to expectations because the nature of engineering at this time was changing in exactly the direction of the shift in the BPR research program. The profession had accelerated its utilization of scientific information and acceptance of scientific methods in the late 19th century, and the highly successful contributions of scientists to such wartime projects as submarine detection and radio created an enormous enthusiasm for scientific research and spurred engineering further in the direction of emulating science. Moreover, such leading figures in the field as Alfred D. Flinn, director of the Engineering Foundation and one of the forces behind creation of the Division of Engineering in the National Research Council,

[2]Nathan Reingold and Arthur Molella, Introduction to "The Interaction of Science and Technology in the Industrial Age" (n. 1 above), p. 628; for additional evidence, see Gernot Bohme, Wolfgang Van Den Daele, and Wolfgang Krohn, "The Scientification of Technology," in *The Dynamics of Science and Technology* (n. 1 above), p. 239. The contradictory Department of Defense studies intended to determine the value of science to technological developments, TRACES and Project Hindsight, offer little guidance in this matter. See Chalmers W. Sherwin and Raymond S. Isenson, "Project Hindsight," *Science* 156 (June 23, 1967): 1571–77; and Illinois Institute of Technology Research Institute, *Technology in Retrospect and Critical Events in Science*, 2 vols. (Chicago, 1968).

encouraged such emulation. An examination of the BPR's investigations of the effect of vehicle impact forces on highways and the behavior of subgrade soils—work that lasted from 1918 until the late 1930s—provides a case study of both this engineering response to scientific research techniques and the problems it presented to one group of engineers.[3]

Federal Highway Research, 1893–1918

At the end of the 19th century, the mass appeal of the bicycle led not only to the first resurgence of interest in highway construction in the United States since the antebellum fad for plank roads but also to congressional creation of an Office of Road Inquiry (ORI) in the Department of Agriculture.[4] The ORI was charged with collecting and disseminating information on road materials and construction methods, an arrangement characteristic of many later Progressive reforms and intended automatically to produce better highways. This purpose was complemented by a pattern developing in other Department of Agriculture agencies, especially the Bureaus of Plant Industry, Animal Industry, and Entomology, which utilized university-trained scientists to solve practical problems through research. As Charles Rosenberg has demonstrated, this orientation toward practicality be-

[3]On science-technology relationships in the 19th century, see Edwin T. Layton, Jr., "American Ideologies of Science and Technology," *Technology and Culture* 17 (October 1976): 688–701; Thomas Parke Hughes, "The Science-Technology Interaction: The Case of High-Voltage Power Transmission Systems," *Technology and Culture* 17: 646–62; David F. Channell, "The Harmony of Theory and Practice: The Engineering Science of W. J. M. Rankine," *Technology and Culture* 23 (January 1982): 39–52; and James Kip Finch, *The Story of Engineering* (Garden City, N.Y., 1960), pp. 387–88. On the changes after World War I, see A. Hunter Dupree, *Science in the Federal Government: A History of Policies and Activities to 1940* (Cambridge, Mass., 1957), pp. 323, 327; Ronald Tobey, *The American Ideology of National Science, 1919–1930* (Pittsburgh, 1971), p. xii–xiii, 35–37, 49–61; Daniel J. Kevles, *The Physicists: The History of Scientific Community in Modern America* (New York, 1978), pp. 117–38; and Alfred D. Flinn, "Research Advances Civil Engineering," *Civil Engineering* 1 (October 1930): 14–16.

[4]The early history of the ORI and American highway development can be traced in Bruce E. Seely, "Highway Engineers as Policy Makers: The Bureau of Public Roads, 1893–1944" (Ph.D. diss., University of Delaware, 1982), chap. 1; see also Phillip Mason, "The League of American Wheelmen and the Good Roads Movement, 1890–1905" (Ph.D. diss., University of Michigan, 1957); George Rogers Taylor, *The Transportation Revolution, 1815–1860* (New York, 1951), pp. 15–31; Thomas H. MacDonald, "The History and Development of Road Building in the United States," *Transactions of the American Society of Civil Engineers* 92 (1928): 1181–206; and U.S. Department of Transportation, Federal Highway Administration, *America's Highways, 1776–1976: A History of the Federal-Aid Program* (Washington, D.C., 1977), pp. 1–63.

came a hallmark of agricultural research situations; after 1900 it also applied to the new National Bureau of Standards.[5]

The Office of Road Inquiry combined the information-gathering and practical research approaches. Initially, the ORI focused its attention on collecting and compiling data on the location of road materials, but rather quickly it began to concern itself with classifying and grading the quality of those resources. Soon the ORI instituted its own research program, beginning with the installation of a physical testing laboratory in 1900 that enabled it to test samples of sand, stone, tar, and cement as a free public service. By 1905, federal engineers at the renamed and reorganized Office of Public Roads (OPR) were considering the optimal use of existing materials, the introduction of new ones, the development of tests for identifying durable specimens, and the effect of the automobile. They also had established a reputation as the leading source of technical expertise on all highway matters.[6]

The OPR investigations relied on a wide variety of full-scale tests and investigations. In many cases, the federal engineers designed tests to compare alternative materials or modes of construction on existing sections of highway. One such study, conducted from 1907 to 1912, involved several procedures for limiting road dust, the test being carried out on public streets in Jamaica, New York. After 1910 the office conducted other investigations at its own experiment station in then-rural Arlington, Virginia, but the basic research approach remained full-scale field studies. Based on this work and results from the free testing program, in 1911 the OPR began issuing typical construction specifications and standard methods of testing materials, and these were widely adopted by local and state road-building agencies.[7]

[5]On science in the Department of Agriculture, see Dupree (n. 3 above), pp. 157–83; T. Swann Harding, *Two Blades of Grass: A History of Scientific Development in the U.S. Department of Agriculture* (Norman, Okla., 1947); and Charles E. Rosenberg, "Science, Technology, and Economic Growth: The Case of the Agricultural Experiment Station Scientist, 1875–1914," *Agricultural History* 45 (January 1971): 1–20. On the Bureau of Standards, see Kevles (n. 3 above), pp. 72–73, 81, 95–96.

[6]These early research efforts can be followed through: U.S. Department of Agriculture, Office of Road Inquiry, *Report of the Agent and Engineer for Road Inquiry* (1893–98); idem, Office of Public Road Inquiry, *Report of the Director of the Office of Public Road Inquiry* (1899–1904); and idem, Office of Public Roads, *Report of the Director of the Office of Public Roads* (1905–18) (hereinafter cited as *OPR Annual Report[s]* with year); see also the large roster of official Department of Agriculture publications prepared by these agencies from 1895–1916.

[7]See *OPR Annual Reports* (n. 6 above), 1905–18; U.S. Department of Agriculture, Office of Public Roads, *Publications of the Office of Public Roads*, Circular no. 88 (August 28, 1908); idem, *Progress Reports of Experiments in Dust Prevention and Road Preservation, 1911*, Circular no. 98 (December 12, 1912); correspondence in Files 10, 59, 335, and 470, General Correspondence, 1893–1916, Records of the Bureau of Public Roads, Record

The record demonstrated that the relatively simple investigations of the OPR prior to World War I provided satisfactory answers to practical problems. The engineers operated in a systematic fashion in both laboratory and field investigations, but they did not consider themselves scientists. They were not concerned with developing theories of material behavior, nor did they express results in the language of mathematics. Instead, they used cut-and-try techniques to find the materials that worked best in actual service. In fact, the office did not even use the term "research" to describe its work until 1916.

Nonetheless, the Office of Public Roads believed that it had built a solid understanding of road construction based on the century-old efforts of such European engineers as Macadam, Telford, and Tresaguet. Federal engineers did not possess all of the answers, but they believed they knew the proper questions, and they were confident of the answers already available. Through an object-lesson construction program of demonstration roads, through models, exhibits, and "Good Roads Trains" that toured the country, through typical specifications issued by the Department of Agriculture, and through programs that sent federal engineers to local road agencies as consultants, the OPR spread its approach to proper road-building procedures. By the start of the war, the OPR claimed that significant problems remained only in the realms of finance and administration, not in technique.[8]

Unfortunately, this confidence was shattered during the winter of 1917–18. When a flood of railroad cars carrying war matériel for France choked freight yards at East Coast ports, government officials suggested that midwestern manufacturers of trucks for the army deliver them to the Atlantic seaboard under their own power and with a

---

Group 30, National Archives, Washington National Records Center, Suitland, Maryland (hereinafter cited as Records of the BPR); U.S. Department of Agriculture, Office of Public Roads, *Methods for the Examination of Bituminous Road Materials*, Bulletin no. 38 (July 27, 1911); idem, *Typical Specifications for the Fabrication and Erection of Steel Highway Bridges*, Circular no. 100 (August 19, 1913); idem, *Typical Specifications for Bituminous Road Materials*, Department Bulletin no. 691 (July 10, 1918); idem, *Typical Specifications for Nonbituminous Road Materials*, Department Bulletin no. 704 (August 30, 1918); and "Object Lessons in Road Building," *Scientific American* 106 (March 16, 1929): 232.

[8]See *OPR Annual Reports* (n. 6 above), 1905–18; correspondence in Files 56, 68, and 369, and "Report of the Work of the 'Good Roads Train' over the Illinois Central Railway System" (1901), File 137, General Correspondence, 1893–1916, Records of the BPR; correspondence in File 430 (for each state), Classified Central File, 1912–50, Records of the BPR; "Help from Uncle Sam," *Colliers* 52 (October 10, 1913): 21–22; "Object Lessons in Road Building" (n. 7 above); the entire collection of bulletins and circulars issued by the road office through the Department of Agriculture; and *America's Highways* (n. 4 above), p. 117.

load of freight. State and federal highway engineers laid out routes to New York, Philadelphia, and Baltimore, often over the newest roads in the country. Much to their consternation, this first large-scale attempt at long-haul trucking precipitated a massive series of road failures. As one federal engineer explained, "Hundreds of miles of roads failed under the heavy motor truck traffic within a comparatively few weeks or months. . . . These failures were not only sudden but also complete, and almost overnight an excellent surface might become impassable."[9]

This disaster caught the Office of Public Roads—soon to be reorganized again as the Bureau of Public Roads (BPR)—totally by surprise, and it could not have happened at a worse time. In 1916, the Federal-Aid Road Act had lodged with the OPR the responsibility for distributing the first federal funds for state road construction, but OPR mechanisms for overseeing this task were quickly fouled in red tape, while the war in Europe imposed limitations on both labor and materials. Together, these factors handcuffed the federal-aid road program and led to numerous complaints regarding the OPR's handling of the funds and the slow rate of actual construction. The road failures appeared to lend credence to those complaints and led to charges of waste and incompetence in the Office of Public Roads.[10]

As part of their defense, BPR engineers urgently launched a series of field investigations to explain the failures. They quickly identified a pair of interlocking factors as culprits: water that froze in the subgrade (soil foundation) and heavy trucks. In the first case, the engineers found that traditional drainage practices, such as Macadam's method of building an impervious layer into the road or the incorporation of tile drains, failed to keep water out of subgrades with clay soils which could be permeated by upward capillary movement. Freezing and thawing could then destroy the road, especially when heavy trucks traversed waterlogged soil foundations. And there were instances where trucks had cracked or crushed surfaces that were not water weakened.[11]

These findings undermined the basic assumptions of highway design. To some degree, however, the situation was not new. After 1905,

[9]*America's Highways* (n. 4 above), pp. 90–100, 103–4, 117–19; *OPR Annual Report* (n. 6 above), 1918, pp. 373–75; "Our Highways and the Burden They Must Carry," *Public Roads* 1, no. 2 (June 1918): 4–29. Quotation from Prevost Hubbard, "Efficiency of Bituminous Surfaces and Pavements under Motor-Truck Traffic," *Public Roads* 1, no. 10 (February 1919): 25.

[10]*America's Highways* (n. 4 above), pp. 103–9; and Seely, "Highway Engineers" (n. 4 above), pp. 160–221.

[11]J. L. Harrison, "Water and the Subgrade," *Public Roads* 1, no. 12 (April 1919): 11–17; W. W. McLaughlin, "Capillary Moisture and Its Effects on Highway Subgrades," *Public Roads* 4, no. 1 (May 1921): 6–8.

larger numbers of automobiles had challenged the durability of macadam surfaces, which utilized earth to bind the stone into an impervious pavement. Logan Page, chief of the OPR and former head of its testing laboratory, found that the powered rear wheels of automobiles loosened the earth binding and sent it flying away in clouds of dust, destroying the road surface. Both this discovery and a variety of solutions to the problem were achieved through the traditional OPR practice of full-scale field studies on existing highways.[12]

In 1918, however, the BPR's reaction to the new problems differed markedly from Page's efforts a decade earlier. This time, bureau engineers ultimately abandoned their emphasis on tests of individual components and full-scale field studies. A researcher explained that the BPR needed to rebuild its understanding in a new way, announcing in 1919 that "With the rational design of road surfaces as a goal, the Bureau of Public Roads has begun experiments to find out something of the fundamentals affecting road design."[13]

The word "fundamentals" signaled the change in research strategies. In scientific circles, fundamental research meant pure research as opposed to applied research directed toward answering some specific problem. The BPR engineer did not mean fundamental in the same way; he was not out to identify ideal laws of nature for the sake of new knowledge as would a scientist. Rather, his concern fitted better into David Channell's description of an engineering science that modified the ideal laws of nature to explain the behavior of actual bodies, thus combining scientific and empirical approaches, with an eye always toward practical utilization.[14] Yet we shall also see the BPR engineers did not entirely reject the scientific connotation of pure research—even as they moved to borrow scientific research methods that high-technology engineering fields had previously found useful.

Fundamental research at the BPR ushered in two significant changes in research method. First, the engineers adopted what they perceived to be the essence of the scientific approach—the acquisition of precise

[12]See *OPR Annual Reports* (n. 6 above), 1905–13; U.S. Department of Agriculture, Office of Public Roads, *Progress Reports of Experiments with Dust Preventives*, Circular no. 89 (April 1908); idem, Circular no. 98 (December 12, 1912); Logan W. Page, "Motor Car and the Road," *Scientific American* 102 (January 15, 1910): 46–47; idem, "Automobiles and Improved Roads," *Scientific American* 109 (September 6, 1913): 178; and idem, "The Selection of Materials for Macadam Roads," in *Proceedings of the Third American Road Congress in Detroit, Michigan, 1913* (Baltimore, 1914), p. 170.

[13]C. A. Hogentogler, "Tests of Impact," *Public Roads* 4, no. 7 (October 1921): 3; see also "Subgrade Investigations Begun by Bureau of Public Roads," *Public Roads* 2, no. 24 (April 1920): 29; and "Bureau of Public Roads Subgrade Study," *Engineering News-Record* 84 (May 27, 1920): 1072–73.

[14]Channell (n. 3 above), pp. 51–52.

quantitative data and the expression of results in mathematical terms. The BPR researchers apparently had come to believe, along with the editor of *Roads and Streets*, that because "the essence of science is definiteness . . . [and] in the most highly developed sciences this means quantification," and because "engineering is the systematic application of science to the solution of economic problems," it followed that "the engineer is fundamentally a quantifier."[15] Second, fundamental research led to an emphasis on developing theoretical understandings of road construction. In both cases, the aim was to replace "mere" empirical rules of behavior and observational methods of research with the mathematical precision credited to science. The upshot of the 1917–18 highway failures, then, was that federal engineers abandoned not only the old rules of construction but also their traditional methods of research.

The BPR targeted several problems for study in the new fashion, including pavement wear and subgrade behavior, beginning in 1918 with work on the effect of impact forces generated by heavy trucks. As the chief of research, A. T. Goldbeck, explained in a 1922 progress report, this was an opportunity to "establish the art of road building as an exact science and to this end let us obtain and use the underlying, fundamental facts."[16] But results of practical utility to highway designers appeared only slowly, if at all, thereby confounding Goldbeck's hopes. The impact study lasted more than twenty years and clearly illustrated both the change in research techniques, especially the emphasis on quantitative data, and the role of those changes in the problems that followed.

Motor Truck Impact Investigations, 1918–1923

With trucks immediately identified by the initial field studies as a primary culprit in the wartime road failures, federal engineers made their first priority the determination of how trucks affected roads. They began in late 1918 by utilizing their traditional research method, placing pressure cells under a full-size concrete slab at the Arlington Experiment Station and driving trucks across the surface to determine the force exerted. The surprised engineers found that the force of impact delivered to the pavement when irregularities in the surface caused the tires to leave the pavement created much more damage than the vehicle's static weight.[17]

[15]"Hoover the Quantifier," *Roads and Streets* 68 (May 1928): 227–28.

[16]A. T. Goldbeck, "Highway Researches and What the Results Indicate," *Good Roads*, 62 (April 19, 1922): 238.

[17]A. T. Goldbeck and E. B. Smith, "An Apparatus for Determining Soil Pressures,"

With this as a guide, the BPR proposed in 1919 to measure the actual magnitude of impact. That decision indicated the first movement toward fundamental research in the bureau. Both the purpose and the design of the experiment appeared more sophisticated and scientific because of the concern for precision measurement. Even so, the researchers retained their old emphasis on full-scale field tests on actual pavement. Their measuring apparatus consisted of a hydraulic jack placed in a box in the pavement so that the bearing plate of the jack was level with the surface. The plate rested on a small copper cylinder which deformed under the impact of 3-ton and 5-ton trucks coming off a 2-inch drop (fig. 1). By comparing the deformation of the copper cylinder with identical cylinders placed in a hydraulic press delivering a known force, the engineers could infer the impact force.[18] Again, the results surprised the investigators, for they observed impact forces that were four to five times greater than the actual weight of the vehicle. The truck's springs caused this phenomenon for, when the vehicle was in the air, these snapped into an unloaded position, accelerating into the pavement the tire with all of the unsprung weight.[19]

Researchers concluded their attempts to understand how impact affected pavement in 1920 by identifying tires as yet another contributing factor. Using the same apparatus of jack and copper cylinders, the engineers drove trucks equipped with worn solid tires, new solid tires, or pneumatic tires over the jack. The tests indicated that worn solid tires created the greatest impact forces. At speeds of 15 MPH, pneumatic tires reduced impact forces by almost 60 percent compared with worn solid tires. Solid tires did not disappear from trucks until the mid-1920s, largely because of the unreliability of the newer pneumatic tires. Even so, an important incentive had appeared for the introduction of new tires.[20]

Proceedings of the American Society for Testing Materials 16, pt. 2 (1916): 306–19; Goldbeck, "Thickness of Concrete Slabs," *Public Roads* 1, no. 12 (April 1919): 34–38.

[18]Some BPR engineers had developed this apparatus in order to test gunpowder during 1917–18. See E. B. Smith and J. T. Pauls, "Preliminary Report of Impact Tests of Auto Trucks on Roads," *Public Roads* 2, no. 15 (July 1919): 8–10. This and all of the tests were discussed in correspondence in File 890 Tests, Motor Trucks, Classified Central File, 1912–50, Records of the BPR.

[19]Smith and Pauls (n. 18 above); correspondence in File 890 Tests, Motor Trucks, Classified Central File, 1912–50, Records of the BPR; "Motor Truck Impact on Roads Five Times Dead Load," *Engineering News-Record* 83 (September 18, 1919): 573–75; A. T. Goldbeck, "The Present Status of Impact Tests on Roadway Surfaces," *Public Roads* 2, nos. 18–19 (October–November 1919): 19; U.S. Department of Agriculture, Bureau of Public Roads, *Report of the Chief of the Bureau of Public Roads* (1920), p. 26 (hereinafter cited as *BPR Annual Report[s]* with year).

[20]Hugh Allen, *The House of Goodyear, a Story of Rubber and Modern Business* (Cleveland,

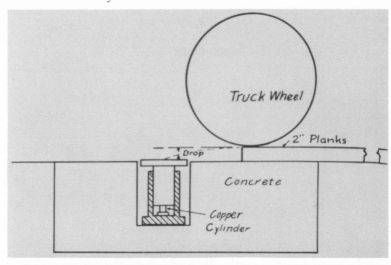

FIG. 1.—Diagram and photograph of apparatus for measuring impact used in initial BPR investigations, 1919–20. (From "Apparatus Used in Highway Research Projects in the United States," *Bulletin of the National Research Council* 6, pt. 4 [no. 35]: 44.)

1943), pp. 30–39, 140–42; "Good Progress in Impact Tests," *Public Roads* 3, no. 27 (July 1920): 18; E. B. Smith, "The How and Why of Motor Truck Impact," *Public Roads* 3, no. 31 (November 1920): 16–18; and correspondence for 1919–22 in File 890 Tests, Motor Trucks, Classified Central File, 1912–50, Records of the BPR.

As BPR engineers completed their initial research program, they had uncovered the factors behind impact damage to highways. A November 1920 report contained a series of practical recommendations for reducing those forces, including the distribution of weight evenly between the front and rear axles, elimination of solid tires, and construction of the smoothest pavements possible. These were commonsense suggestions based on field tests that resembled the BPR's traditional research approach. The goal initially used to justify the research, creation of a generalized approach for the rational design of roads, had not been achieved.[21]

At the end of 1920, however, BPR engineers began again. Even more clearly than before, this second round of research reflected the inroads made by scientific methods, especially the use of quantitative data, into the BPR's research program. The engineers wished to determine the reaction of a slab to impact forces while at the same time measuring the force of impact. Both were impossible if a truck provided the source of impact because the jack-and-cylinder measuring device was separate from the slab. Therefore, BPR researchers replaced trucks with a machine. The device consisted of a loaded box, representing the weight of a vehicle, supported by a truck spring. A plunger below the spring held a wheel that hit the pavement. The engineers could vary the weight in the box and on the plunger, so that dropping the plunger onto the slab could simulate any size truck. A large tower housed this equipment, which tested fifty-six pavement slabs of different materials and thickness at the bureau's Arlington Experiment Station. (See fig. 2.) The soil foundations were identical in order to eliminate another variable.[22]

Through 1921, BPR researchers ran the machine on each slab until severe cracks appeared, determining the most durable pavements and the largest vehicle and height of drop each could sustain. The BPR considered releasing this information in tabular form to guide designers but hesitated for two reasons that reflected the increasing influence of scientific, as opposed to engineering, outlooks in their work. First, the use of only one soil foundation under the slabs limited the applicability of the findings to soils that matched Arlington's. Of all the variables that highway designers had to consider, soil probably varied the most, so the BPR's results had very narrow immediate usefulness.

[21]Smith (n. 20 above).

[22]C. A. Hogentogler, "Tests of Impact on Pavement by the Bureau of Public Roads," *Public Roads* 4, no. 6 (October 1921): 6–18; see also Prevost Hubbard, "Test and Research Investigations of the Bureau of Public Roads," *Public Roads* 2, no. 15 (July 1919): 28–32; "Good Progress in Impact Tests" (n. 20 above); and Earl B. Smith, "Motor Truck Impact Tests of the Bureau of Public Roads," *Public Roads* 3, no. 35 (March 1921): 3–36.

FIG. 2.—Testing tower for supporting the apparatus used in BPR impact investigations, 1921–23. (From *Public Roads* 4 [October 1921]: cover.)

Second, the bureau did not want to release its findings in tabular form because, in the words of one researcher, the "method of selection is not very scientific and can only be considered as a preliminary step."[23]

That this statement reflected significant change in engineering is evident by comparing it with a recent description of the work of Anson Marston, respected dean of engineering at Iowa State University early in the 20th century. Marston's pioneering soil studies led to an exponential equation, still used today, for determining pressure on buried culvert pipe. Yet his publications in Iowa Engineering Experiment Station *Bulletins* from 1913 to 1940 presented a series of design charts, not equations, for Marston believed that practicing drainage engineers could better understand charts. Another engineer noted in 1982 that this choice "had one adverse effect—to give some practicing engineers the impression that the Marston design method is 'empirical' or based on measurements without guiding theory. This of course is not the case."[24] It did not bother Marston, but the younger engineer clearly believed there were few worse situations than being considered an empiricist. This same motive—the fear of appearing "not very scientific"—surely explained the BPR investigators' decision not to publish the results as a table. But the next step demanded production of precise mathematical data and greatly increased the complexity of the experiment. In order to calculate the actual forces of impact, the engineers added to the machine a recording device which plotted on paper the movement of its parts—box, plunger, and slab (fig. 3). From these graphs, calculation of the actual force was possible, but only after two years of effort.[25]

The effect of the change in BPR research methods should be very evident. Emphasizing quantitative data had both increased the complexity of the experiment and delayed results. Moreover, the use of machinery to simulate a truck's impact eliminated from the calculations the factor of vehicle speed. Another problem lay in the application of impact to the center of the slab; the tower was made to straddle each section of pavement, even though the researchers already knew that the greatest impact damage occurred on the edges and corners of the pavement. Although these last two choices increased the artificiality of the experiment, they were accepted because of the desire to express research results in a mathematical, that is, in a more scientifically

[23]Hogentogler (n. 22 above), pp. 3–18, 22, 24; also, *Public Roads* 4, no. 7 (November 1921): 1–18, 27.

[24]R. L. Handy, "ISU Soil Engineering Research," *Marston Muses* 6 (June 1982): 2. See also Anson Marston Papers, Iowa State University Archives, Ames; and Herbert J. Gilkey, *Anson Marston: Iowa State University's First Dean of Engineering* (Ames, Iowa, 1968).

[25]Hogentogler (n. 22 above).

acceptable, fashion. But this ultimately cost the BPR, for even as the engineers struggled through 1922 and 1923 to calculate the impact forces from the graphs, the bureau had to consider the 1921 tests tentative. Unlike the practical suggestions for reducing impact damage produced by the first battery of tests, the second series led only to a call for further research.[26] Once again, a rational design theory for highways had eluded the federal engineers.

The Bates Test Road, 1921–1924

The bureau's initial scientific investigations had not proved especially fruitful. Yet even while federal engineers began tests with the tower apparatus, Illinois State Highway Department engineers inaugurated a study of the same problems—pavement failure, the effect of heavy trucks, and the influence of subgrade soils on road life. A

Fig. 3.—Plunger for delivering impact to the pavement, and recording device for calculating force of impact. (From *Public Roads* 4 [October 1921]: 7.)

[26]See A. T. Goldbeck, "What the Arlington Tests Are Showing," *Engineering and Contracting* 59 (February 7, 1923): 308; *BPR Annual Report* (n. 19 above), 1924, p. 10.

comparison of the Illinois study, known as the Bates Test Road, with the federal research program clearly demonstrates the shift from engineering to scientific methodology and outlook by BPR researchers after World War I.

In 1920, the Illinois State Highway Department constructed a test highway composed of sixty-eight pavement sections, using different specifications, drainage patterns, and soil conditions. The 2-mile oval of brick, concrete, and asphalt slabs was subjected to trucks varying in size from 2,500 to 13,500 pounds. Repeatedly circling the road, these vehicles ultimately destroyed fifty sections of pavement in two research seasons, 1921 and 1922.[27]

Superficially, this test resembled the BPR impact experiments at Arlington. In fact, the bureau provided funds and observers for the Bates Test Road. Illinois highway engineer Clifford S. Older designed and controlled the experiment, however, and, in a startling turnabout from the federal tests, the Bates Road provided information of practical value to highway designers almost immediately. Most significant, the Illinois tests indicated that the correct design for a pavement slab had a cross section with thick edges and a thin center, exactly the reverse of existing design philosophies. Furthermore, the tests suggested the most durable types of construction for a variety of soil conditions. Yet the Arlington Experiment Station could not even confirm the Illinois findings because there was no thickened-edge slab among the pavements being tested there. Even had there been one, the testing apparatus could not have been applied to the edges.[28]

The response to the Illinois findings was immediate and impressive. Almost alone, the Bates Test had removed all doubts about the durability of portland-cement concrete pavement. By the end of 1922, BPR engineers were encouraging state highway departments to adopt the thickened-edge slab design. Illinois, not surprisingly, became the first state to do so, in 1923, and by 1925 thirty-three states had embraced the new specification. Michigan alone estimated it saved $2.2 million in the period 1923–26 because of this change. *Engineering News-Record*

[27]Clifford Older, "Illinois Begins Traffic Endurance Test," *Engineering News-Record* 87 (August 18, 1921): 274–76; Older and H. F. Clemmer, "Preliminary Report on the Bates Road Experiment," *Public Roads* 4, no. 5 (September 1921): 3–11; Illinois Department of Public Works and Buildings, Division of Highways, *Bates Experimental Road; or Highway Research in Illinois*, Bulletin no. 21 (January 1924); and correspondence for 1920–21 in File 407 General and File 407 Illinois, Classified Central File, 1912–50, Records of the BPR.

[28]Goldbeck, "What the Arlington Tests Are Showing" (n. 26 above), pp. 301–12; "Practical Lessons from the Bates Road Tests," *Engineering News-Record* 90 (January 18, 1923): 57–61; "Concrete Pavement Sections with Thickened Edges," *Engineering and Contracting* 60 (July 4, 1923): 20–21.

correctly observed in 1923 that "test roads have seldom given results so positive that a radical change in pavement design followed."[29]

Widely divergent research goals and philosophies largely explained the contrasting results of the Illinois and federal investigations. In a nutshell, Illinois engineers set out to identify the most durable pavements. They built full-size slabs using various materials, construction methods, drainage arrangements, and subgrades, and used real trucks to deliver impact forces to the sample highway. By recreating as many real conditions as possible, they enhanced the validity of the suggestions that they passed on to highway designers. In nearly every way, the state investigators followed BPR's prewar research style. Federal engineers, on the other hand, established the measurement of the actual magnitude of impact forces as their primary goal.They built smaller slabs and used only one soil in order to limit the number of variables. Likewise, they introduced the impact machine to allow exact repetition of the force applied. By isolating each factor in accordance with scientific procedures, the researchers enhanced the acquisition of precise, unequivocal quantitative data on the force of impact. Yet the same choices limited the applicability of the BPR's results.

It was not, however, as if the Illinois experiment completely sacrificed the production of quantitative data. Researchers installed twenty-six BPR pressure cells and a variety of other instruments to measure pavement deflection. They addressed several of the questions considered by federal engineers and even obtained better information on the effect of temperature changes on pavement slabs, the deformation of corners, and the distribution of weight across expansion joints. Nonetheless, BPR engineers considered these full-scale tests insufficiently scientific because they included too many variables.

This attitude showed plainly in their reaction to the Pittsburg Test Road built in 1921 by the California State Highway Department. The experiment followed the Bates model with real trucks running over an oval test track of sample slabs. The BPR again provided observers and furnished war surplus trucks, with California engineers controlling the project. Federal highway engineers were very critical, however. In 1922, the head of research at the BPR complained, "Too much emphasis has been laid on the destruction of slabs and too little on obtaining

[29]Clarkson H. Ogelsby and R. Gary Hicks, *Highway Engineering*, 4th ed. (New York, 1982), p. 732; *BPR Annual Report* (n. 19 above), 1922, p. 471; 1925, p. 30; "Results of Heavy Traffic on Pittsburg Test Road," *Engineering News-Record* 88 (June 29, 1922): 1066–69; A. T. Goldbeck, "Structural Design of Roads," *Engineering News-Record* 89 (November 30, 1929): 942; Michigan State Highway Department, *Eleventh Biennial Report of the State Highway Commission* (1925–26), p. 113; quotation from "Progress Made in Highway Engineering," *Engineering News-Record* 90 (January 18, 1923): 55.

scientific data during the destruction." By "scientific data," he meant quantitative data; he wanted to get measurements on the stress in reinforcing rods, pavement deflection, and bending moments.[30] Yet precisely because they deemphasized such goals, the "empirical" Bates and Pittsburg Test Roads gave engineers clear choices of pavement and serviceable designs under several soil conditions, while the "scientific" tests of the BPR left only graphs of impact forces that required laborious calculations.

Continued BPR Impact Tests, 1924–1940

That a substantial alteration of research methods was taking place at the BPR was apparent by the contrast between state and federal results in 1921–22. But even sharper evidence of the BPR's desire to operate in a scientifically acceptable fashion appeared in a new series of truck-impact investigations begun in 1924. Significantly, these new studies not only took up the call for further research issued by the federal tests concluded in 1923 (a litany that ran through the next fifteen years of BPR annual reports); they also continued the trend of replacing full-scale field studies with complicated mechanical simulators. The equipment provided precise data, but it likewise introduced additional artificiality into research results. Thus the problem first evident in 1921—experimental evidence of limited practicality—not only appeared as a pattern in all later tests, it actually grew worse.

This difficulty was readily apparent in the outcome of the third series of BPR impact investigations. Previous efforts had addressed the reaction of the pavement (1918) and the actual magnitude of impact (1919–23); attention in 1924 focused on both at the same time. The primary obstacle concerned collecting quantitative data. Dissatisfied with the cumbersome system of calculating impact from graphs, the investigators had to devise a means to measure impact without affecting the pavement's reaction. Their creation involved several components, including a truck-mounted accelerometer that recorded the acceleration of the truck's wheels into the pavement and a device that graphed the related movement of the truck's springs. (See figs. 4, 5, and 6.) These two measurements gave the engineers enough data to calculate the force of impact, while a third instrument, a profilometer, measured the pavement imperfections responsible for each impact.[31]

[30]"Results of Heavy Traffic on Pittsburg Test Road" (n. 29 above); and A. T. Goldbeck to Thomas H. MacDonald, June 10, 1922, File 407 California, Classified Central File, 1912–50, Records of the BPR.

[31]Leslie W. Teller, "Impact Tests on Concrete Pavement Slabs," *Public Roads* 5 (April 1924): 4–5; idem, "Accurate Accelerometers Developed by the Bureau of Public Roads," *Public Roads* 5 (December 1924): 1–9; E. B. Smith, "An Accelerometer for Measuring Impact," *Proceedings of the American Society for Testing Materials* 23 (1923): 626–32.

The researchers planned to record a series of impacts delivered to an existing stretch of public highway. Then an improved, less bulky version of the 1921 tower machine would reproduce identical forces on sample slabs at the Arlington Experiment Station (fig. 7). This procedure permitted the engineers to place pressure cells and other apparatus under the sample slabs for measuring deformation and other reactions that could not be determined on actual highways in any mathematically precise manner.[32]

The plan represented an admirable attempt to bridge the gap between the artificiality of the laboratory and the real world, but it proved impossible to implement. Initially, problems with the truck-mounted

Fig. 4.—Apparatus from the BPR investigations of 1924–33. The accelerometer shown here determined the force of impact, while another piece of equipment graphed spring movement; together, they produced a record of the acceleration of the wheels into the pavement. (From *Public Roads* 7 [June 1926]: 71.)

[32]Teller, "Impact Tests on Concrete Pavement Slabs" (n. 31 above), pp. 1–14; J. T. Thompson, "Static Load Tests on Pavement Slabs," *Public Roads* 5 (November 1924): 1–6; C. A. Hogentogler, "Status of the Motor Truck Impact Tests of the Bureau of Public Roads," *Public Roads* 5:11–16; *BPR Annual Report* (n. 19 above), 1924, p. 42.

accelerometers delayed the recording of existing highway surfaces until 1927. Then the researchers struggled for two years, with the help of the National Bureau of Standards, to calibrate the new impact machine. Altogether, eight years were devoted to this experiment, before the *BPR Annual Report* finally announced that its purpose of measuring impact had been fulfilled. The report added that "attention is now turning to the equally important question of the effect on road surfaces of the suddenly-applied forces of motor vehicle impact." Yet this had been the stated purpose of the tests launched in 1924! A proclamation of success could not disguise the complete failure of the attempt to recreate at Arlington the impact forces measured on real roads.[33]

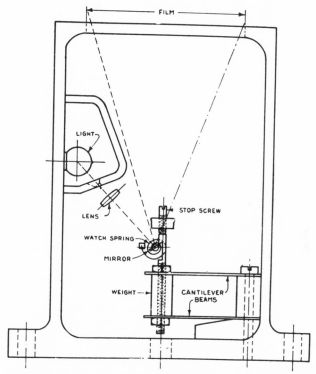

Fig. 5.—Diagram of beam accelerometer for 1924–33 tests. Vibration of the cantilever bearing, recorded on film, enabled BPR engineers to measure impact forces. (From *Public Roads* 5 [December 1924]: 1.)

[33]James A. Buchanan and J. W. Reid, "Motor Truck Impact as Affected by Tires, Other Truck Factors, and Road Roughness," *Public Roads* 7 (June 1926): 69–82; "Motor Truck Impact Tests Now in Progress," *Public Roads* 7 (January 1927): 231; "New

As with the 1921 impact investigation, what lay behind the problems of the 1924–32 tests was the concern for acquiring quantitative data and expressing results in a mathematical form. Bureau engineers had recognized that the 1921–23 research aided very few highway designers because of the narrow range of experimental conditions. But their attempt to correct that deficiency while still maintaining an emphasis on the collection of quantitative data led them beyond the technical capabilities of the day. They all but lost sight of the problem that had justified the project in the first place, getting caught up instead in an intriguing challenge to their ingenuity.

Nor did bureau engineers surrender their quest despite a decade of problems. In 1933 they announced yet a fourth sequence of tests. This time, they abandoned attempts to bring real conditions into the experi-

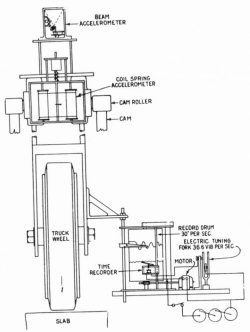

Fig. 6.—Diagram of entire accelerometer experiment, 1924–33. (From *Public Roads* 5 [December 1924]: 4.)

Research Projects Initiated by Bureau of Public Roads," *Public Roads* 8 (August 1928): 124; J. A. Buchanan and G. P. St. Clair, "Calibrations of Accelerometers for Use in Motor Truck Impact Tests," *Public Roads* 11 (July 1930): 81–109; J. A. Buchanan, "Interrelated Effects of Load, Speed, Tires, and Road Roughness on Motor Truck Impact," *Public Roads* 11 (September 1930): 139–52; *BPR Annual Report* (n. 19 above), 1929, p. 48; 1930, p. 48; 1932, p. 40; 1933, p. 42.

ment station and moved entirely into the laboratory. Instead of using an actual road surface, the investigators built a pendulum device to apply impact to a small cantilevered concrete slab (fig. 8). A wheel mounted on the end of the pendulum was equipped with accelerometers to record the impact, while strain gauges on the slab measured the concrete specimen's reaction. The problems of the earlier tests reappeared, for equipment bugs persisted until 1936. Not until the following year could the BPR report, "The work is necessarily slow and painstaking. With the special testing equipment designed and built for this research many thousands of observations have been made."[34]

But that was the last word published on those tests, and the reasons are obvious. The design of this experiment created the most artificial environment yet. Ignoring the cushioning effect of the soil under a slab, the apparatus instead supplied the force to a completely unsupported specimen no more than 3 feet high and of a shape entirely different from a pavement slab. Equally important, application of the force by a pendulum meant ignoring the fact that impact forces had two separate components, sprung and unsprung weights. Finally, this apparatus could not simulate the impact of dual tires, an increasingly common arrangement on trucks by the mid-1930s, nor could it apply force at the edges. All was subordinated to measuring the force of

Fig. 7.—Impact testing machine developed for 1924–33 experiments to replace testing tower. (From "Approaches Used in Highway Research Projects in the United States," *Bulletin of the National Research Council* 6, pt. 4 [no. 35]: 39.)

[34]L. W. Teller, "A Machine for Impact and Sustained Load Tests of Concrete," *Public Roads* 18 (December 1937): 185–94; *BPR Annual Report* (n. 19 above), 1934, pp. 58–59; 1935, p. 57; 1936, pp. 65–66; 1937, p. 68.

330   *Bruce E. Seely*

impact, and, as a result, the tests failed to describe accurately the behavior of highway pavements in actual use. Despite twenty years of research, the BPR had offered little guidance to highway designers attempting to build roads that could stand the impact from heavy trucks.

One might dismiss the BPR's impact investigations as examples of poor research management, and superficially the problem resembled one faced by directors of corporate research and development labora-

Fig. 8.—Some BPR technicians placing a test sample of pavement in the pendulum impact testing machine, used after 1933. (From *Public Roads* 18 [December 1937]: cover.)

tories: How can scientists intrigued with the intricate puzzles of nature be kept working on paths that lead to useful information? Yet few promoters of highway research perceived that this was a question, and no one recognized a problem![35] Indeed, because of the widespread acceptance of the formula that any scientific research eventually produced gains in technical capability, many engineers regarded the impact studies as a model for emulation. T. D. Mylrea, who was head of the Department of Civil Engineering at the University of Delaware and witnessed a test of the pendulum equipment in 1938, wrote to the BPR afterward: "The care with which these tests are being conducted has impressed me very much, and I have come to the conclusion that it would do our Senior students a great deal of good to witness this test procedure."[36]

Mylrea's reaction was understandable, because the last impact test conformed exactly to recommendations made by the director of the Highway Research Board in 1935. Roy W. Crum argued for more research, by which he meant "scientific study rather than experimentation. Much of our present day practice is based upon the experimental method applied in the field, but from now on we must go much deeper into fundamentals."[37] But this produced a curious situation, for a project that must be judged a failure instead emerged as a model for others to follow.

Subgrade Research and Soil Mechanics

The prolonged tests of vehicle impact dramatically illustrated that a scientific research method—the emphasis on quantitative data—failed to aid highway designers, even while older empirical methods reshaped highway construction practices. Bureau engineers also adopted another component of the new methodology, a concern for developing theory. Although none of the resulting projects produced the disastrous results of the impact investigation, there were indications that this shift precipitated problems too. One example was the bureau's celebrated research into soil foundations in the 1920s and 1930s which led to the development of soil mechanics.[38] As with the

[35]On this subject, see Stuart W. Leslie, "Thomas Midgley and the Politics of Industrial Research," *Business History Review* 34 (Winter 1980): 480–503. One of the few to perceive a difficulty was Charles M. Upham; see "Report of the Director," *Proceedings of the Seventh Annual Meeting of the Highway Research Board*, pt. 1 (Washington, D.C., 1927), p. 36.

[36]T. D. Mylrea to Bureau of Public Roads, December 8, 1938, File 890 Tests, Bridge Impact, Classified Central File, 1912–50, Records of the BPR.

[37]Roy W. Crum, "Highway Research," *Engineering News-Record* 114 (January 17, 1935): 75.

[38]The leading soil researcher, Charles Terzaghi, noted in a 1925 paper that the BPR's

impact tests, this project manifested the commitment to fundamental research. As the chief of research explained in 1921, "The day is past for building roads of designs arrived at by rule-of-thumb. Our future designs [of foundations] must be based on sound, scientific, fundamental data."[39] But developing theories of soil behavior created new difficulties for BPR researchers.

Soil behavior was a largely unknown subject before World War I. Early in the 20th century, the Department of Agriculture's Bureau of Soils had started classifying soils, and a very small number of Europeans had begun to consider soil characteristics. But the obstacle to an engineering understanding of this subject lay in the enormous variety of soil conditions; no road builder could expect to encounter uniformity for any distance. This variation had frustrated the few early attempts to develop theoretical schemes for dealing with soils. Only after BPR field studies in 1918 identified clay soils as contributing to the highway failures did federal highway engineers begin a large-scale probe into soil foundation properties.[40]

From the beginning, the concern for fundamental research dominated the BPR soil investigation. One researcher explained early, "At the present time we are lacking in far too many items of cold scientific knowledge," and the project envisioned filling the lacunae before turning to the development of highway design theories.[41] The bureau began by developing laboratory tests to measure soil qualities, a process that required fifteen months. Unfortunately, attempts to determine meaningful values for the properties measured by the tests, such as the water content, clay content, shrinkage, and vertical movement of water, did not proceed at the same brisk pace. The researchers simply had no idea what values were acceptable. Did a clay sample with a 20 percent water content, say, spell future problems? The only solution was to correlate the laboratory values with measurements of identical

work represented "one of the prominent landmarks in the Field of Civil Engineering" ("Concrete Roads—a Problem in Foundation Engineering," in Boston Society of Civil Engineers, *Contributions to Soil Mechanics, 1925–1940* [Boston, 1940], p. 61).

[39] A. T. Goldbeck, "Investigations of Road Subgrades," *Society of Automotive Engineers Journal* 8 (April 1921): 346.

[40] See C. A. Hogentogler, *Engineering Properties of Soils* (New York, 1937); Karl Terzaghi, *From Theory to Practice in Soil Mechanics* (New York, 1960), pp. 4–5, 59–60, 62–63; *BPR Annual Report* (n. 19 above), 1920–38; and *America's Highways* (n. 4 above), pp. 324–25.

[41] J. L. Harrison, "Water and the Subgrade: Subgrade Investigations Begun by the Bureau of Public Roads," *Public Roads* 2, no. 24 (April 1920): 29; quotation from *BPR Annual Report* (n. 19 above), 1926, p. 33.

soil under roads that had failed and others that had not. These values, called soil constants, were the focus of the BPR's research.[42]

The long, tedious effort to connect laboratory and reality apparently encouraged BPR researchers to model their purpose, as well as their methods, on those of basic research scientists. The BPR reported in the mid-1920s that "those who are working in this important field are urged less by the hope of finding a final solution than by the passion of the scientists for facts."[43] And a breakthrough did not appear until 1925, when immigrant German scientist Charles Terzaghi introduced a theory of soil consolidation that enabled the BPR to develop a soil classification system (fig. 9). While investigators made progress in the identification of constants for moisture retention and flow characteristics of soil after Terzaghi began to cooperate with the BPR in 1926, not until the late 1920s did researchers tentatively address the question of modifying soil behavior, or soil mechanics. By 1931, the BPR was assisting state highway departments in establishing soil laboratories using equipment and methods developed by federal investigators. Even so, researchers continued wrestling with the difficulties of laboratory values, which may explain why the bureau did not bring soil mechanics to the general attention of the National Research Council's Highway Research Board until 1936 and 1937. And only after World War II did soil mechanics become an integral part of American highway construction practice.[44]

This record may read like an endorsement of the fundamental

[42]A. T. Goldbeck and F. H. Jackson, "Tests for Subgrade Soils," *Public Roads* 4, no. 3 (July 1921): 15–21; J. L. Harrison, "The Effect of Soil Moisture on Highway Design," *Municipal and County Engineering* 62 (April 1922): 122–25; A. C. Rose, "Practical Field Tests of Subgrade Soils, *Public Roads* 5 (August 1924): 10–15; J. R. Boyd, "Procedure for Testing Subgrade Soils," *Public Roads* 6 (April 1925): 93–101; idem, "The Present Status of Subgrade Studies," *Public Roads* 6 (September 1925): 137–62.

[43]*BPR Annual Report* (n. 19 above), 1926, p. 33.

[44]Charles Terzaghi, *Principles of Soil Mechanics* (New York, 1926); idem, "Simplified Soil Tests for Subgrades and Their Physical Significance," *Public Roads* 7 (October 1926): 153–63, 170; idem, "Principles of Final Soil Classification," *Public Roads* 8 (May 1927): 41–53; C. A. Hogentogler, Charles Terzaghi, and A. M. Wintermeyer, "Present Status of Subgrade Soil Testing," *Public Roads* 9 (March 1928): 1–8; correspondence for 1920–37 in File 407 General, Classified Central File, 1912–50, Records of the BPR; C. A. Hogentogler, A. M. Wintermeyer, and E. A. Willis, "Subgrade Soil Constants, Their Significance, and Their Applicability in Practice," *Public Roads* 12 (June 1931): 89–108; *BPR Annual Report* (n. 19 above), 1932, p. 44; National Academy of Sciences, Highway Research Board, *Proceedings of the Annual Meeting*, vols. 15–25 (1935–45); "Use of Soil Studies in Road Work Emphasized at Research Meeting," *Engineering News-Record* 117 (December 3, 1936): 791–93; "Road Research at High Level," *Engineering News-Record* (December 9, 1937): 918; and Clarkson H. Ogelsby, *Highway Engineering*, 3d ed. (New York, 1975), pp. 454–59.

research process, for the slow acquisition of knowledge eventually did lead to theoretical explanations. But there are flaws in this success story related directly to the BPR researchers' emulation of scientists. A goal of the BPR soil research effort had been the replacement of empirical rules and judgment based on experience with the scientific precision of laboratory tests—as evidenced by devoting the entire attention of the program from 1922 until 1935 to determining laboratory constants. Yet the gains offered to highway designers by this strategy were not always obvious; laboratory tests were exceedingly uncertain, and, as several engineers noted in 1933, "Examination of failures had probably done more than anything else to improve methods of design."[45]

Specific criticism came from a few engineers who felt that the entire emphasis on soil constants was incorrect. D. P. Krynine, a noted Russian soil researcher who emigrated and joined the Yale faculty in the late 1920s, complained that the emphasis on soil constants produced too many complex mathematical theories that had not been tested

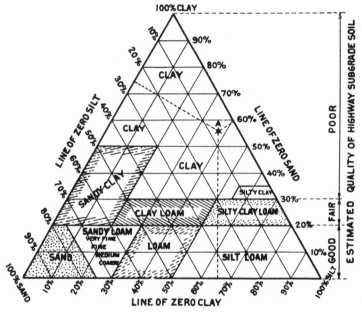

FIG. 9.—Trilinear soil classification scheme developed by the BPR in the 1920s. (From *Public Roads* 5 [August 1924]: 13.)

[45]See *BPR Annual Reports* (n. 19 above), 1920–38; quotation from paper by R. G. C. Baston, reported in "Public Works, Roads and Transport Congress," *Engineering* 136 (November 24, 1933): 569; see also "Doubtful Value of Laboratory Work," *Chemical Age* (London) 29 (November 18, 1933): 459.

against "sound engineering judgment based on experience which cannot be replaced by theories." Krynine urged the preparation of a simple manual containing direct applications of research for engineers in the field.[46] Two British engineers had more fundamental doubts, however, noting in a 1949 text that "many of these [soil] constants bear no simple relation to the physical and mechanical properties which engineers encounter when dealing with solid materials. The difficulty of understanding the significance of such tests tended to make many engineers regard soil mechanics with disfavour." Soil researchers slowly came to the same realization, for the British engineers reported a change in focus in their 1953 revised text. They pointed out that the difficulty in obtaining representative samples for testing "had encouraged a trend towards testing *in situ* instead of in a laboratory."[47]

Significantly, Charles Terzaghi, the father of soil mechanics, also ridiculed the optimistic emphasis on laboratory tests in a 1936 paper that blamed the problem on engineering education:

> The major part of the college training of civil engineers consists in the absorption of the laws and rules which apply to relatively simple and well-defined materials, such as steel or concrete. This type of education breeds the illusion that everything connected with engineering should and can be computed on the basis of *a priori* assumptions. As a consequence, engineers imagined that the future science of foundations would consist in carrying out the following program: Drill a hole in the ground. Send the soil samples obtained from the hole through a laboratory with standardized apparatus served by conscientious human automatons. Collect the figures, introduce them into the equations and compute the result. . . .

Terzaghi's conclusion was that, unfortunately, "no such formulas can possibly be obtained except by ignoring a considerable number of vital factors."[48]

Terzaghi repeatedly stressed that soils are too complex for theoretical and mathematical treatment. He advocated semiempirical methods based on the clinical approach of medicine, utilizing tests only as a tool. That remained the approach; as a 1958 civil engineering text cautioned students, although soil mechanics is an essential tool for

[46]"Progress in Highway Research Reported at Baltimore Meeting," *Engineering News-Record* 127 (December 18, 1941): 887–88. On Krynine, see Winifield Scott Downs, ed., *Who's Who in Engineering*, 3d ed. (New York, 1931), p. 747.

[47]P. Leonard Capper and W. Fisher Cassie, *The Mechanics of Engineering Soils* (New York, 1949), p. 26; ibid., 2d rev. ed. (1953), p. 27.

[48]Quoted in Karl Terzaghi, *From Theory to Practice* (n. 40 above), p. 63.

engineers, "the fact remains that they practice an art that is not wholly based on science. Therefore the soundness of many of their decisions must depend largely upon experiences and good judgment."[49] Since the BPR goal of replacing experience by laboratory tests could not work in soil mechanics, it is not surprising that its system required extensive modification before it could be adopted. Viewed as engineering, then, the soil research had a fundamental flaw. Hardy Cross, a noted civil engineering educator, pointed out the basis for this flaw when he wrote: "Experience is a guide which may be miscellaneous, fragmentary, unsatisfactory, often second-hand, frequently inaccurate, but no engineer will discount its tremendous importance as evidence.

"All nature is trying to tell something of how its forces act. . . . Men may try to duplicate her phenomenon in a laboratory, but we never exactly reproduce the true natural problem, never fully ferret out her secrets."[50]

Other Scientific Investigations in the BPR

The vehicle impact and subgrade studies were only two of the BPR's research efforts during the period between the world wars. Its investigators also examined numerous other questions, in many instances following the same pattern of borrowing scientific research methods. Special attention was devoted to developing rational, that is, mathematical, methods of bridge design, with the most notable experiment involving the destructive testing of a full-size concrete arch bridge in North Carolina. Investigators had the opportunity to observe how closely theoretically predicted performance matched actual load-bearing ability. Other experimenters developed a mathematical analysis for banked curves and trigonometric formulations for designing ideal intersections. The BPR also underwrote the first theoretical analysis of the stresses in concrete slabs by H. H. Westergaard at the University of Illinois.[51]

---

[49]Robert W. Abbett, *American Civil Engineering Practice*, vol. 1 (New York, 1958), p. 7-02; also, Glennon Gilboy, "Soil Mechanics Research," *Proceedings of the American Society of Civil Engineers* 55 (October 1931): 1165.

[50]Hardy Cross, *Engineers and Ivory Towers*, ed. Robert C. Goodpasture (New York, 1952), p. 105.

[51]On bridge research, see correspondence for 1925–29 in File 810.2 North Carolina, Classified Central File, 1912–50, Records of the BPR; *BPR Annual Report* (n. 19 above), 1925, p. 30; 1927, p. 35; 1928, pp. 41–42; Albin L. Gemeny and W. F. Hunter, "Loading Tests on a Reinforced Concrete Arch," *Public Roads* 9 (December 1928): 185–208; Clyde T. Morris, "Yadkin River Bridge Tests," *Proceedings of the American Society of Civil En-*

Many of these projects reached more successful conclusions than the two programs discussed above. In some cases this may have stemmed from the dimensions of the topic. To guide the researchers in the bridge studies, for instance, there was the nature of bridges as artificial structures as well as a century of work in bridge design theory. Hence not every attempt to borrow scientific methods failed. But neither should the problems encountered by the impact and subgrade researchers be classed as unfortunate exceptions to the rule. This becomes apparent by comparison of those efforts with four of the most important highway investigations conducted in the late 1940s and 1950s. Significantly, all of these were test roads. The Hybla Valley Road Test (1944–54) at Arlington, Virginia, specifically studied asphalt pavements, while three others—Road Test One in Maryland (1949), the WASHO Road Test in Idaho (1952–54), and the AASHO Road Test in Illinois (1958–60)—considered a broader range of materials. All copied the Bates Test Road in testing a variety of sample pavements under different soil conditions with real trucks of varying weight. Significantly, every program produced information immediately useful to designers. And the reason for this outcome had not changed since the Bates Test in 1922—the experiments had as their goal identification of the best pavements and construction methods by considering the interaction rather than the separation of variables.[52]

The bureau's return to this method did not indicate a total shift in its research work. In fact, the focus on fundamental research remained very much alive—in 1967, almost $4.5 million of the $30.6 million for

---

gineers 55 (March 1929): 608–17; on other projects, see T. F. Hickerson, "Laying Out Circular Curves by Deflections from the P.I.," *Public Roads* 3, no. 20 (October 1920): 13–18; A. L. Luedke and J. L. Harrison, "Superelevation and Easement as Applied to Highway Curves," *Public Roads* 3, no. 31 (November 1920): 3–12; Joseph Barnett, *Transition Curves for Highways* (Washington, D.C., 1938); "Methods of Computing the Intersection of a Line with a Spiral and Any Curves Parallel to the Spiral," *Public Roads* 22 (May 1941): 63–66; H. M. Westergaard, "Stresses in Concrete Pavements Computed by Theoretical Analysis," *Public Roads* 7 (April 1926): 25–35; and idem, "Mechanics of Progressive Cracking in Concrete Pavements," *Public Roads* 10 (June 1928): 65–71.

[52]On the various road tests, see National Academy of Sciences—National Research Council, Highway Research Board, *Road Test One–Md.*, Special Report no. 4 (1952); idem, *The WASHO Road Test*, pt. 1, Special Report no. 18 (1954), and pt. 2, Special Report no. 22 (1955); and idem, *The AASHO Road Test: History and Description of Project*, Special Report no. 61A (1961). See also Fred Burggraf, "Full-Scale Road Tests Produce Data for Sound Highway Development," *Roads and Engineering Construction* 92 (December 1954): 117–18; Burggraf and W. B. McKendrick, Jr., "AASHO Road Test," *Civil Engineering* 30 (January 1960): 3–5; U.S. Department of Commerce, Bureau of Public Roads, *Annual Report of the Chief of the Bureau of Public Roads* (Washington, D.C., 1947–59); and *America's Highways* (n. 4 above), pp. 333–35.

highway research was spent on theoretical investigations.[53] Yet the fact is that, as the BPR geared up for the interstate program in the immediate postwar period, it returned to observation-oriented full-scale field tests to guide its construction work. This indicates that scientific research efforts failed to provide the rational design method initially envisioned after the First World War.

Conclusion

The Bureau of Public Roads' discouraging experience in highway research from 1918 to 1940 leaves at least two related questions to be answered. Why were federal engineers so much more eager to adopt scientific, quantitative methods than state investigators? And why did they retain those methods despite apparent limitations in their utility for solving practical problems?

In addressing the first question, we must combine several elements. At first glance, it would appear that the BPR chose the only proper course. The highway failures of 1917–18 indicated that empirical research and construction approaches had failed too. The newer scientific methodology promised to rectify the problems. Just as the engineering leaders of the period did, most current engineers probably would judge the bureau's methods superior to the state projects precisely because of their quantitative focus and theoretical aspirations. But closer study shows that the older research techniques had not failed so much as construction practice had been undone by the heavy motor truck. The answers to the problems of 1918 could still be found by the older research style, as both the Illinois researchers and those in the 1950s demonstrated.

Nonetheless, the BPR chose to move in new directions, and several factors explain this choice. Above all other considerations was the public perception of science and scientists as the heroes of World War I. Engineers hoped that copying scientific techniques could connect engineering to the new enthusiasm. This attitude was plainly evident in the activities of Thomas H. MacDonald, chief of the Bureau of Public Roads from 1919 until 1953. MacDonald had long been a proponent of highway research, both as chief highway engineer in Iowa and later as head of the bureau. He had been a student of Anson Marston, the Iowa State pioneer in soil studies. Marston, in turn, was a Cornell graduate who had worked with Robert Thurston and passed on to MacDonald Thurston's emphasis on a scientific approach to engineering. Thus the

[53]U.S. Department of Transportation, Federal Highway Administration, *Highway Research and Development Studies Using Federal-Aid Research and Planning Funds: Fiscal Year 1968* (Washington, D.C., 1967).

postwar enthusiasm for research struck a responsive chord in the BPR's chief.[54]

The status of engineers was a vital consideration in MacDonald's views of research, as evidenced in a 1923 paper titled "The Objectives of Highway Research" in which he asked why engineers had not "been accorded more readily and more consistently leadership and the rewards, not necessarily material, which generally accrue to recognized authority." The answer, he felt, was "the lack of engineering research." He added that "if the [engineer] is making progress, if he is succeeding, if he finally proves beyond all doubt the ability of the engineer to master and thus lead in a new and major translation in our national life, it will be through research."[55] Not surprisingly, MacDonald played a leading role in the establishment and early operation of the Highway Research Board in the National Research Council's Division of Engineering. If based within the National Academy of Sciences in this way, MacDonald believed that highway engineers could share the prestige with which that body emerged from the war. Indeed, this viewpoint was shared by the engineers, such as Anson Marston, who had pushed for the creation of a Division of Engineering in the National Research Council.[56]

Status comparisons between engineers and scientists were not, however, the only explanation for MacDonald's embrace of scientific research methods; an additional factor was the prestige of the bureau itself. Long recognized as the pacesetter in highway engineering practice, it was incumbent on the BPR to remain on the cutting edge. Such a sentiment certainly motivated the federal investigator who urged in 1927 that local requests for research on practical matters such as testing materials be turned down on the grounds that "the energies of

[54]On MacDonald and Marston, see Gilkey (n. 24 above), pp. 13–15. Correspondence in the BPR files regarding the HRB shows that contact between the two men continued over the years. See BPR File 001.11, Classified Central File, 1912–50, Records of the BPR.

[55]Thomas H. MacDonald, "Objectives of Highway Research," paper presented to the Advisory Board for Highway Research, November 9, 1923, pp. 2–3 (copy in Thomas H. MacDonald Reports, Box 84-N, Bentley Library, University of Michigan, Ann Arbor).

[56]On MacDonald and the Highway Research Board, see Seely, "Highway Engineers" (n. 4 above), chap. 5; and National Academy of Sciences, Highway Research Board, *Ideas and Actions: A History of the Highway Research Board, 1920–1970* (Washington, D.C., 1971). On the motives of engineers in creating a Division of Engineering, see correspondence among MacDonald, Marston, and A. N. Talbot, August–December, 1919, File 001.11, Classified Central File, 1912–50, Records of the BPR; Anson Marston, "A National Program for Highway Research," File 1919, Marston Papers (n. 24 above); and Alfred D. Flinn, "Engineering Foundation: Division of Engineering of National Research Council: Their Origin, Work, Plans, Needs," *Transactions of the American Society of Civil Engineers* 86 (1923): 1286–91.

the Bureau, except in rather exceptional cases, should be expended on basic research."[57] This proposal would have reversed the entire justification for the BPR research effort—one that had stood for more than thirty years—and it never was adopted as official policy. Yet the fact that state highway engineers, in the words of the director of the Highway Research Board, "do not often have time for long range scientific inquiry," further encouraged the BPR to conduct fundamental research.[58]

The final component in the BPR's shift to scientific research was the criticism levied in the period 1918–21 regarding alleged inefficiency in the handling of the Federal-Aid System, inefficiency which the road failures seemed to prove. Again, the adulation of scientific methodology provided an apparent answer to a problem, for, by wrapping its research efforts in the cloak of fundamental research, BPR investigators believed they could prove their critics wrong.[59]

These factors explain why the BPR abandoned its full-scale field investigations as not scientific enough. But they do not account for the dogged adherence to the new methodology. That explanation lies in two other factors. First, the engineers were caught up in challenging problems requiring sophisticated equipment to gain data. In other words, they were having fun. But more important, every field of engineering was following the same steps which highway researchers took. Indeed, the history of many areas of engineering over the first fifty years of the 20th century, especially after World War I, records a scramble to adopt the powerful problem-solving tools of science and to emulate the spectacular successes earlier gained by electrical and chemical engineering.[60] In engineering education, James Kip Finch noted that "the engineering graduate of the early 1900s found the largely descriptive and rule-of-thumb instruction of pre–World War I days of little use in mastering the involved technology of later en-

[57]E. F. Kelley to A. T. Goldbeck, September 29, 1927, File 001.11, Classified Central File, 1912–50, Records of the BPR.
[58]R. W. Crum, "Highway Research in the United States," *Chemistry and Industry* 57 (July 23, 1938): 697.
[59]On the legislative wrangles and their resolution, see Seely, "Highway Engineers" (n. 4 above), chap. 2.
[60]For evidence of engineers encouraging scientific research, see Bohme et al. (n. 2 above), and a series of contemporary articles, including: F. C. Lea, "Science and Engineering," *Engineering* 128 (August 16, 1929): 188; Mervyn O'Gorman, "Bringing Science into the Road Traffic Problem," *Engineer* 159 (January 11, 1935): 53; A. A. Potter, "Scientific Research and Future Road and Street Building Programs," *Roads and Streets* 77 (March 1934): 94; Alfred D. Flinn, "Research Advances Civil Engineering," *Civil Engineering* 1 (October 1930): 15; F. M. Dawson, "The Role of Fundamental Research," *Journal of Engineering Education* 39 (December 1948): 195.

gineering practice." This situation led to greater emphasis on mathematics and physics in the classroom.[61] A. Hunter Dupree and Daniel J. Kevles have discussed the explosion of scientific research which saw the number of corporate laboratories increase from 300 in 1925 to 1,625 only a decade later. The nature of engineering was so clearly changing that in 1935 Earle B. Norris, the dean of engineering at Virginia Polytechnic Institute, approvingly explained, "The scientist is pretty much of an engineer and the engineer must be a fair sort of scientist."[62]

All of these changes reflected an assumption that quantification of data was the essence of science. Certainly they show actions in accordance with the sentiment of the mission chief scientist of the 1980 *Voyager* flyby of Saturn, who reiterated the assertion that "'until you have numbers, you don't have a science.'"[63] This definition of science is vastly inadequate; certainly Galileo, Faraday, and Darwin had not relied on mathematics. Yet, however we may judge that statement, BPR researchers assumed that the precision of mathematical expression was the hallmark of scientific research. And the current hierarchy of the sciences exhibits the same assumption, for it clearly places theoretical physics near the pinnacle because it is based, more than other branches, on mathematics. It should not be surprising that engineering has continued this movement toward scientific methodology, as evidenced not only by the adoption of a similar hierarchy but also by the continued placement of greater emphasis on mathematics in engineering education.[64]

There is no doubt that this adoption of scientific methods and movement toward science in other ways have proved essential to advances in many modern engineering fields. Still, the Bureau of Public Roads' experience in highway research demonstrates that there is nothing automatically beneficial about adopting the methods of science. The key point was well made by L. C. Laming, a mechanical

[61]Finch (n. 3 above), pp. 387–88; also idem, *Trends in Engineering Education: The Columbia Experience* (New York, 1948), pp. 5–6.

[62]Dupree (n. 3 above), pp. 323–27; Kevles (n. 3 above), pp. 117–38; and Earle B. Norris, "Research as Applied to Engineering," *Civil Engineering* 5 (May 1935): 408.

[63]Rick Gore, "Saturn: Riddle of the Rings," *National Geographic* 160 (July 1981): 24.

[64]Evidence of the growing mathematization of engineering can be found in those books promoting engineering as a career, e.g., Phillip Pollock, *Careers and Opportunities in Engineering* (New York, 1967), pp. 54–56; and in Finch, *Trends in Engineering Education* (n. 61 above). See also Eugene S. Ferguson, "The Mind's Eye: Nonverbal Thought in Technology," *Science* 197 (August 26, 1977): 827–36. Engineers today certainly are aware of this change; see, e.g., Henry D. Claybrook, "Readers Write," *Civil Engineering* 52 (May 1982): 26; and John Huston, "Stop the World—I Want to Get Off," *Civil Engineering* 51 (December 1981): 66–67.

engineer at the Imperial College of Science and Technology in London, who explained, "Until a graduate has the attitude that analysis is a tool for him to use, and the laws of science are the rules of the game, but not the game itself, we have failed to produce an engineer."[65] Federal highway engineers, for a variety of reasons, apparently lost sight of this and mistakenly believed their new tools were "the game itself."

The most disquieting lesson about the Bureau of Public Roads' experience, however, is that no federal researcher even recognized the problems that this attitude created for BPR research efforts. After all, not only were their actions in conformity with the recommendations of the leading figures in the field, but they also knew that everyone in highway engineering looked to them for technical leadership. The assumption shared by everyone, both then and now, was that scientific research inevitably led to improvements in practice. One expects to find difficulties of translation between science and engineering, but this author was unprepared for the unsettling results of the BPR's research program. When an initial hypothesis—that the BPR's problems stemmed from poor research management—proved unsatisfactory and gave way to the conclusion that the attempts to emulate scientists were at fault, there followed an identification with a character in a children's fairy tale. The character was the child who realized, even though everyone around him ostensibly did not notice, that the emperor had no clothes. This leads to the final question: If the most respected highway engineers in the interwar period, obviously no dupes, were unable to see through the mystique of science, how many others have also fallen into the same trap?

[65] L. C. Laming, "Engineering Science versus Technology?" *Engineering* 216 (May 1976): 352–54; see also Douglas Lewin, "The Relevance of Science to Engineering—a Reappraisal," *Radio and Electronic Engineer* 49 (March 1979): 119–124.

Institutions

DEFINING PROFESSIONAL BOUNDARIES: CHEMICAL ENGINEERING IN THE EARLY 20TH CENTURY

TERRY S. REYNOLDS

Several studies of the engineering profession in America have suggested that one of the critical elements in the process of professionalization has been the definition of boundaries and relationships between engineering disciplines and their border areas, particularly science and business.[1] The best and most comprehensive of these studies, Edwin T. Layton's *The Revolt of the Engineers: Social Responsibility and the American Engineering Profession*, argues that the tensions created in drawing the boundaries between engineering and science and engineering and business were central to the entire history of the engineering profession in the United States.[2] To support this thesis, Layton drew heavily on the evolution of the civil, mining, mechanical, and electrical engineering professions and on the histories of their professional societies.

Layton did not deal with chemical engineering, probably because chemical engineering was a very small profession during the period that formed the heart of his study, 1890–1930. Chemical engineering, however, deserves attention for several reasons. First, chemical engineering today is one of the "big four" engineering fields, along with civil, mechanical, and electrical engineering; and second, chemical engineering faced unique problems in delineating its boundaries with

TERRY REYNOLDS is professor of history and head of the Department of Social Sciences at Michigan Technological University. He has authored several books, including a history of water power and a history of the American Institute of Chemical Engineers. He is currently working with Bruce Seely on a history of American engineering education.

[1]E.g., Herbert A. Shepard, "Engineers as Marginal Men," *Journal of Engineering Education* 47 (1957): 536–42; George Strauss, "Professionalism and Occupational Associations," *Industrial Relations* 2, no. 3 (May 1963): 11, 22–29; David Noble, *America by Design: Science, Technology, and the Rise of Corporate Capitalism* (New York, 1977), pp. 3–66, passim; and Edwin T. Layton, Jr. *The Revolt of the Engineers: Social Responsibility and the American Engineering Profession* (Cleveland, 1971 [and Baltimore, 1986]).

[2]Layton, *Revolt of the Engineers*, esp. chap. 1.

This essay originally appeared in *Technology and Culture*, vol. 27, no. 4, October 1986.

its associated science, chemistry. Moreover, the unique nature of the problems faced in delimiting boundaries with chemistry led chemical engineers to create a new "engineering science" and to accept a much closer initial relationship to business than any of the major engineering disciplines with the possible exception of mining engineering.

In this article, I will explain why chemical engineering faced a more serious problem than the older engineering disciplines in marking out its frontiers with its associated science, why chemical engineers wished to create a professional identity independent of chemists, how they established themselves as independent of chemistry, and how the problems of delineating the frontiers between chemical engineering and chemistry led chemical engineering to closely embrace business management.

Chemical Engineering and Chemistry

The major engineering disciplines that professionalized before chemical engineering seem to have had only limited problems in defining their relationships with science. Mechanical and civil engineering, both of which evolved from craft traditions, went through a period in which practically trained and academically trained engineers argued over the emphasis that should be put on scientific or theoretical training.[3] But even in this period, the arguments were over degree of emphasis, not over the importance of at least some science. Moreover, before the end of the 19th century, both civil and mechanical engineers were able to make use of science as a tool, not only to improve engineering practice but also to distinguish their emerging professional fields from related, but lower-status, craft groups that performed or were capable of performing some of the same tasks as engineers. For civil engineers, these groups included surveyors, carpenters, and building contractors, as well as decorative-oriented architects.[4] For mechanical engineers, these groups included mechanics and machinists.[5] Thus, in

---

[3]E.g., see Daniel H. Calhoun, *The American Civil Engineer: Origins and Conflict* (Cambridge, Mass., 1960), pp. 137–38, 178–80; Monte A. Calvert, *The Mechanical Engineer in America, 1830–1910: Professional Cultures in Conflict* (Baltimore, 1967), pp. 63–106; Noble, *America by Design*, pp. 27–29.

[4]For the use of "science" by civil engineers to distinguish themselves from lower-status groups see, as examples, Thomas C. Clarke, "The Education of Civil Engineers," American Society of Civil Engineers, *Transactions* 3 (1874): 261, in conjunction with his "Science and Engineering," ibid. 35 (1896): 508–19; see also Calhoun, *American Civil Engineer*, pp. 57, 139, 178–79; William H. Wisely, *The American Civil Engineer, 1852–1974: The History, Traditions and Development of the American Society of Civil Engineers* (New York, 1974), pp. 298–301; and Richard M. Levy, "The Professionalization of American Architects and Civil Engineers, 1865–1917" (Ph.D. diss., University of California, Berkeley, 1980), pp. 87, 89, 291–92, and elsewhere.

[5]Calvert, *The Mechanical Engineer in America*, pp. 162–67, for instance.

the late 19th century the "applied" or "engineering" sciences that civil and mechanical engineers drew on or had developed had proved an aid, not a hindrance, to the professionalization process.

Electrical engineering's relationship to science was somewhat different. Electrical engineering had emerged more directly from an academic science (physics)—unlike civil and mechanical engineering, whose origins lay in craft traditions. But electrical engineering, like civil and mechanical engineering, did not face serious difficulties in delineating its relationship with physics. Physicists offered no more opposition to the emergence of electrical engineering as a distinct profession than they had to the emergence of civil or mechanical engineering, both of which had either created new branches of physics (thermodynamics) or had drawn from selected branches of physics (mechanics) in the process of profession formation. In the late 19th century the American physics community was quite small and largely imbued with the idea of pure science. Physicists either ignored or welcomed the emergence of electrical engineering as a distinct discipline; they did not openly resist it.[6]

The early relationship between chemical engineering and its science, however, was much more difficult. There were good reasons for this. Unlike the other major engineering disciplines, chemical engineering grew largely out of a well-established science, not a craft tradition or a professionally weak science. By 1900 chemistry was already a relatively large profession in America with somewhere between 5,000 and 10,000 practitioners and a tradition of industrial employment.[7]

Thus, many of the leaders of the American chemical community did not view the creation of a new and distinct engineering discipline deriving from their science with favor. At a 1908 meeting organized to discuss the need for a distinct society for chemical engineers, William McMurtrie, a former president of the American Chemical Society, declared: "'Chemist' is a good enough term for me, and if the chemists stand up for their names and their interest as well as the engineers do,

[6]Robert Rosenberg, "Test Men, Experts, Brother Engineers, and Members of the Fraternity: Whence the Early Electrical Work Force?" *IEEE Transactions on Education* E-27 (1984): 207–9, and "American Physics and the Origins of Electrical Engineering," *Physics Today*, October 1983, pp. 53–54.

[7]William McMurtrie, "The Condition, Prospects, and Future Educational Demands of the Chemical Industries," *Journal of the American Chemical Society* 22 (1901): 80, says more than 5,000 chemists were at work in the United States, more than 80% in industry; Arthur D. Little, "The Chemist and the Community," *Science* 25 (1907): 651, noted that the 1900 census listed 8,847 chemists in the United States. See also Arnold Thackray et al., *Chemistry in America, 1876–1976: Historical Indicators* (Dordrecht, 1985), pp. 16–18, 20–21, 108, and elsewhere.

you do not need any society to speak for you. Stick to what you have. I do not think we have any need to form any other organization. . . . Stick to the term 'chemist' and don't get away from it."[8]

At that same meeting, M. T. Bogert, the incumbent president of the American Chemical Society, also attempted to persuade chemical engineers not to create a separate disciplinary identity.[9]

Two factors reinforced the opposition of the leaders of the American Chemical Society to the creation of chemical engineering as a distinct discipline. First, the American Chemical Society was in the process of creating professional divisions and journals to serve the professional needs of its industrial members.[10] Second, scaling up chemical work from the laboratory to the industrial level, the forte of chemical engineers, did not necessarily require the creation of a separate discipline. Another model for the scaling-up process already existed, a model that did not threaten the professional unity of chemistry.

This latter point requires further explanation. The rapid growth of the American chemical industry in the late 19th and early 20th centuries had created a need for a means of bridging the gap between laboratory processes and full-scale chemical production. A model for doing this had been created late in the 19th century. In Germany, research chemists who developed new products or processes were themselves placed in charge of scaling the processes up to a semi-industrial level, with mechanical or civil engineers and other technical personnel put at their disposal to handle the engineering aspects of the projects. In other words, the German model called for retaining traditional disciplinary boundaries and the traditional division of labor between chemist and engineer and for relying on a team that included both chemist and engineer for scale up, instead of using chemical engineers, that is, individuals with special training in both chemistry

[8]McMurtrie's comments are contained in "Proceedings of a Meeting Held at Parlor A, Belmont Hotel, Saturday Evening, January 18, 1908, for the Purpose of Discussing the Advisability of Forming a Society of Chemical Engineers," privately printed (1908), quoted in F. J. Van Antwerpen and Sylvia Fourdrinier, *High Lights: The First Fifty Years of the American Institute of Chemical Engineers* (New York, 1958), pp. 19–20.

[9]Van Antwerpen and Fourdrinier, *High Lights*, pp. 12–13, 15, 17, 21. Dr. Hugo Schweitzer, another prominent American chemist, had stated in 1904 that he was "absolutely against" the introduction of chemical engineering in the education of chemists and argued for restricting chemists to pure chemistry. See J. B. F. Herreshoff, "The Chemical Engineer," *Electrochemical Industry* 2 (1904): 172–173 (comment of Schweitzer).

[10]Charles A. Browne and Mary E. Weeks, *A History of the American Chemical Society* (Washington, D.C., 1952), pp. 83–85, 89; Herman Skolnik and Kenneth M. Reese, *A Century of Chemistry: The Role of Chemists and the American Chemical Society* (Washington, D.C., 1976), pp. 12–15, 309.

and engineering.[11] Carl Duisberg, one of the creators of the German scale-up model, argued in 1896 that this approach was quite appropriate to American, as well as German, conditions. He pointed out that there was extensive experience in American manufacturing in the division of labor and that the German approach to handling scale-up problems with teams of chemists and engineers was but another application of this method of work organization.[12] The large number of prominent American chemists educated in Germany in the late 1800s meant, of course, that American chemists were quite familiar with the German model.[13]

In summary, the emergence of a distinct professional identity for chemical engineering in America was fraught with difficulties and by no means an inevitable consequence of the growth of large-scale chemical production. A separate identity for chemical engineering was opposed by prominent leaders in the American chemical community, leaders who preferred to see chemical engineering remain a branch of chemistry. American chemists involved in industrial work already had in theoretical chemistry a means of distinguishing themselves from factory hands in chemical plants and had, within the existing chemical profession, a strong disciplinary community with organizations and journals prepared to serve their professional needs. There was, in addition, an alternative model for solving the scale-up problems involved in translating a laboratory process to the plant level, a model that did not require the creation of a new profession.

Why, then, did chemical engineering emerge as a distinct profession in America? Three factors contributed: the nature of the American chemical industry around 1900, concern over the role and status of chemists within that industry, and the belief of some of these chemists

[11]Karl Schoenemann, "The Separate Development of Chemical Engineering in Germany," in *History of Chemical Engineering*, ed. William Furter (Washington, D.C., 1980), pp. 249–71; Jean-Claude Guédon, "Conceptual and Institutional Obstacles to the Emergence of Unit Operations in Europe," ibid., pp. 62–70; and Klaus Buchholz, "Verfahrenstechnik (Chemical Engineering)—Its Development, Present State and Structure," *Social Studies of Science* 9 (1979): 33–62.

[12]Guédon, "Conceptual and Institutional Obstacles," p. 67, who cites Carl Duisberg, "The Education of Chemists," *Journal of the Society of Chemical Industry*, July 1931, pp. 173–74. For an American chemist who seems to have favored adopting the German model, see M. T. Bogert, "The Chemical Engineer," *Electrochemical Industry* 2 (1904): 173; see also W. D. Richardson, "The Usefulness of Chemistry in the Industries," *Science* 27 (1908): 805, who noted that it might ordinarily be necessary to have the chemical engineer "embodied" in two individuals—one a chemist, the other an engineer.

[13]Thackray et al., *Chemistry in America*, pp. 189–91, indicates that in the period between 1876 and 1905 over half of the presidents of the American Chemical Society had been trained in Germany, most at Göttingen or Leipzig.

that border engineering disciplines posed a threat to their perceived role.

The American chemical industry around 1900 differed from its German counterpart in several respects. The American industry focused on bulk production of a few simple, widely used chemicals, such as sulfuric acid, alkali, and phosphates. The chemical processes involved in the production of these chemicals were generally well known and uncomplicated. The mechanical elements of these processes, on the other hand, were complex, especially since American mass production required handling huge volumes of raw materials, intermediate chemicals, and final products. The German chemical industry, by contrast, focused on the production of much smaller quantities of much more complex and costly chemicals, particularly dyes and pharmaceuticals. Production machinery and mechanical processing were thus less important, and the chemistry of the production processes more important, in Germany than in America.[14] These conditions made separating chemical engineering from chemistry easier in America than it would have been in Germany.

The status anxieties of those American industrial chemists who were involved in designing, constructing, and managing full-scale chemical production processes and their perception of external threats to their positions, however, also contributed significantly to the birth of chemical engineering in America. An explanation of these phenomena requires a review of the role and status of chemists in American industry around 1900.

By 1900, chemists were being employed in growing numbers in American chemical and metallurgical works.[15] Generally, these industrial chemists fell into three groups: analytical chemists, research chemists, and, for lack of a better term, production chemists. Let us briefly look at each of these three groups.

Analytical chemists worked almost entirely in laboratories, where they carried out tests of raw materials coming into the plant, the

[14]L. F. Haber, *The Chemical Industry 1900–1930* (Oxford, 1971), pp. 14–17, 26–29, 61, 108–34, 173–84; Gaston Du Bois, "The Development of Our Chemical Industries," *The Chemical Engineer* 20 (1914): 244–45; "National Faults in Manufacturing," *Metallurgical and Chemical Engineering* 9 (1911): 5; "Prof. F. Haber on Electrochemistry in the United States," *Electrochemical Industry* 1 (1902–3): 350; B. Rassow, "What Did We Chemists Learn in America?" *Industrial and Engineering Chemistry* 6 (1914): 33–35.

[15]William H. Nichols, "The Efficiency and Deficiencies of the College-trained Chemist When Tested in the Technical Field," *Industrial and Engineering Chemistry* 1 (1909): 103; Arthur Comey, "Certain Phases of Technical Chemical Research," *The Chemical Engineer* 15 (1912): 109; McMurtrie, "Condition, Prospects, Educational Demands" (n. 7 above), pp. 79–80.

intermediate compounds produced, and the final products sold by the plant. Their primary responsibility is best described as quality control. Of the three groups, analytical chemists were by far the most numerous. For example, a 1907 survey of University of Illinois graduates in chemistry indicates that, of those employed in industrial positions, around 70 percent were either clearly or probably employed in analytical laboratories.[16] Charles Burgess, an early promoter of chemical engineering education at the University of Wisconsin and later a chemical manufacturer, commented in 1911 that "fully 90 per cent" of college graduates who entered industrial chemistry started in analytical chemistry.[17]

Research chemists, the second type of chemist working in industry, had job responsibilities which, like those of analytical chemists, were laboratory focused. But they were concerned with the development of new chemical products and reactions, not routine quality-control tests. Research chemists were the focus of industrial chemistry in Germany, where their success in producing a variety of new synthetic dyestuffs had made them nearly indispensable and given them, and other chemists indirectly, considerable status.[18] In the United States, however, industrial research chemists were rare before 1910 and had not made a significant impact on the American chemical industry.[19] Widespread use of chemists in American industrial research laboratories was largely a post–World War I phenomenon.

The impetus for the creation of a distinct identity for chemical engineering came from the concerns of the third major group of

[16]University of Illinois, Department of Chemistry, "The Study of Chemistry at the University of Illinois, Urbana, Illinois, 1907" (Urbana, 1907), pp. 27–28. Of the 149 graduates surveyed, around fifty were no longer practicing chemistry. They were either working in non-chemical-related businesses, working as physicians, working as pharmacists, or deceased. I classified as probably engaged in analytical work the following classes: "railroad laboratories," "packing companies," "metallurgy/metallurgical chemistry," "fuel testing/research," and "analytical chemists or chemical experts" (the single biggest category). My thanks to P. Thomas Carroll of Rensselaer for providing me with this information.

[17]Charles F. Burgess, "The Efficiency of College Training of Men for the Chemical Industries," *Journal of the American Chemical Society* 33 (1911): 611.

[18]A British Chemical Mission to Germany after World War I reported that in German chemical works it was "usual for the chemist rather than the engineer to have the determining voice" (cited in William J. Hale, "The Immediate Needs of Chemistry in America," *Industrial and Engineering Chemistry* 13 [1921]: 463); and Victor Cambon, a French observer of the German chemical industry reporting in 1908, compared the research chemists in one German plant to popes set within sanctuaries (quoted in Guédon, "Conceptual and Institutional Obstacles" [n. 11 above], p. 64).

[19]I know of no hard quantitative data on the number of chemists involved in industrial research in the United States ca. 1900–10. But statements from a number of contempor-

chemists employed by American industry in the early 20th century—production chemists. Production chemists were made up of two subgroups. One subgroup consisted primarily of chemists employed as plant superintendents and plant managers. Their primary responsibility was production. They supervised the operation of chemical plants, sometimes designed and constructed new apparatus for the plants in their charge or new plants, and attempted to improve the operation and efficiency of their systems. The second subgroup consisted of chemical consultants involved in chemical plant design, construction, and troubleshooting. Most members of this subgroup had had previous experience in midlevel chemical management positions involving supervision of chemical production or chemical plant construction before setting up consulting practice.

The focus of the concerns felt by both subgroups of production chemists was the low regard in which manufacturers held analytical chemists, by far the largest of the groups of chemists working in American industry. For instance, Charles Burgess, a Wisconsin chemical engineer, complained in 1905 about a manufacturer who did not distinguish between a chemistry graduate, a former office boy, and a man recently taken from a labor gang, all of whom he had working at similar tasks in his analytical laboratory.[20] William H. Walker, an MIT chemical engineer and an early contributor to the development of chemical engineering science, similarly noted that many industrial chemists were "little more than testing machines; men who possess the ability to do nothing more than the most strictly routine analysis."[21] And W. D. Richardson, chief chemist for Swift & Company, observed in 1908 that chemical analysis had come to be regarded as "a more or

---

ary sources indicate that chemists involved in industrial research were relatively recent in America and that American manufacturers generally did not yet know how to use them properly. See, for example, "Prof. F. Haber on Electrochemistry in the United States" (n. 14 above), p. 350; Robert Duncan, "Excerpts from Addresses on Certain Problems Connected with the Present-Day Relation between Chemistry and Manufacture in America," *The Chemical Engineer* 13 (1911): 90–91; L. H. Baekeland, "Science and Industry," *Metallurgical and Chemical Engineering* 8 (1910): 336–37, and "What's the Matter with the American Chemist," *The Chemical Engineer* 18 (1913): 129–31; Harvey J. Skinner, "The Debt of the Manufacturer to the Chemist," *The Chemical Engineer* 14 (1911): 307; and C. F. Burgess, "Research in Chemical Industry," *Metallurgical and Chemical Engineering* 13 (1915): 921. See also Edward H. Beardsley, *The Rise of the American Chemistry Profession*, University of Florida Monographs, Social Sciences, no. 23 (Gainesville, 1964), p. 65.

[20]Burgess, "Efficiency of College Training" (n. 17 above), pp. 612–13.

[21]William H. Walker, "Some Present Problems in Technical Chemistry," *Electrochemical and Metallurgical Industry* 3 (1905): 61.

less perfunctory and disagreeable duty" which could often be taught to boys of no special education.[22]

Other chemists similarly believed that analytical chemists were held in low regard. Archibald Craig, with Ledoux and Company of New York, noted that analytical chemists were regarded as being of the same grade as machinists, draftsmen, and cooks.[23] George Auchy, who had experience working in the metallurgical industry, asserted earlier that "employers were almost unanimous" in believing that chemical analysis was "quick and easy work."[24] And numerous other chemists complained that manufacturers did not recognize that chemists were more than just laboratory animals.[25]

The low wages paid to analytical chemists in the early 1900s further suggest that they were not held in high regard by manufacturers. The salary of a skilled artisan in 1905 was about $1,200 annually. An academically educated analytical chemist could expect to start at a much lower salary, slightly over $700 per year, with increases, according to one commentator, "not very frequent."[26]

To make matters worse, the status of the analytical chemist, already low, seemed to be in danger of dropping even lower as a result of new

---

[22]W. D. Richardson, "The Improvement of Analytical Processes," *Industrial and Engineering Chemistry* 2 (1910): 99.

[23]Archibald Craig, "Status of the Analytical Chemist," *Chemical Age* 28 (1920): 474.

[24]George Auchy, "An American Institute of Chemistry," *Industrial and Engineering Chemistry* 1 (1909): 65.

[25]"Why Not 'The American Society of Chemical Engineers'?" *The Chemical Engineer* 2 (1905): 392; Baekeland, "Science and Industry," p. 336; Oskar Nagel, *The Mechanical Appliances of the Chemical and Metallurgical Industries* (New York, 1908), p. 301; Walker, "Some Present Problems in Technical Chemistry" (n. 21 above); Alex Silverman, "The Chemist and the Glass Manufacturer," *The Chemical Engineer* 12 (1910): 130–31; and "Applied Chemistry," *Metallurgical and Chemical Engineering* 13 (1915): 677. In "Factory or Laboratory," *The Chemical Engineer* 14 (1911): 349–50, the editor advised young chemical graduates that if they had to choose between employment in a laboratory or as a common laborer in a chemical plant to choose the latter. The author argued that it was "infinitely easier" to move into positions of responsibility by starting as a laborer than by starting in the laboratory because of the tendency on the part of manufacturers and supervisory people to regard laboratory help as impractical in managing plant operations. See also Burgess, "Efficiency of College Training" (n. 17 above), pp. 611–18, and Rollin G. Myers, "A Plea for a Chemists' Protective Association," *Industrial and Engineering Chemistry* 7 (1915): 798–99.

[26]"Are Analytical Chemists Poorly Paid?" *The Chemical Engineer* 4 (1906): 85–86. For additional material on the wages of analytical chemists, see H. B. Pulsifer, "Hygiene in Lead Smelting," ibid. 20 (1914): 65; Myers, "Plea"; Burgess, "Efficiency of College Training," p. 611; Duncan, "Excerpts" (n. 19 above), pp. 90–91; and Skinner, "Debt of the Manufacturer" (n. 19 above), p. 308. In 1921 chemists were complaining that they were paid little more than stenographers. See Walter Ferguson, "Compensation of Chemists and Salesmen," *Chemical and Metallurgical Engineering* 24 (1921): 460.

developments, particularly the growing standardization of analytical tests and the commercialization of analytical testing. As tests were standardized and commercial labs took in large quantities of analyses on a low-bid basis, some chemists feared that intelligence and professional discretion were no longer to be key ingredients in the industrial analytical laboratory. To run an analysis, one would simply follow, "parrot-like," the directions laid down by a standard test manual instead of exercising judgment about the best test to apply. Speed would become more important than any other factor. The editor of *The Chemical Engineer* feared in 1908, for example, that the emphasis on speed in making standard analyses on a commercial basis would mean that any "Hungarian boy" could be taught to make twenty to thirty determinations for silicon a day, and he would cost his employer no more than $350 a year. The result would be the decline of analytical chemistry from a profession to a craft, or worse.[27] Charles Burgess at Wisconsin noted "a tendency for the college chemist to regard his technical brother as one immersed in routine work and to whom chemistry has become a trade rather than a profession."[28] Other observers also expressed fears that analytical chemistry was fast becoming a realm where quick and dirty work was replacing intellectual discrimination.[29]

Since analytical chemists were the most widely used chemists in the American chemical industry, production chemists seem to have feared that the declining status of analytical chemists was affecting or would affect them as well. For instance, several early chemical engineers expressed concern over the tendency of manufacturers to confuse analysts with other types of chemists. L. H. Baekeland, developer of several new chemical products and a prominent early chemical engineering consultant, noted the "widespread belief" that the main occupation of a chemist was to analyze substances and, at another time, noted that it was a "common mistake" to imagine that the main work of

[27]"The Future of Analytical Chemistry," *The Chemical Engineer* 9 (1909): 23–24. See also "Standard Methods of Analysis," ibid. 12 (1910): 27; W. F. Hillebrand, "The Present and Future of the American Chemical Society," *Science* 25 (1907): 86–88, 94; and "A Question That Will Not Down," *Chemical and Metallurgical Engineering* 24 (1921): 193.

[28] Burgess, "Efficiency of College Training," p. 617 (see also pp. 611–13).

[29]For instance, Auchy, "An American Institute of Chemistry" (n. 24 above); W. Murray Sanders, "The Chemical Engineer," *Electrochemical Industry* 2 (1904): 292; W. D. Richardson, "The Improvement of Analytical Processes," *Industrial and Engineering Chemistry* 2 (1910): 99; Walker, "Some Present Problems" (n. 21 above); Herreshoff, "The Chemical Engineer" (n. 9 above), p. 171; "Are Analytical Chemists Poorly Paid?" (n. 26 above), p. 86; A. D. Little, "The Chemists' Place in Industry," *Industrial and Engineering Chemistry* 2 (1910): 64; and Myers, "Plea" (n. 25 above), pp. 798–800.

the chemist was confined to performing chemical analysis.[30] Similarly, J. S. Brogdon, a chemist working in the fertilizer industry, noted in 1912 that fertilizer managers had, for the most part, risen from the sales force and tended to "confuse the services of an analyst with those of a chemist."[31] And W. D. Richardson noted in 1908 that many manufacturers dealt only with the most common type of industrial chemist—the analyst—and that this influenced their entire view of the profession.[32]

The production chemists' status anxieties in the early 20th century are further evidenced by the discussions that accompanied the formation of a separate institutional home for chemical engineers in 1907 and 1908. In 1907, at a meeting of the American Society for Testing Materials, a number of production chemists met to discuss whether to form a society with strict membership standards which would be composed of chemists involved in designing, constructing, and operating chemical plants. One of the primary arguments raised at that meeting in favor of the creation of such a society was that it "would tend to raise the standing of chemists among manufacturers."[33] In June 1908, at the organizational meeting of the American Institute of Chemical Engineers (AIChE), Charles McKenna, one of the new society's officers, delivered an address in which he sought to justify the creation of a separate organization for chemical engineers. In that address McKenna argued that one of the primary functions of the new society should be to secure greater recognition for the chemical engineer, since manufacturers and the general public seemed ignorant of his existence and expertise and tended to assume that chemists were unable to deal with things outside of the laboratory.[34] Finally, the institute's constitution in 1908 listed as the organization's second objective, after advancing applied chemical science, giving the profession of chemical engineering "such standing before the community as will justify its recognition by Municipal, State, and National authorities in public works."[35]

[30]L. H. Baekeland, "Applied Chemistry," *Metallurgical and Chemical Engineering* 13 (1915): 677, and "Science and Industry," *Metallurgical and Chemical Engineering* 8 (1910): 336.

[31]J. S. Brogdon, "The Analyst versus the Chemist," *Industrial and Engineering Chemistry* 4 (1912): 684–85.

[32]W. D. Richardson, "The Analyst, the Chemist and the Chemical Engineer," *Science* 28 (1908): 399. See also Richard K. Meade, "Research Work," *The Chemical Engineer* 2 (1905): 328–30, and Myers, "Plea" (n. 25 above), p. 798.

[33]"Preliminary Discussion on the Advisability of Organizing the American Institute of Chemical Engineers," *AIChE Transactions* 1 (1908): 2.

[34]Charles F. McKenna, "The Justification of the American Institute of Chemical Engineers," *AIChE Transactions* 1 (1908): 10–11.

[35]"Constitution," *AIChE Transactions* 1 (1908): 21.

Production chemists' status concerns were exacerbated by the related fear that mechanical engineers were usurping their domain of chemical plant design, construction, and management. It is impossible, because of lack of data, to determine quantitatively how serious the perceived mechanical engineering penetration of the chemical industry really was. But some production chemists certainly regarded this as a real and serious threat. Richard Meade, the chemical engineering journalist who first called for the creation of a professional society for chemical engineers in 1905, commented, in making that call, that, because of manufacturers' ignorance of the chemical engineer's potential, many plants were being built by mechanical engineers.[36] That same year MIT's William H. Walker noted "recent attempts on the part of certain mechanical engineers to appropriate the term [chemical engineer] to themselves."[37] In 1908 Richard McKenna, the second president of the American Institute of Chemical Engineers, in his address justifying the formation of the institute, commented that the new organization was needed to counter beliefs like those held by the president of one of the largest chemical manufacturing companies, who had publicly declared that he would rather employ an engineer with a smattering of chemistry than a chemist with a smattering of engineering.[38] A year earlier, in 1907, Joseph Richards, a chemical engineer and the founder of the American Electrochemical Society, noted that mechanical engineers, not chemists, were often serving as the "physicists" of large chemical works.[39]

Moreover, there is evidence that some mechanical engineers considered chemical plant design as their area. For example, Charles Lucke, a mechanical engineering professor at Columbia, saw the mechanical engineer, not the chemical engineer, as following the chemist in the process of scaling up chemical reactions and pointed out that the equipment used in chemical plants largely consisted of standard mechanical appliances.[40]

[36]"Why Not 'The American Society of Chemical Engineers'?" (n. 25 above).

[37]William H. Walker, "What Constitutes a Chemical Engineer," *The Chemical Engineer* 2 (1905): 1.

[38]McKenna, "The Justification of the American Institute of Chemical Engineers," p. 10.

[39]Joseph W. Richards, "Rule of Thumb vs. Engineering," *Electrochemical and Metallurgical Industry* 5 (1907): 5. For other examples, see "Industrial Chemistry and Industrial Fellowships," *The Chemical Engineer* 13 (1911): 171; "Room for Research Work in Chemical Engineering," ibid. 7 (1908): 31; "A Proposed Society of Chemical Engineers," ibid. 6 (1907): 44; C. F. Burgess, "Electrochemistry as an Engineering Course," Society for the Promotion of Engineering Education, *Proceedings* 10 (1902): 128; H. M., "Industrial Chemistry and Industrial Fellowships," *The Chemical Engineer* 13 (1911): 171; and W. D. Richardson, "Analyst, Chemist and Chemical Engineer" (n. 32 above), p. 401.

[40]Charles Lucke, "Chemical Industry and the University," *Industrial and Engineering*

The perceived threat from mechanical engineering was, of course, related to the production chemists' parallel concerns over the low status of chemists working in industry. If "chemist" implied to the manufacturer the low-paid drudge working in his laboratories for a pittance and believed to be totally incapable of contributing outside of the laboratory, the manufacturer would be inclined to turn to mechanical or even civil engineers for advice in new plant design and in the modification of existing plants and to self-trained foremen, not to chemical engineers, for management personnel.

The membership requirements of the American Institute of Chemical Engineers, the society founded in 1908 by production chemists to represent their interests, reflected both the fears of mechanical engineering's expanding domain and the status concerns that had created the desire of production chemists to build a fence between themselves and other chemists. To become an "active" member of the AIChE one had to be at least thirty years old and proficient in "chemistry *and* in some branch of engineering as applied to chemical problems" (my italics). In addition, active members were required, at the time of election, to be actively engaged in work involving the application of chemical principles to practice. Further, members were required to have ten years of practical experience in chemical manufacturing if they had no academic degree, five years of practical experience with an academic degree.[41] Moreover, the institute's membership committee tended to interpret these already-strict requirements as the minimum expected and not necessarily as sufficient to achieve membership.[42]

These elitist membership standards served several functions. First, they minimized the possibility of direct conflict with the well-established American Chemical Society (ACS), whose leaders, as previously noted, opposed the creation of the new organization. The requirements were too rigid to draw many chemists from the general pool of ACS membership and seriously threaten the older society's growth and direction.[43] The insistence on expertise in both chemistry

---

Chemistry 7 (1915): 1012–14. AIChE president David Wesson asserted in 1920 that mechanical engineers still felt they could do all the engineering necessary in the chemical industry. See "The New Orleans Meeting of the American Institute of Chemical Engineers," *Chemical and Metallurgical Engineering* 23 (1920): 1119.

[41]"Constitution" (n. 35 above).

[42]Martin H. Ittner, "The Membership Committee," *AIChE Transactions* 28 (1932): 317–18. M. T. Bogert, president of the American Chemical Society, accused the founders of AIChE of seeking to create an engineering elite among chemists employed by industry (Van Antwerpen and Fourdrinier, *High Lights* [n. 8 above], pp. 15, 20).

[43]Terry S. Reynolds, *75 Years of Progress—a History of the American Institute of Chemical*

and engineering also excluded from membership most mechanical engineers as well as practically all analytical chemists. The strict requirements also served to provide those production chemists and consultants who formed or were soon admitted to the new organization with some of the status or recognition that the declining position of the analytical chemists had threatened. The ACS membership had conferred neither status nor recognition, since the ACS had admitted to full membership anyone interested in the advancement of chemistry.[44] The rigid membership standards also had other secondary benefits. They proved attractive to some chemical engineering executives. Although chemical consultants spearheaded the creation of the institute, making up more than 40 percent of its charter members, around 15 percent of early AIChE members were chemical executives (general managers, treasurers, vice-presidents, presidents, secretaries of corporations or companies). (See table 1.) Their presence in the institute further increased its status and prestige.

In summary, anxiety over the declining status of analytical chemists and fears that mechanical engineers were assuming roles rightfully belonging to chemical engineers stimulated production chemists to attempt to define their profession in a manner that distinguished them from other chemists.

However, status concerns, fears of mechanical engineering invasion, and even the subsequent creation of an elite society to represent production chemists were not sufficient to create a new profession. A profession-to-be, in addition to establishing an institutional base, faces the problem of defining its boundaries in the realm of knowledge in a manner distinct from its potential rivals and doing so in such a way as to secure recognition of these boundaries. Defining the cognitive realm of the new "profession" of chemical engineering proved to be a very difficult problem. The basis of the production chemists' claim to superiority over factory hands who had worked their way up to plant superintendency and over mechanical engineers engaged in chemical plant design lay in their knowledge of theoretical chemistry. Yet, this provided no means of clearly distinguishing chemical engineers from other chemists, particularly analytical chemists.

For almost a decade after production chemists formed the American Institute of Chemical Engineers in 1908, there remained no clear

---

Engineers 1908–1983 (New York, 1983), pp. 6–7; Jean-Claude Guédon, "Il progett dell'ingegneria chimica: L'affermazione delle operazioni di base negli Stati Uniti," *Testi & Contesti* 5 (1981): 14–16 also emphasizes the importance of the AIChE's restrictive membership requirements in limiting conflict with the ACS.

[44]Browne and Weeks, *History of the American Chemical Society* (n. 10 above), p. 88.

TABLE 1
CLASSIFICATION OF EARLY MEMBERS OF THE
AMERICAN INSTITUTE OF CHEMICAL ENGINEERS

	1908 Charter Members (40)	1909 Members (101)
Executives (presidents, vice-presidents, general managers, treasurers, etc., of chemical companies)...............	6 (15%)	13 (13%)
Consultants/members of consulting firms	16 (40%)	28 (28%)
Midlevel management (laboratory heads, superintendents, managers, chemical directors, chief engineers, works managers, etc.)....................	9 (22.5%)	31 (31%)
Technical positions (chemical engineers, chemists, metallurgists, research chemists, etc.).....................	3 (7.5%)	17 (17%)
Academic positions (figures in brackets are those academics also indicating consulting work)	4 (10%) [2][5%]	8 (8%) [3][3%]
Government positions	1 (2.5%)	3 (3%)
Unknown............................	1 (2.5%)	1 (1%)

SOURCE—"List of Members, June 1909," *AIChE Transactions* 1 (1908): 46–50. Charter members are listed in Terry S. Reynolds, *75 Years of Progress—a History of the American Institute of Chemical Engineers 1908–1983* (New York, 1983), p. 184.

conception of how to define chemical engineering's relationship to chemistry or to engineering. The AIChE's membership criteria did little to clarify matters; they were sufficiently strict to keep analytical chemists, shop foremen, and most mechanical engineers out of the organization. But they did not really define chemical engineers as more than chemists who had, through practical experience, acquired a smattering of engineering. They did not secure recognition of chemical engineering as a distinct branch of knowledge from other professions, nor did they dispel the feeling in some circles that chemical engineering was an "unholy" alliance of two radically different beasts.[45] Chemical engineering, in other words, did not at first have a recognizable, distinct, cognitive base in science to give it legitimacy in scientific, engineering, industrial, or academic circles as a separate branch of knowledge.

As a result, the "profession" of chemical engineering existed precari-

[45]Walker, "What Constitutes a Chemical Engineer" (n. 37 above), p. 2, and Charles H. Benjamin, "Balance of Courses in Chemical Engineering," *The Chemical Engineer* 14 (1911): 439.

ously. In 1908, the year chemical engineers created their first professional organization, the chemical profession strongly reasserted its claim to the field of production chemistry by creating within the American Chemical Society a division of Industrial Chemistry and Chemical Engineering and recruited Arthur D. Little, hitherto active in the creation of AIChE, to head it.[46] Neither did chemical engineering receive recognition in engineering circles. The AIChE was omitted consistently from key joint engineering committees in the 1910s, with the American Chemical Society often invited to participate instead.[47] In 1920 one AIChE member pithily commented: "We are not even able to persuade the engineers that we are engineers."[48]

Only in the period 1915–25 were American chemical engineers able to devise a means of defining themselves in a way that distinguished chemical engineering from both chemistry and mechanical engineering. The key concept, implicit for some decades but first explicitly stated by Arthur D. Little in 1915, was "unit operations."[49] In his 1922 report to the AIChE on chemical engineering education, Little used the concept as a way of distinguishing chemical engineers from chemists and mechanical engineers. He stated:

> Chemical engineering as a science . . . is not a composite of chemistry and mechanical and civil engineering, but a science of itself, the basis of which is those unit operations which in their proper sequence and coordination constitute a chemical process as conducted on the industrial scale. These operations, as grinding, extracting, roasting, crystallizing, distilling, air-drying, separating, and so on, are not the subject matter of chemistry as such nor of mechanical engineering. Their treatment in the quantitative way . . . and . . . the materials and equipment concerned in them is the province of chemical engineering. It is this selective emphasis on the unit operations themselves in their quantitative aspects that differentiates chemical engineering from industrial chemistry, which is concerned primarily with general processes and products.[50]

[46]Skolnik and Reese, *Century of Chemistry* (n. 10 above), pp. 309–10.

[47]*AIChE Bulletin*, no. 13 (1916): 31–32; *Chemical and Metallurgical Engineering* 21 (1919): 1199; and "New Orleans Meeting" (n. 40 above).

[48]*AIChE Bulletin*, no. 24 (1921), p. 53.

[49]For the background of the unit operations concept, see Guédon, "Il progett dell'ingegneria chimica" (n. 43 above), pp. 5–27; Martha Moore Trescott, "Unit Operations in the Chemical Industry: An American Innovation in Modern Chemical Engineering," in *A Century of Chemical Engineering*, ed. William Furter (Washington, D.C., 1982), pp. 1–18, and *The Rise of the American Electrochemicals Industry, 1880–1910* (Westport, Conn., 1981), pp. 135–70.

[50]"Report of Committee on Chemical Engineering Education of the American Institute of Chemical Engineers, 1922," n.p., unpaginated.

What were the key elements involved in the use of unit operations as a means of defining the borders of the chemical engineering profession? First, unit operations focused on the physical or mechanical operations involved in chemical manufacture, not on the chemical reactions or chemical products that concerned the traditional industrial chemist. Second, unit operations were conceived as processes that operated on an industrial scale, not a laboratory scale. Scale provided a second means of defining chemical engineering in a manner distinct from chemistry. Third, unit operations created a new "engineering science" by focusing on the quantitative study of classes of man-made objects like distillation towers, evaporators, and agitators, instead of on the natural objects of the chemists, like molecules and compounds.[51] Fourth, and perhaps most important, unit operations provided a means of rationally delimiting the intellectual sphere in which chemical engineering operated and thus provided a means of organizing academic instruction for chemical engineers and of justifying the creation in academia of departments of chemical engineering distinct from chemistry.

Chemical engineers quickly recognized the power of unit operations as a tool for sharply discriminating between their profession and chemistry, particularly analytical chemistry. In December 1919 the American Institute of Chemical Engineers asked its Committee on Chemical Engineering Education to begin considering steps the organization could take to improve education in the discipline. Significantly, A. D. Little, popularizer of unit operations as a means of organizing the scientific base of chemical engineering, was named as chairman.[52] In 1922 Little's committee presented to the institute a comprehensive report on chemical engineering education in America, a report that recommended using unit operations as a means of defining the profession's boundaries and of judging the quality of educational programs in the field.[53]

Little's report was accepted enthusiastically. Institute president David Wesson called it "monumental," and Ralph McKee, one of the institute's directors, observed: "The Committee have written a prescription, and it is our duty to see that the prescription is filled at the

[51]Edwin Layton, "Mirror-Image Twins: The Communities of Science and Technology in 19th-Century America," *Technology and Culture* 12 (October 1971): 565–67, identified laws relating to man-made objects as a characteristic of engineering science. The adoption of unit operations, which emphasized the mechanical nature of chemical manufacture, matched the nature of the American chemical industry in the early 20th century.

[52]*AIChE Bulletin*, no. 20 (1919): 50–53.

[53]See n. 50 above.

drug store, and given to the patients. . . ."[54] The institute, seeing in unit operations a means of systematizing and standardizing the chemical engineer's knowledge base along scientific lines and of thus legitimizing the profession's distinctiveness, quickly undertook a program to codify "proper" educational standards for chemical engineering programs, standards in large part based on the use of unit operations to distinguish between chemistry and chemical engineering.[55] So effective was this program that by 1928 Alfred Holmes White, a prominent educator and a future (1929–30) AIChE president, was able to state that "almost all schools which teach chemical engineering" had come to recognize unit operations as providing "the framework for the engineering side of chemical engineering."[56]

The acceptance of unit operations as a means of defining chemical engineering's intellectual boundaries left the profession with several legacies. First, the movement toward mechanical engineering entailed by the adoption of unit operations enabled production chemists to expand their position in chemical plant management at the expense of self-trained plant foremen and managers who had worked their way up from the labor force. The self-trained manager competing in engineering know-how against the ordinary industrial chemist could often more than hold his own. But he was increasingly at a disadvantage in competing against chemical engineers trained in unit operations.

Second, by concentrating on the physical processes involved in industrial-scale chemical manufacture, unit operations drew chemical engineering steadily away from chemistry and ultimately moved the new branch of knowledge to some point on the spectrum of knowledge between chemistry and physics. Increasingly, chemical engineers' exposure to advanced chemistry came primarily in the form of physical chemistry. This movement permitted chemical engineers to claim— and it has become an enduring part of the chemical engineering ideology—that chemical engineering is the broadest and best of the engineering fields because it draws from three related disciplines

[54]*AIChE Bulletin*, no. 24 (1921): 31 (for Wesson's comment), 34 (for McKee's).

[55]Reynolds, *History of the American Institute of Chemical Engineers* (n. 43 above), pp. 13–15.

[56]A. H. White, "Chemical Engineering Education in the United States," *AIChE Transactions* 21 (1928): 65. White used the term "unit processes" instead of "unit operations." Unit processes later came to mean something completely different than unit operations, but in the 1920s the two terms were often used interchangeably, and the context of White's comments makes it clear that he was referring to Little's unit operations, not what later came to be called unit processes.

(mathematics, physics, and chemistry), while all other engineering fields draw from only two (mathematics and physics).[57] The professional self-satisfaction that this belief has provided may in part account for the high proportion—60–75 percent—of chemical engineers who associate themselves with the American Institute of Chemical Engineers. This is a higher proportion of practitioners than the other "founder" societies can claim for their disciplines.[58] At the same time, however, the positioning of chemical engineering between chemistry and physics has created an enduring fear within the profession that the move away from chemistry has been too great and that some movement back toward chemistry is needed.[59]

The primary concern that led chemical engineers to adopt unit operations as a means of defining the domain of chemical engineering—the desire to distinguish chemical engineers from chemists—also played a major role in setting the attitude chemical engineers took toward their relationship to business management.

Chemical Engineering and Management

Early in the process of professionalization most of the other major engineering fields seem to have had some difficulty in defining their relationship with business management. Some only slowly accepted corporate management as a logical part of an engineer's career path. For example, as late as 1909 the president of the oldest major engineering professional society, the American Society of Civil Engineers, cautioned civil engineers against embracing business ideals too closely,

[57]For expressions of the belief that chemical engineering is the broadest of the engineering fields, see, e.g., M. C. Whitaker, "Some Professional Obligations," *The Chemical Engineer* 20 (1914): 1; Robert L. Pigford, "Chemical Technology: The Past 100 Years," *Chemical and Engineering News* 54 (April 6, 1976): 190; William Furter, "Chemical Engineering and the Public Image," in Furter, ed., *History of Chemical Engineering* (n. 11 above), pp. 393–99; and Ralph Landau, "The Chemical Engineer—Today and Tomorrow," *Chemical Engineering Progress* 68 (June 1972): 14.

[58]Gloria M. Lambson, "Forecasting Employment Demand for Chemical Engineers" (M.B.A. thesis, New York University, 1976), p. 19, estimated that 69.1% of all chemical engineers belonged to AIChE in 1970; Landau, "The Chemical Engineer," p. 9, gives figures suggesting between 51% and 60% belong. Current AIChE officials estimate between 50% and 75%. For the other "founder societies" the figures may be as low as 30%–40% if one compares membership to census returns. Somewhat higher figures are provided by *Engineering Manpower: A National Problem or a National Resource?* (New York, 1975), p. 49, citing "memo to ASCE Board of Directors, E. Zwoyer, File 1-7-1, Apr. 1975." This memo gave 62.5% professional society membership to chemical engineers, 55% to civil, and 50% to electrical and mechanical engineers.

[59]E.g., Pigford, "Chemical Technology," pp. 192, 200, 202, and Landau, "The Chemical Engineer," p. 19.

warning that engineering was becoming the "tool of those whose aim it is to control men and to profit by their knowledge."[60] Frederick Taylor's scientific management attracted a significant following among mechanical engineers at the turn of the century partly because it promised them an autonomous power base within the emerging corporations.[61] Even electrical engineering, which was practically born within the large corporation, went through a period of conflict over the relationship between scientific professionalism and business, a struggle that led, in 1912, to a nasty palace revolution within that discipline's leading professional society, the American Institute of Electrical Engineers.[62]

Chemical engineering, however, went through no such struggles. From the beginnings of the profession, chemical engineers, despite their ties to the science of chemistry, embraced management within the company or corporation as an essential part of what it meant to be a chemical engineer. For example, in 1907, Richard Meade, editor of *The Chemical Engineer* and prominent in the founding of the American Institute of Chemical Engineers, defined "chemical engineer" as "one who manages, hence designs, operates or directs chemical enterprises. . . ."[63] M. C. Whitaker, a Columbia chemical engineering professor and former general superintendent of the Welsbach Company and an active early member of the AIChE, in 1911 defined the area in which the chemical engineer worked as the "organization, operation, and management of existing or proposed processes."[64] That same year the president of Brooklyn Polytechnic Institute saw the chemical engineer as a "managing chemist" who remained in the laboratory only long enough to get his business bearings. He argued that among the principal characteristics of the successful chemical engineer was "a natural aptitude for administration and management."[65] John Olsen, for nearly two decades the executive secretary of the AIChE, declared in 1908 to the committee considering the formation of the institute: "I think this committee should draw a sharp line of distinction, that in this movement it is the men in responsible charge of the operations of the

[60]"Discussion," American Society of Civil Engineers, *Transactions* 64 (1909): 573 (Onward Bates).

[61]Layton, *Revolt of the Engineers* (n. 1 above), pp. 134–53.

[62]Ibid., pp. 80–93. See also A. Michal McMahon, *The Making of a Profession: A Century of Electrical Engineering in America* (New York, 1984), chap. 4.

[63]"What Is a Chemical Engineer?" *The Chemical Engineer* 6 (1907): 40.

[64]M. C. Whitaker, "The Training of Chemical Engineers," *The Chemical Engineer* 13 (1911): 1, and *Metallurgical and Chemical Engineering* 9 (1911): 9.

[65]Fred. W. Atkinson, "The Development of the Chemist as an Engineer," *The Chemical Engineer* 13 (1911): 9.

country, and that the proposition up to date has not included all of these men in the industry, but merely the men in charge [*sic*]."[66]

Other prominent figures also closely linked chemical engineering to the management of chemical facilities. David Wesson, 1920 AIChE president, declared that the chemical engineer "in the broadest sense of the word" was "a chemical manager, that is, one who manages enterprises requiring the knowledge and application of chemistry."[67] Harry McCormick, the second editor of *The Chemical Engineer*, argued in 1914 that chemical engineers needed to be able to design chemical apparatus and plants *and* superintend the construction and operation of these plants.[68] Finally, the central importance of management to early chemical engineers was expressed somewhat more crudely by Milton C. Whitaker in 1911: "Of what consequence is the chemical reaction which depends upon labor to make it work, if the chemist does not know how to make a 'Dago' work?"[69]

Not surprisingly, in light of these views from early spokesmen for the chemical engineering profession, the AIChE incorporated management responsibilities into its membership criteria when it was founded in 1908. At least five years "in responsible charge of operations requiring the elaboration of raw materials, the design of machinery involving chemical processes, or the application of chemistry to industry" were required for full membership.[70] Moreover, the early AIChE was dominated by chemical engineers with supervisory responsibilities and management associations. Around 75 percent of the members listed in AIChE's 1909 directory identified themselves with groups that had or had had management responsibilities (chemical executives, mid-level chemical managers in industry or government, and chemical consultants). (See table 1.) The remaining 25 percent of AIChE's early membership was divided between academics and chemists or chemical engineers with no readily apparent supervisory responsibilities, although the sketchiness of the listings in the AIChE's directory does not preclude the possibility that many of these men were, or had been, in positions involving some supervisory responsibilities.

Why did chemical engineering at its inception choose to so closely

[66]Van Antwerpen and Fourdrinier, *High Lights* (n. 8 above), p. 15, citing "Proceedings of a Meeting." See also Benjamin, "Balance of Courses" (n. 45 above).

[67]David Wesson, "What Is a Chemical Engineer?" *Chemical and Metallurgical Engineering* 23 (1920): 50.

[68]"Do Business Men Know What a Chemist Really Is and What He Can Do for Them?" *The Chemical Engineer* 19 (1914): 90.

[69]Duncan, "Excerpts" (n. 19 above), pp. 93–94 (comment by Whitaker).

[70]"Constitution" (n. 35 above), p. 22.

embrace management? The management duties that came as a natural part of much of production chemistry no doubt influenced this close association, but other factors certainly contributed. For example, if ties to management could be developed, chemical engineers could avoid the status losses that they feared were overtaking analytical chemists. Moreover, supervisory responsibilities at the design, construction, and operation levels provided a clear functional difference between chemical engineering and chemistry. In the absence of a good, cognitive base for distinguishing between chemical engineering and chemistry before the emergence of unit operations, chemical engineers could embrace management responsibilities as the key to defining the profession's boundaries with chemistry.

The inclusion of management functions within the very definition of what it meant to be a professional chemical engineer in the formative stages of the profession established a tradition which has continued to dominate the profession throughout this century. Even in 1973, at the end of the first great employment crisis faced by chemical engineers since the founding of the profession, one which might have shaken ties to management, only 12 percent of chemical engineers polled felt that they were part of the "rank and file."[71]

The very early close association of management responsibilities with professional goals among chemical engineers suggests that, for chemical engineering, the tension between science and business has not been as central an issue in professional development as Edwin Layton has suggested it was for the four older engineering groups (civil, mining, electrical, and mechanical engineering).[72] In chemical engineering the central tension has not been the balancing of scientific professionalism with business; it has been the drive of chemical engineers to persuade executive management that chemical engineers should be treated as part of the management team and be the main focus of recruitment when executive positions open up.

In summary, the chemical engineering profession's definitions of its relationships to the critical border areas of science and business during this century were strongly conditioned by the perceived status of analytical chemists around 1900. Because production chemists feared that analytical chemists were losing status and that this loss of status would carry over to them and encourage absorption of the areas of chemical plant construction, design, and supervision by mechanical

<hr />

[71]Chemical engineers, unlike most other engineers, were not seriously affected by the Great Depression of the 1930s. Reynolds, *History of the American Institute of Chemical Engineers* (n. 43 above), p. 27. "Issues of Professionalism and Employment," *Chemical Engineering Progress* 69 (November 1973): 41.

[72]Layton, *Revolt of the Engineers* (n. 1 above), pp. 1–24 and elsewhere.

engineers or even scientifically untrained foremen, production chemists sought definitions for chemical engineering that would sharply distinguish chemical engineers from analytical chemists. These definitions moved the profession cognitively away from chemistry and toward physics and engineering and functionally placed its aspirations, from the very inception of the profession, firmly in the direction of management.

ACADEMIC ENTREPRENEURSHIP AND ENGINEERING EDUCATION: DUGALD C. JACKSON AND THE MIT-GE COOPERATIVE ENGINEERING COURSE, 1907–1932

W. BERNARD CARLSON

Between 1880 and 1930, the American engineering profession underwent a remarkable transformation. During these years, the number of practicing engineers grew from 7,000 to 226,000, an increase of over 3,000 percent.[1] In 1880 there were only three established engineering fields, civil, mechanical, and mining; the next five decades saw the creation of specialties ranging from aeronautical and metallurgical engineering to sanitation and heating-refrigeration. Representing these new engineers were a variety of professional societies that set standards for education and conduct as well as stimulating the diffusion of new engineering knowledge through journals and meetings.[2]

Why did engineering grow so rapidly and emerge as a profession between 1880 and 1930? Although the development of the engineering profession can be linked to the general process by which the American middle class shaped a number of occupations into professions, the

W. BERNARD CARLSON teaches the history of technology in the School of Engineering and Applied Science at the University of Virginia. He is the author of *Innovation as a Social Process: Elihu Thomson and the Rise of General Electric, 1870–1900* (Cambridge University Press, forthcoming). He received the 1989 SHOT-IEEE Life Members Fund Prize for his essay in this volume.

[1] Edwin T. Layton, Jr., *The Revolt of the Engineers: Social Responsibility and the American Engineering Profession* (Cleveland, 1971; Baltimore, 1986), p. 3.

[2] The rise of professional engineering in America has been extensively explored by historians. Among the major studies are Daniel H. Calhoun, *The American Civil Engineer: Origins and Conflict* (Cambridge, Mass., 1960); Monte A. Calvert, *The Mechanical Engineer in America, 1830–1910: Professional Cultures in Conflict* (Baltimore, 1967); Raymond H. Merritt, *Engineering in American Society, 1850–1875* (Lexington, Ky., 1969); David F. Noble, *America by Design: Science, Technology, and the Rise of Corporate Capitalism* (New

This essay originally appeared in *Technology and Culture*, vol. 29, no. 3, July 1988.

expansion of engineering has also been attributed to major changes occurring in American business and industry during this period.[3] As business firms used larger and more complex technological systems such as electric power grids, assembly lines, and continuous-flow chemical processes, they employed a growing number of engineers to design, supervise, and maintain these systems.[4] Because such systems were often central to the operation of these firms, engineers frequently assumed managerial positions and applied their expertise not only to designing machinery but also to the human and financial aspects of industrial organizations. Successful in the application of their methodology to both technical and organizational problems, engineers grew in social and economic status. By the early 20th century, American engineers had come to be regarded as the creators of a richer, more rational civilization, and it is hardly surprising that social commentators such as Thorstein Veblen should have perceived them as the natural leaders of the future.[5]

Engineering education was intimately involved in the transformation of the engineering profession. Schools of engineering grew rapidly from 1880 to 1930 in order to provide engineers who could build complex technological systems and participate in the business hierarchy. Whereas approximately 226 students were awarded engineer-

York, 1977); Bruce Sinclair, *A Centennial History of the American Society of Mechanical Engineers* (Toronto, 1980); Terry S. Reynolds, *75 Years of Progress—a History of the American Institute of Chemical Engineers* (New York, 1983); and A. Michal McMahon, *The Making of a Profession: A Century of Electrical Engineering in America* (New York, 1984).

[3]On the American middle class and the rise of the professions, see Burton Bledstein, *The Culture of Professionalism: The Middle Class and the Development of Higher Education in America* (New York, 1976), pp. 80–128, and Robert H. Wiebe, *The Search for Order, 1877–1920* (New York, 1967), pp. 111–21.

[4]Thomas P. Hughes, *Networks of Power: Electrification in Western Society, 1880–1930* (Baltimore, 1983); David A. Hounshell, *From the American System to Mass Production, 1800–1932* (Baltimore, 1984); and Alfred D. Chandler, Jr., *The Visible Hand: The Managerial Revolution in American Business* (Cambridge, Mass., 1977).

[5]During the first half of the 20th century, nearly one-half of all of the engineers trained at the Massachusetts Institute of Technology (MIT) went into business or managerial careers; see John B. Rae, "Engineering Education as Preparation for Management: A Study of M.I.T. Alumni," *Business History Review* 29 (March 1955): 64–74. For a discussion of the image of engineers in American literature and popular culture in the early 20th century, consult Cecelia Tichi, *Shifting Gears: Technology, Literature, Culture in Modernist America* (Chapel Hill, N.C., 1987), pp. 97–170. See also Thorstein Veblen, *The Engineers and the Price System* (New York, 1921), and, for a discussion of Veblen's interest in engineers, see Edwin T. Layton, Jr., "Veblen and the Engineers," *American Quarterly* 14 (Spring 1962): 62–72.

ing degrees in 1880, by 1930 over 11,000 students received such degrees.[6] Prior to the 1890s, engineering education tended to be practical and highly specific, emphasizing shop work and hands-on experience with state-of-the-art machinery. In the early 20th century, however, engineering education became increasingly based on theoretical science, mathematics, and laboratory instruction that provided the student with concepts applicable to a variety of problems. To supplement scientific training, the engineering student also took management and economics courses that provided insight into the business world. Finally, in order to furnish engineers capable of undertaking the research desired both by universities and by business firms, a few engineering schools in the early 20th century established graduate degree programs. Through these curriculum changes, engineering educators played a prominent role in the transformation of the engineering profession.[7]

Historians have interpreted these changes in the engineering profession and engineering education as part of the larger trend toward the integration of science and technology into American business and industry. In so doing, they have generally adhered to one of two perspectives. First, scholars such as David F. Noble have argued that, soon after major corporations rationalized their internal operation and organization, they turned their attention to controlling and shaping the external political and social environment. Noble has suggested that a new elite, the corporate engineers, "ventured to design

[6]For statistics on engineering graduates, see Dugald C. Jackson (hereafter cited as DCJ), *Present Trends of Engineering Education in the United States* (New York: Engineers' Council for Professional Development, 1939), p. 18, and Douglas L. Adkins, *The Great American Degree Machine: An Economic Analysis of the Human Resource Output of Higher Education* (Berkeley, 1975), pp. 217, 287. I am grateful to Terry S. Reynolds for helping me locate these statistics.

[7]On the history of engineering education, see Society for the Promotion of Engineering Education (SPEE), *Report of the Investigation of Engineering Education, 1923–1929*, 2 vols. (Pittsburgh, 1930), 1:541–49; L. P. Grayson, "A Brief History of Engineering Education in the United States," *Engineering Education* 68 (December 1977): 246–64; James G. McGivern, *First Hundred Years of Engineering Education in the United States (1807–1907)* (Spokane, Wash., 1960); Frederick E. Terman, "A Brief History of Engineering Education," *IEEE Proceedings* 64 (September 1977): 1399–1407; Robert Rosenberg, "The Origin of EE Education: A Matter of Degree," *IEEE Spectrum* 21 (July 1984): 60–68; and Jeffrey K. Stine, "Professionalism vs. Special Interest: The Debate over Engineering Education in Nineteenth Century America," *Potomac Review* 27 (1984–85): 72–94.

a new (yet old) social order, one dominated by the private corporation and grounded upon the regulated progress of scientific technology."[8] Corporate engineers accomplished this by exploiting the patent system, appropriating universities for industrial research, and reorganizing the workplace through scientific management. They were especially successful in transforming engineering education. According to Noble, through the creation of corporate training programs, cooperative courses between universities and companies, and new national agencies for standardizing technical education, corporate engineers secured a dependable supply of subordinates and potential successors. These new programs ensured that the corporate hierarchy would be filled with technically proficient individuals habituated to corporate life and prepared to assume managerial responsibility. At the heart of this perspective are two assumptions: first, that science and technology were redesigned so as to serve the needs of corporate capitalism, and second, that the process of transformation took place smoothly and with a minimum of conflict.

In contrast to Noble's perspective, other historians have suggested that the integration of science and technology into American business was hardly smooth and orderly. Rather, the integration occurred only gradually, marked by the clash of goals, values, and professional cultures. Within firms and professional societies, various groups of engineers and scientists frequently debated not only technical issues but also what their role was to be in American society. In chronicling the unsuccessful design and testing of the copper-cooled engine at General Motors during the 1920s, Stuart W. Leslie found that the project's failure could be attributed to the inability of engineers in the research and production divisions to agree on a common design for the engine. Similarly, George Wise and Leonard S. Reich have observed that a clash of values between scientists and managers informed the creation of research laboratories at General Electric and AT&T; while scientists sought the freedom to pursue "pure" research, managers were continually seeking products and patents to improve their firms' marketing position.

Beyond the individual firm, Terry S. Reynolds has demonstrated that chemical engineers struggled to create a disciplinary niche for themselves between the fields of chemistry and mechanical engineering, both of which claimed hegemony over industrial chemistry. Like-

---

[8]Noble (n. 2 above), p. xxiv. In a similar vein, E. Richard Brown has argued that doctors and corporate interests reshaped the medical profession in the early 20th century to serve their common goals; see *Rockefeller Medicine Men: Medicine and Capitalism in America* (Berkeley, 1979).

wise, Monte Calvert has argued that mechanical engineering was shaped by a struggle between shop- and college-trained engineers. Whereas the former perceived the field as a group of self-made men known for their entrepreneurial skills and machine shop expertise, the latter sought to define the field in terms of average men who received a college education in science and would serve in the industrial bureaucracy. Calvert saw this struggle as a clash of cultures, and this viewpoint was extended beyond a single field to the entire engineering profession by Edwin T. Layton, Jr. In *The Revolt of the Engineers,* Layton observed that engineers are caught between the business culture of their corporate employers and the scientific culture of their professional colleagues. Reviewing events in the history of engineering organizations from 1900 to 1940, Layton suggested that this tension was fundamental in shaping the engineering profession. Most recently, Peter Meiksins has reviewed the history of the American Association of Engineers and proposed that the engineering profession has been shaped alternately by conflict and cooperation among a traditional pro-business elite, another elite that wished to establish the engineer as an independent professional, and the rank and file who were concerned with employment issues such as salaries and promotions. Emerging from all these studies is a perspective that the integration of science and technology into American business was a process that was marked by tension and conflict, and that change occurred only as different groups reconciled divergent goals, values, and cultures.[9]

In this article, I wish to continue the tradition of viewing the transformation of the engineering profession as a clash of values and ideas. In particular, the aim is to investigate how the evolution of engineering education may be viewed from this perspective by examining Dugald

[9]For an overview of the recent historiography of professionalization as the clash of values and ideas, see Louis Galambos, "Technology, Political Economy, and Professionalization: Central Themes of the Organizational Synthesis," *Business History Review* 57 (Winter 1983): 471–93, esp. 489–90. See also Stuart W. Leslie, "Charles F. Kettering and the Copper-cooled Engine," *Technology and Culture* 20 (October 1979): 752–76; George Wise, "A New Role for Professional Scientists in Industry: Industrial Research at General Electric, 1900–1916," *Technology and Culture* 21 (July 1980): 408–29; Leonard S. Reich, "Industrial Research and the Pursuit of Corporate Security: The Early Years of Bell Laboratories," *Business History Review* 54 (1980): 503–29; Terry S. Reynolds, "Defining Professional Boundaries: Chemical Engineering in the Early 20th Century," *Technology and Culture* 27 (October 1986): 694–716; Calvert (n. 2 above); Layton, *The Revolt of the Engineers* (n. 1 above); Peter Meiksins, "Professionalism and Conflict: The Case of the American Association of Engineers," *Journal of Social History* 19 (1986): 404–21; and Meiksins, "The 'Revolt of the Engineers' Reconsidered," *Technology and Culture* 29 (April 1988): 219–46.

C. Jackson's efforts to create a cooperative course for electrical engineering students at the Massachusetts Institute of Technology (MIT) and the General Electric Company (GE) from 1907 to 1932. Located between academia and industry, this course reveals how the needs and expectations of business were merged with the values and ideals of engineering schools. Moreover, the MIT-GE course suggests that there was an ongoing debate over the exact role that the college-trained engineer should play in American business: was he to be a factory supervisor, a highly paid technician, a scientifically trained designer, or a corporate executive?

I also wish to supplement the conflict view of the integration of science, technology, and business by arguing that Jackson shaped engineering education by functioning as an academic entrepreneur. Like entrepreneurs in the business world, he creatively matched the supply and demand forces of the marketplace. Through his efforts at MIT, Jackson coordinated the demands of industry for trained personnel with the supply of engineers produced by universities. To do so was a risk-taking enterprise, demanding the invention of new institutional arrangements and curricula and with no guarantee that either business or academia would be satisfied with the outcome. Nevertheless, as the MIT-GE course illustrates, Jackson persevered and forged an institutional link between academia and industry that permitted him to implement his distinctive vision of the engineer as corporate leader. In serving as an academic entrepreneur, Jackson was by no means unique; other scholars have demonstrated how William A. Henry at the University of Wisconsin, George Ellery Hale at Cal Tech, Roger Sherman at the University of Illinois, and Frederick E. Terman at Stanford University played similar entrepreneurial roles that helped join science, technology, and business.[10]

[10]I am grateful to Stuart W. Leslie for suggesting that I view Jackson as an entrepreneur. For a discussion of the role of the entrepreneur in technological history, see his essay, "Whatever Happened to Entrepreneurial History?" forthcoming in *Business History Review*. For a conceptualization of the entrepreneur, see C. Joseph Pusateri, *A History of American Business* (Arlington Heights, Ill., 1984), pp. 3–12, and Joseph Schumpeter, *Theory of Economic Development* (Cambridge, Mass., 1934). For other examples of academic entrepreneurs, consult Charles E. Rosenberg, *No Other Gods: On Science and American Social Thought* (Baltimore, 1976), pp. 160–62; Robert H. Kargon, "Temple to Science: Cooperative Research and the Origin of the California Institute of Technology," *Historical Studies in the Physical Sciences* 8 (1977): 3–31; P. Thomas Carroll, "Academic Chemistry in America, 1876–1976: Diversification, Growth, and Change" (Ph.D. diss., University of Pennsylvania, 1982); McMahon (n. 2 above), pp. 181–87; Effie G. Bryson, "Frederick E. Terman: Educator and Mentor," *IEEE Spectrum* 21 (March 1984): 71–73; and Stuart W. Leslie and Bruce Hevly, "Steeple Building at Stanford: Physics, Electrical Engineering, and Microwave Research," *IEEE Proceedings* 73 (July 1985): 1169–80.

To develop this view of Jackson as academic entrepreneur, I shall first provide information on his career and suggest how his experiences led him to develop his view of the professional engineer. A second section will briefly review the origins of cooperative engineering education and GE's interest in training engineers. With this background material in place, I shall then address the formation and development of the course, emphasizing Jackson's educational philosophy, his desire to build a strong academic department, and General Electric's indifference toward the course.

Jackson's Early Career and His Vision of the Professional Engineer

Jackson's family background, education, and early career prepared him well for assuming a prominent position of leadership in the electrical engineering profession. He was born in 1865 to a family of educators and engineers. His father Josiah taught mathematics at Pennsylvania State College, and two of his brothers became electrical engineers. Older brother John served for many years as head of the electrical engineering department at Penn State, and brother William worked with Dugald in the consulting engineering firm of Jackson and Moreland.[11]

As an engineering student at Penn State, Jackson became curious about the new field of electrical technology and spent the summer of 1884 working with the inventor William Stanley, who later perfected one of the first alternating current systems. After receiving a bachelor's degree in civil engineering in 1885, Jackson spent two years studying electrical engineering with Professor William A. Anthony at Cornell University. At Cornell, Jackson was impressed by Anthony's efforts to involve students in research. In 1887, he joined two Cornell alumni, J. G. White and Harris J. Ryan, in organizing the Western Engineering Company in Lincoln, Nebraska. This firm constructed electric lighting and street railway systems throughout the Midwest. Becoming especially interested in street railway motors, in 1889 Jackson joined the Sprague Electric Railway and Motor Company, which was soon absorbed by the Edison General Electric Company. At Edison

---

[11]See James E. Brittain, entry for DCJ, *Dictionary of American Biography*, ed. John A. Garraty (New York, 1977), Suppl. 5, pp. 354–55; Michael Bezilla, *Engineering Education at Penn State: A Century in the Land Grant Tradition* (University Park, Pa., 1981), pp. 25–26, 31–32; Vannevar Bush, "A Tribute to Dugald C. Jackson," *Electrical Engineering* 70 (December 1951): 1063–64. Jackson wrote several textbooks with his brother John, including *An Elementary Book on Electricity and Magnetism and Their Applications* (New York, 1902).

General, he rose to the rank of Chief Engineer for the Central District, supervising projects in thirteen midwestern states.[12]

Because of his extensive practical experience, Jackson was selected in 1891 to head the newly formed electrical engineering department at the University of Wisconsin. At Wisconsin Jackson blossomed as an engineering educator. Since the College of Engineering was supported by a significant portion of the land grant funds as well as 1 percent of the state's railroad license tax, Jackson was able to acquire quickly a departmental laboratory and to hire several new instructors. Concerned that the citizens of Wisconsin might see the newly appointed engineering professors as too theoretical and not practical, Jackson encouraged his colleagues to join him in consulting work. By skillfully combining new facilities, new professors, and a novel emphasis on consulting, Jackson succeeded in establishing a thriving department at Wisconsin.[13]

Much of Jackson's work at Wisconsin centered on developing a new undergraduate curriculum. Drawing on his practical experience, he railed against electrical engineering curricula based exclusively on physics and mathematics. This situation had arisen because many of the first electrical engineering programs had been established within physics departments, but, according to Jackson, students needed a balance of theory and practice. "While theory alone . . . cannot make a practical man," he wrote in 1892, "it is the one who can follow the guide of theory along the paths of practical work and experience, who makes the fully developed engineer." Jackson's curriculum included two years of basic science and mathematics followed by two years of courses covering the theory and practice of electrical engineering. To teach students about business affairs, Jackson added economics, sociology, and composition. Wherever possible, he had the students follow their classroom lectures with laboratory exercises that encouraged them to think for themselves. Jackson also advocated that bright students undertake their own research projects, arguing that

[12]H. H. Norris, "Dugald C. Jackson, President of the American Institute of Electrical Engineers," *Engineering News* 65 (February 2, 1911): 135–36.

[13]DCJ, "Some Highlights in the Evolution of Electrical Engineering Education," *Electrical Engineering* 58 (April 1939): 165–68; Thomas J. Higgins, "A Resourceful College of Engineering," in *A Resourceful University: The University of Wisconsin in Its 125th Year* (Madison, 1975), pp. 37–53; and Storm Bull, "Technical Education at the University of Wisconsin," *Wisconsin Engineer* 3 (January 1899): 1–17. I am grateful to Terry S. Reynolds for sharing his notes on this last article with me.

this independent work further stimulated their enthusiasm for electrical engineering.[14]

In developing the electrical engineering program at Wisconsin, Jackson formulated a vision of the role that the engineer should play in society. The electrical engineer should not so much be a technician or designer as he should be a leader, manager, and executive. The electrical engineer, he often said, must be a man "competent to conceive, organize, and direct extended industrial enterprises of a broadly varied character." In viewing his students as potential leaders, Jackson differed significantly from other engineering educators, who believed that their students should become scientifically trained, creative designers, and from industrial leaders, who sought college-educated engineers for the routine tasks of manufacturing, testing, selling, and installing products.[15]

Jackson's vision of the electrical engineer as a leader possessing scientific knowledge and practical judgment reflects his personal, professional, and cultural contexts. Like other teachers, Jackson wanted to train students in his own image and likeness. Since his own industrial and consulting work centered on the development of large-scale electrical systems, he trained his students to be leaders in managing such systems. In fact, according to his colleagues Vannevar Bush and Karl Wildes, much of Jackson's later teaching focused on the management of large utility systems.[16]

From a professional standpoint, Jackson emphasized science and theory as the basis of the engineer's expertise out of a desire to help establish electrical engineering as a legitimate and respected field. As

[14]See Robert Rosenberg, "American Physics and the Origins of Electrical Engineering," *Physics Today* 36 (October 1983): 48–54. The quote is from DCJ, "The Technical Education of the Electrical Engineer," *AIEE Transactions* 9 (1892): 476–86, esp. 479–80. For a view of how Jackson taught electrical engineering as a balance of theory and practice, see the two college textbooks he wrote: *A Text-Book on Electro-Magnetism and the Construction of Dynamos* (New York, 1893) and, with John Price Jackson, *Alternating Currents and Alternating Current Machinery* (New York, 1896).

[15]Quote is from DCJ, "Methods of Teaching Electrical Engineering," paper presented before the Second Pan American Scientific Conference, Washington, December 27, 1915–January 8, 1916 (Washington, D.C., 1917), p. 2, on file in General Electric Transfile (hereafter cited as GE Trans.) 1915–23, MIT, Elihu Thomson Papers, Library of the American Philosophical Society, Philadelphia (hereafter cited as Thomson Papers). See also DCJ, "The Typical College Course Dealing with the Professional and Theoretical Phases of Electrical Engineering," *AIEE Transactions* 22 (1903): 599–607, esp. 601, 600. For another interpretation of Jackson's philosophy of engineering education, see McMahon (n. 2 above), pp. 70–77.

[16]Vannevar Bush, *Pieces of the Action* (New York, 1970), p. 254; Karl L. Wildes and Nilo A. Lindgren, *A Century of Electrical Engineering and Computer Science at MIT, 1882–1982* (Cambridge, Mass., 1985), p. 44.

Burton Bledstein has observed, many professional groups in 19th-century America based their expert knowledge on scientific principles since science was seen as a source of transcendent and objective authority in a democratic society. Including science in his curriculum was also highly desirable from the standpoint of academic politics since it helped quell the fears of nonengineering faculty, who sometimes opposed including engineering in the university, arguing that it lacked a true intellectual basis.[17]

In the broadest sense, one can relate Jackson's view of the engineer as a scientifically trained leader or professional with the Progressive Era's celebration of the expert. Although Progressives promoted many different reforms, one of their pervasive beliefs was that society would be saved by college-trained experts who would apply science to all problems. Jackson himself expressed this belief in a 1920 speech in which he proclaimed, "The redemption of this country . . . depends upon the even-tempered fairness of mind and clearness of thought, outspokenly expressed and supported by action, of the tens of thousands of college-trained men (and notable [sic] those trained in science) whose training mark them for leadership."[18] Most frequently associated with Progressivism are specially trained city managers, social scientists, and public health officials, but one could easily add engineers who improved the city, built the roads for the new automobiles, designed the new regional power systems, and generally directed technology toward satisfying the needs of the emerging mass consumer society. As much as any profession in early 20th-century America, engineers were expected to bring about improvement and reform.[19]

At first glance, it may seem surprising to consider Jackson in terms of Progressivism since he was an opponent of Morris L. Cooke, the well-known liberal reformer. In 1914 Jackson and Cooke fought a

[17]Bledstein (n. 3 above), pp. 88–92. For an unsuccessful effort to use science to develop an engineering field, see Bruce E. Seely, "The Scientific Mystique in Engineering: Highway Research at the Bureau of Public Roads, 1918–1940," Technology and Culture 25 (October 1984): 798–831. On the opposition of nonengineering faculty to engineering in the university, see Bruce Sinclair, "Inventing a Genteel Tradition: MIT Crosses the River," in New Perspectives on Technology and American Culture, ed. Bruce Sinclair (Philadelphia, 1986), pp. 1–18, esp. 3–6.

[18]DCJ, "College Men Necessary to Save Nation," Boston American (March 7, 1920), Dugald C. Jackson Biographical File, MIT Museum, Cambridge.

[19]On the engineer as Progressive Era expert, see Henry S. Pritchett, "The Relation of Educated Men to the State," Science, new ser. 12 (November 2, 1900): 657–66; Bruce E. Seely, "Engineers and Government-Business Cooperation: Highway Standards and the Bureau of Public Roads, 1900–1940," Business History Review 58 (Spring 1984): 51–77; and Donald Stabile, Prophets of Order: The Rise of the New Class, Technocracy and Socialism in America (Boston, 1984).

nasty battle in the Philadelphia Electric Company's rate case in which Jackson was the utility company's expert witness and Cooke was Philadelphia's chief municipal engineer. Given his conservative political opinions (particularly his faith in big business), Jackson was hardly a reformer in the sense that Cooke was. Whereas Cooke wanted engineers to be more responsive to social concerns and change big business from without, Jackson believed that engineers should support corporations and work to improve business from within. Yet what these two men shared was a belief in the potential of the engineer as expert; they differed only on the organizational context in which the engineer should bring about change. Consequently, although he favored big business, Jackson did not simply accept the status quo of corporate America; rather, he believed that engineers should become the leaders of industry in order to improve its efficiency and its responsiveness to the needs of society. This vision of the engineer as leader and expert was the essence of Jackson's engineering philosophy, and, as we shall see, he pursued it even when it ran counter to what big business wanted or needed.[20]

Impressed with his engineering philosophy and his successful department at Wisconsin, MIT hired Jackson in 1907 to rebuild its electrical engineering program. MIT had established this program in 1882 (the first in America), but Course VI, as it was known in MIT parlance, had fallen into decline by the early 1900s.[21] Following a strategy similar to the one he had used at Wisconsin, Jackson immediately campaigned for additional laboratories, offices, and equipment. He reorganized the undergraduate curriculum and added a graduate program that awarded its first doctorate in 1910. Jackson actively promoted industrially sponsored research, knowing it would provide additional financial support as well as prestige for the department. To handle this work, Jackson created the Division of Elec-

---

[20]Jackson outlined his vision of the electrical engineer as conservative reformer in his presidential address before the American Institute of Electrical Engineers; see "Electrical Engineers and the Public," *AIEE Transactions* 30, pt. 2 (1911): 1135–42. On Jackson, Cooke, and the Philadelphia Electric Company case, see Layton, *The Revolt of the Engineers* (n. 1 above), pp. 163–68; Kenneth E. Trombley, *The Life and Times of a Happy Liberal: A Biography of Morris Llewellyn Cooke* (New York, 1954), pp. 35–70; Jean Christie, *Morris Llewellyn Cooke: Progressive Engineer* (New York, 1983), pp. 28–29; and Nicholas B. Wainwright, *History of the Philadelphia Electric Company, 1881–1961* (Philadelphia, 1961), pp. 113–21.

[21]Wildes and Lindgren (n. 16 above), pp. 16–30; Samuel C. Prescott, *When M.I.T. Was "Boston Tech" 1861–1916* (Cambridge, Mass., 1954), pp. 118 and 184–85; DCJ to R. C. Maclaurin, September 24, 1912, Records of the Office of the President, 1897–1932 (AC 13) (hereafter cited as MIT President Records), Box 4, Fol. 324, Institute Archives and Special Collections, Massachusetts Institute of Technology, Cambridge.

trical Engineering Research in 1913, and it was soon conducting research for GE, Stone and Webster, and American Telephone and Telegraph. With these reforms, Jackson succeeded in rapidly expanding the department; when he arrived the department had only forty-six full-time students, but by 1909 it had over 200.[22]

In restructuring the curriculum, encouraging graduate study, and establishing a program of commercial research, Jackson functioned as an academic entrepreneur. Through these innovations, he capitalized on opportunities available in industry to build a strong electrical engineering program at MIT. Such efforts were not, however, simply for self-aggrandizement or to serve industry. Rather, Jackson created a strong institutional base because he sincerely wanted to train electrical engineers according to his distinctive conception. He believed that electrical engineers were the future leaders of American business and society, and he continually sought better ways to train engineers to fulfill his vision.

The Origins of Cooperative Engineering Education and GE's Interest in Engineering Education

Jackson's vision of the engineer can be most clearly seen in his efforts to establish a cooperative course at MIT, but, to understand this program, it is useful to review the origins of cooperative engineering education. Cooperative courses are programs arranged between engineering schools and business firms in which students alternate between taking university courses and working in industrial shops and offices. The first cooperative course had been introduced by Herman Schneider at the University of Cincinnati in 1906. In teaching engineering, Schneider had noticed that students lacked any practical experience of working with men and machines, experience necessary for them to become effective factory supervisors. To remedy this, Schneider devised a six-year program in which the students alternated weekly between the university and manufacturing shops. At the university, the co-op students studied the same subjects as regular engineering students, but their lectures were arranged so that they

[22]See Wildes and Lindgren (n. 16 above), pp. 42–49; Untitled Memo, ca. 1908, Thomson Papers, GE Trans. 1905–11, M–N; DCJ to E. Thomson, March 11, 1908, Thomas Papers, GE Trans. 1905–11, M–N; and "Report of the Visiting Committee," 1909, Dugald C. Jackson Papers (MC 5) (hereafter cited as Jackson Papers), Box 4, Fol. 250, MIT Archives. Statistics for 1907 are from "President's Report," January 1908, p. 54, MIT Archives, and those for 1909 are from "Report of the Visiting Committee Assigned to the Electrical Engineering Department," February 27, 1912, Thomson Papers, GE Trans. 1915–23, MIT.

directly related to the work the students were doing concurrently in the shops. At the factory, the students learned about manufacturing techniques by working in different departments; in a machine-tool plant, for instance, students worked in the foundry, pattern shop, machine shop, drafting room, and engineering offices. Students who enrolled in the course found that the work was greasy and hard, that shop foremen were gruff, and that the wages were quite low. Consequently, in the first year of the Cincinnati course, one-half of the students dropped out. The cooperating firms found, however, that the program provided a dependable supply of high-caliber shop apprentices, and they were excited by the long-term prospect of acquiring engineers with shop experience.[23]

The cooperative idea was soon taken up by other engineering schools, and by the mid-1920s it had been adopted by sixteen institutions, about 10 percent of all engineering schools.[24] (See table 1.) In all likelihood, cooperative education was considered desirable since it supplemented the students' theoretical training by adding a spatial or visual dimension to their understanding of technology and by exposing them to the social relations of the shop floor.[25] From its inception, the cooperative engineering course was a persuasive and popular idea, and it is still regarded as an important way to train students; the electrical engineering department at MIT today maintains a cooperative program with GE and other companies.[26]

The first to recognize the possibility of a cooperative program between MIT and GE was not Jackson but Magnus Alexander, chief design engineer at GE's Lynn, Massachusetts, plant.[27] GE and its predecessors had a long-standing interest in training engineers at its factories; since 1885 the Thomson-Houston Electric Company had

[23]Herman Schneider, "The Cooperative Course in Engineering at the University of Cincinnati," *SPEE Proceedings* 15 (1907): 399–411; Herman Schneider, "Notes on the Co-Operative System," *SPEE Proceedings* 18 (1910): 395–423; John T. Faig, "The Effect of Cooperative Courses upon Instructors," *SPEE Proceedings* 20 (1912): 97–106; Clyde W. Park, *Ambassador to Industry: The Idea and Life of Herman Schneider* (Indianapolis, 1943), pp. 45–98.

[24]Bulletin no. 12, "A Study of the Cooperative Method of Engineering Education," in SPEE, *Report on Engineering Education* (n. 7 above), 1:558–625.

[25]On the role of visual and spatial thinking in engineering, see Eugene S. Ferguson, "The Mind's Eye: Nonverbal Thought in Technology," *Science* 197 (1977): 827–36, and Brooke Hindle, *Emulation and Invention* (New York, 1981), pp. 133–38.

[26]For the history of cooperative education in the electrical engineering program at MIT, consult Wildes and Lindgren (n. 16 above), pp. 124–37, and *Hand in Hand: On the Occasion of Fifty Years of Industry-aided Selective Cooperative Education* (Medford, Mass., 1958).

[27]On Magnus Alexander's career, see H. M. Gitelman, "Management's Crisis of Confidence and the Origin of the National Industrial Conference Board, 1914–1916," *Business History Review* 58 (Summer 1984): 153–77, esp. 157–59.

TABLE 1

ENGINEERING SCHOOLS WITH COOPERATIVE PROGRAMS IN 1925

School	Year Founded	Co-op Enrollment
University of Cincinnati.	1906	1,144
Northeastern University	1909	1,032
University of Pittsburgh	1910	469
University of Georgia.	1912	330
University of Akron	1914	146
University of Detroit	1915	400
Massachusetts Institute of Technology.	1917	224
Marquette University	1919	447
Newark Technical School	1919	175
Evansville College (Indiana).	1920	95
New York University	1921	111
Detroit Institute of Technology.	1921	88
University of North Carolina.	1922	188
Southern Methodist University	1925	95
University of Louisville	1925	58
Drexel Institute.	1925	88

SOURCE.—Society for the Promotion of Engineering Education, *Report of the Investigation of Engineering Education, 1923–1929*, 2 vols. (Pittsburgh, 1930), 1:563–67.

NOTE.—In addition to the schools listed above, Cleveland YMCA School of Technology, Antioch College, and Worcester Polytechnic Institute offered cooperative engineering courses, but information is not available on their founding date and enrollment. Two engineering schools, at Lafayette College and Harvard University, experimented with cooperative courses in the early 1920s but abandoned them after a few years.

maintained the test course that introduced engineers to the firm's product line by having them conduct quality-control tests on equipment. Such testing required careful measurement and control yet was boring and repetitive; consequently, it fell on novice engineers who could be trusted but who lacked seniority to qualify them for more interesting work. After two years "on Test," the new engineers advanced to positions in sales, design, or manufacturing. The test course served GE's needs through the 1890s, but after that the firm needed more engineers than the course could produce. In response, Alexander investigated various schemes for recruiting and training new engineers. In 1907, after visiting the University of Cincinnati to observe the cooperative course, he proposed a similar program between MIT and the Lynn works. A cooperative course, Alexander advocated, would produce young engineers with "an excellent preparation for high grade engineering work" and would teach them "to realize the full importance of the economics of production in the design and manufacture of machines."[28]

[28]See George Wise, " 'On Test': Postgraduate Training of Engineers at General Electric, 1892–1961," *IEEE Transactions on Education* E-22 (November 1979): 171–77. Also useful is E. E. Boyer, "Practical Shop Training for Technical Graduates," January 3,

Early Efforts to Establish the MIT-GE Cooperative Course,
1907 and 1917

On hearing of Alexander's proposal for a cooperative course, Jackson responded enthusiastically. As a new innovation in engineering education, a successful cooperative course would attract attention to his department. It would obviously provide closer ties between the department and industry, leading to additional financial support and research opportunities. And naturally, it would be an ideal mechanism for matching the supply of students Jackson was training with the needs of industry.[29] But to ensure that the course served his vision of the engineer, Jackson sought at the outset to alter its purpose. Alexander had suggested that the course should emulate the Cincinnati course and produce both design engineers and factory supervisors. Instead, Jackson advocated that the MIT cooperative program should produce men "of large vision and finer training . . . for the distinctively higher executive positions," and not just "better $2000 to $3000 men."[30] Jackson was not interested in training technicians; he wanted his students to be leaders of industry. Even though this view did not correspond exactly with GE's need for engineers, Alexander apparently accepted this view, and all subsequent descriptions of the course mentioned the managerial potential of the graduates. From the outset, Jackson strove to shape the course to reflect his vision of the engineer rather than simply serve GE's manpower needs.

Through 1907 and 1908, Jackson worked closely with the MIT faculty in formulating a curriculum. Along with concerns about the level of scientific instruction and the number of vacations, the faculty was worried that the students might be exploited; thus, they had GE agree that the students' work was to be conducted "with direct (but not sole) reference to its educational value."[31] In general, the faculty thought that a cooperative program should produce better engineers by integrating theory with practice. Yet, they were not enamored with either the idea of shop training for the students or the notion of engineers as managers, as some faculty members still believed that they should teach students how to be scientifically minded designers.

1907 (typescript) in John W. Hammond File, General Electric Company, Schenectady, New York (hereafter cited as Hammond File). The quote is from M. W. Alexander to Henry Pritchett, June 10, 1907, Jackson Papers, Box 2, Fol. 157.

[29]"Cooperative Course in Electrical Engineering," ca. 1907, Jackson Papers, Box 2, Fol. 157, and H. S. Pritchett to M. W. Alexander, September 24, 1907, MIT President Records, Box 8, Fol. 11-1.

[30]DCJ to H. Schneider, November 12, 1907, Jackson Papers, Box 4, Fol. 245.

[31]"Revised Draft of a Plan for a Course in Electrical Engineering for Engineer Apprentices," March 1908, Thomson Papers, GE Trans. 1908–13, M–N.

In their report on the proposed course, a faculty committee warned: "There is danger that students who are devoting so large a portion of their time to manual work may acquire a narrow, mechanical point of view, which will tend to prevent them from becoming engineers of a high originative type, even though they may be exceptionally well-trained as engineers for merely executive positions."[32]

The concerns of the faculty were integrated into the plan of the course by a joint MIT-GE committee, and in April 1908 the faculty approved the course. The proposed six-year course required selected students to rotate between MIT and the Lynn plant every six months. The program was expected to start the following fall, and the necessary arrangements were begun.[33] At General Electric, however, Alexander found it difficult to obtain support for the course. During 1907 an economic recession curtailed industrial production and GE executives became reluctant about initiating the new course. As long as demand was down for their products, GE officials probably did not want to make an investment in engineers they might not need. In May 1908, Alexander regretfully reported that the course would have to be abandoned temporarily "on account of a lack of enthusiasm on the part of some of the General Electric officials."[34]

Significantly, GE management was not alone in its lack of enthusiasm for a cooperative course. A number of engineers and educators voiced similar concern in the discussion following a paper on cooperative education given by Alexander at the 1908 meeting of the American Institute of Electrical Engineers (AIEE). B. A. Behrend, the leading

[32]"Report of the Committee of the Faculty on the Proposed Cooperative Course in Electrical Engineering," ca. December 1907, Jackson Papers, Box 4, Fol. 247.

[33]The joint MIT-GE committee decided that the course would have between forty and sixty students who would receive the same theoretical preparation as other electrical engineering students. During the six-year program, the students would spend the first six months at the Lynn plant and the next four and one-half years in rotation between MIT and Lynn. In the sixth year of the course, they would attend MIT full-time to write theses. While at the Lynn plant the students would work fifty-five hours per week in the shops or forty-nine hours per week when assigned to the engineering offices. Alexander was confident that the students would produce a large amount of good commercial work that would more than compensate GE for the student wages and the cost of special supervisors for the course. See "Report of the Faculty," 1907 (n. 32 above); H. W. Tyler to M. W. Alexander, February 18, 1908, MIT President Records, Box 8, Fol. 116; Karl L. Wildes, "Electrical Engineering at MIT," manuscript, MIT Archives, pp. 4–5; "Revised Draft," 1908 (n. 31 above); M. W. Alexander to F. P. Fish, December 22, 1907, MIT President Records, Box 2, Fol. 157; "Co-Operative Engineering Course," n.d., Jackson Papers, Box 4, Fol. 249; and M. W. Alexander, "The New Method of Training Engineers," *AIEE Transactions* 27, pt. 2 (1908): 1459–71, esp. 1468–69.

[34]M. W. Alexander to DCJ, May 21, 1908, Jackson Papers, Box 4, Fol. 249.

engineer at Allis-Chalmers, argued that in alternating between the classroom and the shop a boy would not be able to learn anything properly and concluded that the cooperative plan was "the most vicious educational innovation that has been proposed in years." Engineering educators pointed out that a cooperative program would eliminate elective courses, that it would not permit colleges to develop a student's character and self-discipline, and that engineering curricula were already overcrowded and could not be restructured to allow sufficient time for factory work. Out of the twelve engineers who took part in the discussion, only Jackson and Elihu Thomson, GE's preeminent inventor, spoke out in favor of the course.[35] Clearly, as this discussion reveals, not all engineers and educators favored cooperative education. If he was going to utilize cooperative education to implement his vision of the engineer, Jackson realized that he was going to have to work hard to overcome objections.

In spite of General Electric's initial rejection and the criticism they received from their professional colleagues, neither Alexander nor Jackson gave up the idea of a cooperative course. Instead, they contemplated other ways of bringing academia and industry closer together, and they waited for an opportune time to resubmit the cooperative scheme to corporate headquarters. Over the next few years, Jackson expanded his commercial research program and promoted graduate education. Alexander held a conference at the Lynn plant to acquaint engineering professors with the test course and to discuss the best ways to train men for important engineering positions.[36]

[35]See "Discussion on 'A New Method of Training Engineers' and 'The Relation of the Manufacturing to the Technical Graduate,'" *AIEE Transactions* 27, pt. 2 (1908): 1480–97. In the fall of 1907, Jackson wrote to leading engineers and MIT alumni to solicit their opinions about cooperative education, and in return he received a number of replies criticizing the proposed co-op plan. Several engineers objected to the course because they felt it took away from the theoretical preparation for which MIT was known. Others, such as Arthur D. Little, were ambivalent; although Little acknowledged the gap between the needs of industry and engineering education, he thought that a co-op course would overspecialize the students. G. M. Basford, president of the American Locomotive Company, opposed cooperative education because he believed that the student engineer would not learn anything with the thoroughness that factory operations required. Furthermore, Basford was concerned that once the student left the course and became a factory supervisor he would have difficulty earning the respect of his superiors and the men working under him. For samples of the letters that Jackson sent out, see DCJ to C. L. Edgar, November 1, 1907, DCJ to C. A. Stone, November 1, 1907, and DCJ to F. B. Abbott, November 5, 1907, Jackson Papers, Box 4, Fol. 245. For the critical replies, see H. V. Hayes to DCJ, November 6, 1907, Jackson Papers, Box 4, Fol. 245; E. B. Raymond to DCJ, December 4, 1907, A. S. Baldwin to DCJ, December 4, 1907, and A. D. Little to E. Morss, December 3, 1907, all in Jackson Papers, Box 4, Fol. 246; G. M. Basford to E. Morss, November 21, 1907, MIT President Records, Box 2, Fol. 157.

[36][DCJ]] to M. W. Alexander, December 4, 1908, Jackson Papers, Box 4, Fol. 249 and M. W. Alexander to DCJ, January 15, 1909, Jackson Papers, Box 4, Fol. 250; M. W.

The next chance to introduce the cooperative course came with the industrial boom of World War I. In April 1917 Alexander reported that the president of GE, Edwin W. Rice, Jr., favored the plan. At MIT, under the guidance of Jackson and President Richard C. Maclaurin, a faculty committee formulated a new version of the cooperative plan. This new plan, which became known as Course VI-A, called for a five-year sequence in which selected juniors alternated between MIT and GE for three years and then received master's degrees. The master's degree level was chosen because Jackson now insisted that the students undertake an independent research project. The course was approved by the faculty in May 1917, but in so doing they expressed concern about who would pay the extra expenses. Because the new program did not match the regular MIT academic calendar, it required additional instructors for summer courses. To remedy this difficulty, President Maclaurin personally took charge of the matter.[37]

Maclaurin was a hard-driving New Zealander who in his first seven years as president of MIT had raised over $4 million to begin building MIT's new Cambridge campus.[38] Hence, he was the ideal person to approach General Electric about contributing to the support of Course VI-A. Maclaurin estimated that the extra cost would be $7,000 for the first year and $19,000 per year once three classes of students were enrolled. When GE officials proved unsympathetic to his request for funds, a faculty committee restructured the program to reduce its cost. Finally, after protracted negotiations, GE reluctantly agreed to contribute $3,500.[39] The difference was presumably absorbed by Jackson's department. GE approved of the course but did not want it badly enough to underwrite even one-half of the start-up costs. In contrast, Jackson appears to have wanted the course badly enough that he was willing to absorb a significant portion of the expenses.

Alexander to R. C. Maclaurin, January 22, 1910, MIT President Records, Box 2, Fol. 11-1; M. W. Alexander to R. C. Maclaurin, December 28, 1912, MIT President Records, Box 8, Fol. 11-1.

[37]Duncan Davis to DCJ, March 31, 1917, Jackson Papers, Box 4, Fol. 251; M. W. Alexander to R. C. Maclaurin, April 27, 1917, MIT President Records, Box 8, Fol. 11-1; R. C. Maclaurin to M. W. Alexander, April 30, 1917, MIT President Records, Box 8, Fol. 11-1; R. C. Maclaurin to DCJ, May 9, 1917, Jackson Papers, Box 4, Fol. 251; and R. C. Maclaurin to E. W. Rice, Jr., May 9, 1917, Thomson Papers, GE Trans. 1915–23, MIT.

[38]Henry Greenleaf Pearson, *Richard Cockburn Maclaurin* (New York, 1937), pp. 97–161; Prescott (n. 21 above), pp. 247–64; Wildes and Lindgren (n. 16 above), pp. 49–52; and Sinclair, "Inventing a Genteel Tradition" (n. 17 above), pp. 11–17.

[39]R. C. Maclaurin to E. W. Rice, Jr., May 9, 1917, and R. C. Maclaurin to E. W. Rice, Jr., June 5, 1917, Thomson Papers, GE Trans. 1915–23, MIT; R. C. Maclaurin to F. P. Fish, May 23, 1917, MIT President Records, Box 9, Fol. 212-1; and DCJ to Administrative Committee, April 6, 1920, MIT President Records, Box 4, Fol. 326.

Although World War I permitted the introduction of the cooperative course, ironically it also prevented the course from being fully implemented. Two classes enrolled in the program in 1917 and 1918, but they were forced to withdraw with the establishment of the Student Army Training Corps. With the entry of the United States into the war, all engineering students were liable to the draft except for those attending designated schools with instruction in military affairs. Hence, the first co-op students returned to MIT and joined the student corps. For the moment the cooperative course was halted, and neither of the first two classes completed the program.[40]

The Cooperative Course in the 1920s

Following World War I, both MIT and GE were anxious to reestablish the cooperative course and capitalize on its benefits. At MIT, Jackson recruited William H. Timbie from Pratt Institute to supervise the course on a day-to-day basis.[41] In July 1919 it was announced that juniors would be admitted to the cooperative program the following fall. General Electric began interviewing candidates and selected twenty-eight men on the basis of their academic record, past working experience, conduct, initiative, self-reliance, and general appearance. To help finance the course, GE now promised MIT an annual contribution of $5,000.[42]

Course VI-A in the 1920s differed significantly from the original 1907 plan. The course now reflected Jackson's vision of the engineer as expert and executive; it was to develop "the highest scientific, engineering, and administrative capacities of those who desire to become

[40]Because of the war and because the course had not been approved until after the spring term had ended, Jackson was in the peculiar situation of having to look for students to take the course. Since only a few electrical engineering juniors were interested in taking the course, Jackson hastily recruited transfers from other colleges in order to have eighteen students to begin the course in 1917. See DCJ to Harold Pender, June 7, 1917, Jackson Papers, Box 4, Fol. 251. Jackson sent similar letters asking for students to C. F. Scott, M. C. Beebe, J. H. Parker, and M. MacLaren, all engineering educators at other universities. See also M. W. Alexander to DCJ, January 30, 1918, MIT President Records, Box 8, Fol. 11-1; DCJ to Administrative Committee, April 6, 1920, MIT President Records, Box 4, Fol. 326, and Wildes and Lindgren (n. 16 above), pp. 57 and 60–62. For an informed discussion of the SATC, see Noble (n. 2 above), pp. 217–23.

[41]On William H. Timbie, see Wildes and Lindgren (n. 16 above), p. 129, and *Hand in Hand* (n. 26 above).

[42]See Wildes (n. 33 above), pp. 4–32, 4–33, 4–37, 4–38, and William H. Timbie, "Cooperative Course in Electrical Engineering of the Massachusetts Institute of Technology," *AIEE Journal* 44 (June 1925): 613–17, esp. 615. I was unable to find any information in the MIT or GE records about why GE willingly gave $5,000 at this time.

leaders in manufacturing industries."[43] Hardly anything was now said about the importance of giving the student practical shop training in order to make him a good factory supervisor. Instead, the curriculum emphasized theoretical knowledge and exposure to corporate culture. Students entered the course at the start of their junior years because their first two years at MIT were devoted to basic science and mathematics courses. To ensure that the students possessed a thorough mastery of electrical engineering science, they spent the fifth year of the program doing research at both MIT and GE.[44] In terms of purpose and curriculum, the cooperative program served Jackson's vision of the engineer rather than GE's need for test engineers or factory supervisors.

During their fourteen-week terms at General Electric's Lynn plant, the students performed manual labor but, more important, studied in detail the problems related to the administration of large manufacturing enterprises and the implementation of engineering projects. Beginning in the machine shop, they subsequently worked in the armature winding department, drafting office, testing departments, and general factory production sections. The students regularly attended talks by GE department heads on manufacturing techniques and recent engineering developments.[45]

As the students worked by day in the plant, by night they took college courses. As Jackson explained, these courses were to keep the students "wide awake and in touch with their Technology studies." Four nights each week, MIT instructors went to Lynn and taught electrical engineering, accounting, literature, business psychology, and composition. Significantly, the organizers of VI-A considered these evening classes a means of disciplining the student; through them, the student acquired "the habit of working days and studying nights, a habit which experience proves is hard to break in later years."[46]

[43]M. W. Alexander and DCJ, "Requirements of the Engineering Industries and the Education of Engineers," *Mechanical Engineering* 43 (June 1921): 391–95.

[44]M. W. Alexander, "Co-operative Engineering Course Conducted Jointly by Massachusetts Institute of Technology and General Electric Company, Lynn Works," June 18, 1917, Thomson Papers, GE Trans. 1915–23, MIT.

[45]See Timbie, "Cooperative Course in Electrical Engineering" (n. 42 above), pp. 614–16; Timbie, "A Co-Operative Course in Electrical Engineering," 1920, Jackson Papers, Box 4, Fol. 255. For memorabilia and details about student life at the Lynn works, see "VI-A Album," Historical Collections, MIT Museum.

[46]First quote is from DCJ to R. C. Maclaurin, October 22, 1919, Jackson Papers, Box 4, Fol. 252. Second quote is from Alexander and DCJ, "Requirements of the Engineering Industries" (n. 43 above), p. 397. See also Timbie, "Cooperative Course in Electrical Engineering" (n. 42 above), pp. 613–14, 616. Notably, of the sixteen colleges having cooperative engineering courses, only the MIT program required students to take courses during their working periods; see SPEE, *Report on Engineering Education* (n. 7 above), 1:576.

Not just in the evening classes, but throughout the cooperative program, emphasis was placed on shaping the student's character and behavior. In 1908, Alexander had expressed the ideal of discipline when he stated that "under such a plan, the freedom enjoyed by students during the college career is happily interrupted by the stern discipline that must prevail in a business organization."[47] Although this was still the goal in the 1920s, it was now phrased in terms of teamwork, company loyalty, and human relations. During their co-op assignments, the students were expected to enter heartily into the spirit of the company. Working in the shop and engineering office, the students learned teamwork and cooperation, and, most important, they learned about the relationship between managers and workers:

> The students soon come to understand the sterling qualities of the men with whom they are working and study to adapt themselves to the personal characteristics and eccentricities of the various foremen under whom they are working, an experience which will be of the utmost importance to them when they have arrived at a position where they are directing the work of others. So during their period at the plants they secure experience not only in Electrical Engineering, but also in what is called for lack of a better name, "Human Relations."[48]

Thus in Course VI-A, the students were not only given technical training but taught how to behave within the corporate hierarchy.

By 1920, Course VI-A appeared to be a successful program. Fifty-two students were enrolled and Jackson reported: "The men are full of enthusiasm, are gaining a great deal out of their shop experiences and are working at their studies with a vigor that is very gratifying and reassuring."[49] Yet at the same moment, Jackson was called on to defend the program. In the years immediately following World War I, MIT suddenly found itself in a serious financial crisis. Although they devised the "Technology Plan" to raise money, MIT administrators sought other ways to cut expenses.[50] Under these conditions,

[47]Alexander, "New Method of Training Engineers" (n. 33 above), p. 1465.

[48]Timbie, "Cooperative Course in Electrical Engineering" (n. 42 above), pp. 614, 616, 617.

[49]Quote is from DCJ to Administrative Committee, April 6, 1920, MIT President Records, Box 4, Fol. 326.

[50]In 1917 MIT lost its share of the McKay Endowment, which it had been sharing with Harvard and which had been used primarily to support the electrical engineering department. This was followed by a refusal of the Massachusetts legislature to continue its annual appropriation of $100,000 for MIT as it had done since 1911. In response to this crisis, MIT organized the "Technology Plan," a scheme whereby money was

Jackson was asked in March 1920 whether the costs of the cooperative course could be reduced. He adamantly refused to make any changes, arguing that the quality of scientific training should not be compromised in order to save money. As the administration pressed the issue, Jackson defended the course by revealing that it had become profitable. Between GE's annual contribution of $5,000 and the students' tuition, the program generated not only sufficient revenue to pay the extra expenses of evening and summer classes but also some surplus. This seemed to satisfy the administration, and the course was left intact.[51]

From this incident, it is apparent that MIT as a whole did not necessarily regard the cooperative course as an integral part of its institutional strategy. On one hand, the cooperative course complemented the missions of applied science and industrial support being promoted at MIT by William H. Walker. On the other hand, the course flew in the face of the vision of MIT as a center for pure research that was promoted by Maclaurin and Arthur A. Noyes.[52] Given the ambiguity about MIT's mission in the early 1920s, all one can conclude is that the top administration had no clear interest in the cooperative course; rather, it was a component of Jackson's departmental strategy. Through the course, Jackson strengthened his department by expanding enrollments, acquiring prestige, and even making some profit. But most important, the course permitted him to train engineers who fulfilled his vision.

Initially, General Electric seems to have been satisfied with Course VI-A. Bright and hard-working, the students showed promise of becoming good managers and engineers. Hoping to increase the benefits

raised by offering the services of MIT's personnel, laboratories, and libraries for the conduct of industrial research. See Sinclair, "Inventing a Genteel Tradition" (n. 17 above), pp. 6–7; Noble (n. 2 above), pp. 141–43; Wildes and Lindgren (n. 16 above), pp. 62–63; and John W. Servos, "The Industrial Relations of Science: Chemical Engineering at MIT, 1900–1939," *Isis* 71 (1980): 531–49, esp. 539. See also C. H. Warren to DCJ, March 19, 1920, and DCJ to C. H. Warren, March 23, 1920, Jackson Papers, Box 4, Fol. 253.

[51]DCJ to Administrative Committee, April 6, 1920, and April 8, 1920, MIT President Records, Box 4, Fol. 326, and E. B. Wilson to Administrative Committee, December 4, 1920, MIT President Records, Box 4, Fol. 324. That the cooperative course was profitable for MIT is not particularly surprising; in their study of engineering education in the mid-1920s, the SPEE found that cooperative courses were often economically advantageous for engineering schools since they permitted a school to accommodate more students without enlarging its physical facilities (thus lowering the capital expenditure for training each student), while at the same time larger student enrollments increased tuition income. See SPEE, *Report on Engineering Education* (n. 7 above), 1:609–11.

[52]Servos (n. 50 above) and Sinclair, "Inventing a Genteel Tradition" (n. 17 above), pp. 11–12.

of the course, GE officials allowed the 1920 class to be 80 percent larger than the previous one. In addition, to assist the students in paying their living expenses while working at the plant, the company raised their wages from $15 to $21 per week.[53]

But these positive steps were short-lived, revealing that GE was ambivalent about the course. Just as business conditions in 1907–8 had led GE to reject the original cooperative plan, so a recession in 1921 prompted a cutback on the course. GE management reduced student wages to their original rates and limited the incoming class to forty. Jackson probably concluded that he had no choice but to let GE make these cuts.[54] At the same time, though, he compensated for GE's reductions by placing a few co-op students with the Edison Electric Illuminating Company of Boston and the Boston Elevated Railway Company.[55] Proceeding in this way, Jackson acted as a good academic entrepreneur; he not only maintained the momentum of the program but used this opportunity to diversify it.

Not only did the management of GE cause difficulties for Jackson, but students did so as well. In December 1921, a student, Paul Kellogg, complained about a reprimand he had received from a company supervisor, after he had missed several shop lectures: "We don't give a damn about any of you fellows. You go to any damn lecture that's given, whether you've heard it before or not, or get out, every one of you."[56] Upset by this scolding, Kellogg as well as several other VI-A students suspected that GE was not really interested in the course.

While GE officials immediately assured Kellogg that the company had every intention of standing behind the course, Jackson responded by calling Kellogg to his office (see fig. 1) to explain to him the purpose

---

[53]"Report of the Electrical Engineering Department," ca. 1920, Thomson Papers, GE Trans. 1915–23, MIT; C. K. Tripp to DCJ, June 22, 1920, and DCJ to C. K. Tripp, June 26, 1920, both in Jackson Papers, Box 4, Fol. 253. See also Alexander and DCJ, "Requirements of the Engineering Industries" (n. 43 above), p. 395.

[54]In early 1921 there was some talk on the part of GE about dropping the entire course; Jackson, however, reminded GE that, when the course was first arranged, it was agreed that it would run for at least five years and that afterward either side could withdraw only on giving one year's notice. Furthermore, he maintained that it was important to keep the program going even in a dull period in order to ensure that a supply of trained men would be available in prosperous times. See DCJ to R. H. Rice, January 13, 1921, R. H. Rice to DCJ, January 10, 1921, and W. H. Timbie to DCJ, March 22, 1921, all in Jackson Papers, Box 4, Fol. 254.

[55]DCJ to Administrative Committee, October 3, 1921, Thomson Papers, GE Trans. 1915–23, MIT; "Program for Visiting Committee," ca. 1922, MIT President Records, Box 2, Fol. 128; and Massachusetts Institute of Technology, *President's Report, January 1922* (Cambridge, 1922), pp. 67–68.

[56]P. M. Kellogg to DCJ, December 16, 1921, Jackson Papers, Box 4, Fol. 255.

Fig. 1.—Dugald C. Jackson in his office at MIT in the late 1930s. (Historical Collections, MIT Museum.)

of the course and what was expected of him. He reminded Kellogg that the primary interest of MIT and GE was that he should become "a sound-minded co-operative man with the scientific training which will enable him to become a creative leader in the manufacturing industries."[57] To do this, Kellogg had to learn to use tact and reasonable compromise. As an engineering educator, Jackson spoke to the student in this manner to ensure that he learned to conform in order to become a good corporate engineer. As an academic entrepreneur, though, Jackson treated Kellogg as he did probably because he wanted no one to raise doubts about the course, doubts that could only weaken the overall position of his department.[58]

[57]E. Thomson to P. M. Kellogg, December 20, 1921, Thomson Papers, 1921 Letterbook, and P. M. Kellogg to DCJ, December 31, 1921, Jackson Papers, Box 4, Fol. 255. To assure Kellogg that both GE and MIT had only his best interests in mind, Jackson told Kellogg that neither institution profited from the course; as he later wrote to F. P. Fish: "In respect to the expense of the course, I pointed out to him that the Institute and the General Electric Company are spending twice as much money per annum on his education as he contributes in fee, so that neither the Institute nor the General Electric Company are [sic] making a profit out of him as a student." See DCJ to F. P. Fish, January 5, 1922, Jackson Papers, Box 4, Fol. 255.

[58]Jackson's final comment on the incident was "I think the young man received what he deserved and that it will do him good." See DCJ to E. Thomson, January 5, 1922,

Jackson had no sooner dealt with Kellogg than a more serious issue arose with the cooperative program, that of employing graduates of the course at GE. This reveals the differences between Jackson's vision of the engineer and GE's needs and preferences. As early as March 1922, Jackson inquired about what arrangements were being made for hiring the first graduating class of cooperative students. Not only was it inherently prestigious to secure good jobs for his students, but Jackson was anxious that his students assume the executive positions for which he had trained them. Initially, the company had contemplated a special hiring procedure for the MIT co-op students, but then decided to follow its long-standing recruitment policy for new engineers. This meant that the co-op students hired would start at the low salary of $27 per week and would be assigned to the Schenectady testing department (see fig. 2), where they would work until selected by another department. Jackson vehemently objected to this. In addition to their undergraduate degree, he argued, the cooperative graduates had worked in industry for eighteen months and had done the advanced work required for a master's degree. Jackson obviously thought these students deserved special treatment.[59]

Jackson's objections fell on deaf ears, however. GE management argued that favoring the MIT graduates would create discontent among the ranks of their student engineers who came from other engineering schools and the test course. If MIT co-op men proved superior to those from other schools, then they would be given higher starting salaries in the future. For the moment, as Elihu Thomson, GE's distinguished inventor and acting president of MIT, put it, "the Company must follow the course which seems to it the most conducive, not only to the satisfaction of all its men, but which accords with its own best interest."[60] In the face of this opposition, Jackson apologized for in-

---

Jackson Papers, Box 4, Fol. 255. Kellogg apparently learned his lesson and was later hired by AT&T. See W. H. Timbie to DCJ, October 16, 1923, Jackson Papers, Box 4, Fol. 256.

[59]The policy of having newly hired engineers go through the Schenectady testing department dates back to 1901; see E. W. Rice, Jr., Circular Letter A-209, Re: Additions to Engineering Force, January 28, 1901, Hammond File, Item L-5706. See also C. E. Arvidson, "G-E Testing Department Assists Graduate Engineers in Securing Positions after Giving Them Practical Training," [GE] *Schenectady Works News* (January 4, 1924), Hammond File, L-991; DCJ to F. C. Pratt, March 7, 1922; and W. H. Timbie, Statement, July 1, 1922, both in Jackson Papers, Box 4, Fol. 255.

[60]E. Thomson to F. C. Pratt, March 16, 1922, and F. C. Pratt to E. Thomson, March 14, 1922, both in Jackson Papers, Box 4, Fol. 255; F. C. Pratt to DCJ, Thomson Papers, GE Trans. 1915–23, MIT.

FIG. 2.—Student engineer conducting tests on large rotating electrical equipment at General Electric's Schenectady works, 1931. He is standing next to the test and control bench and is probably testing the electrical machine at the right. (Ellenberger Collection, IEEE Center for the History of Electrical Engineering. Thanks to Joyce Bedi for alerting me to this photo.)

terfering with the internal affairs of the company; his only interest, he explained, was in "increasing the intimacy between the industries and the engineering schools."[61] Yet he was plainly disappointed that the first cooperative students would not assume the management positions that he hoped they would attain.

By July 1922, the result of General Electric's hiring policy was apparent: only five of the first twenty-five graduates of the cooperative course chose to stay with GE. Although most of the students indicated that they preferred to work for GE, they left as other companies made them better offers. By neglecting to create a procedure by which cooperative graduates were brought into the company, GE failed to reap the full benefits of the course.[62]

The problem of placing graduates of the MIT-GE cooperative program was not fully resolved in later years. Throughout the 1920s and

[61]DCJ to F. C. Pratt, March 23, 1922, and DCJ to E. Thomson, March 23, 1922, both in Thomson Papers, GE Trans. 1915–23, MIT. In writing Pratt, Jackson commented, "I recognize that even so great a corporation as the General Electric Company cannot do the co-operative work for all the industries, but of course you and I would like to see it taking leadership."

[62]W. H. Timbie, Statement, July 1, 1922, Jackson Papers, Box 4, Fol. 255.

1930s GE hired the MIT co-op students in fits and starts, guided by the economic conditions and their immediate manpower needs. In some years they took as many as two-thirds of a class, whereas at other times they took as few as one-tenth. (See table 2.) In all likelihood, the reason why GE did not hire more of the course's graduates was because the course did not precisely fulfill the firm's specific needs for technical manpower. GE needed engineers, to be certain, but did it want young engineers who had been told by Jackson that they were destined to be managers and executives? How useful were MIT graduates with their strong theoretical preparation in performing the routine jobs of factory supervision and technical sales? From years of experience with the test course, GE officials knew that investing heavily in the training of individuals was a risky enterprise; to get one good engineer one frequently had to train one or two others who did not meet company standards.

Given these problems GE managers were cautious about investing too much in the MIT course and may have preferred to use their own in-house training programs. Notably, GE supplemented its test course in 1923 with an advanced engineering course that prepared engineers for key technical and managerial positions; it is likely that this program preempted the role of the MIT course and received more attention from GE executives. Finally, the co-op course could provide only a small fraction of the entry-level engineers required annually by GE. During the 1920s it supplied no more than 10 percent of the total of engineers hired each year.[63] In summary, although GE officials showed some interest in the cooperative course, they were not enthusiastic about it because it failed to provide the quality or quantity of engineers they needed.

The experience of not being able to place the co-op graduates soured Jackson on the GE cooperative course. Rather than actively improve the GE program, during the next few years Jackson diversified the overall course and continued to pursue his vision of the

[63]On the advanced engineering course, see Wise, " 'On Test' " (n. 28 above), p. 175. For statistics on entry-level engineers and cooperative graduates, see W. H. Timbie to DCJ, October 16, 1923, Jackson Papers, Box 4, Fol. 256, DCJ to G. Swope, March 4, 1929, Jackson Papers, Box 4, Fol. 260, and A. B. Caine to PRO Exempt Staff and Program Partners on "Entry Level Acceptances," April 5, 1979, Applicant Referral Center, General Electric Company, Schenectady, New York. I am grateful to George Wise for providing me with this document, which lists the number of entry-level engineers hired by the company from 1919 to 1978. From 1924 to 1929, GE hired between 328 and 432 new engineers annually; during the same period, the course graduated between thirty and fifty students, and GE took no more than one-half of these graduates in any year.

TABLE 2

PLACEMENT OF MIT COOPERATIVE PROGRAM STUDENTS
IN ELECTRICAL ENGINEERING, 1921–1941

	1921	1923	1925	1927	1929	1931	1933	1935	1937	1939	1941
Students who co-oped with General Electric	11	9	10	6	6	5	3	7	7	7	3
Graduates hired by General Electric	4	1	5	3	4	0	1	2	2	4	1
Students who co-oped with other firms*	0	1	4	3	8	10	11	4	3	4	0
Graduates hired by other firms	0	0	1	2	3	0	1	0	1	3	0
Total respondents†	11	10	14	9	14	15	14	11	10	11	3
Total graduates	25	30	45	38	35	40	28	15	42	48	19

SOURCE.—*Hand in Hand: On the Occasion of Fifty Years of Industry-aided Selective Cooperative Education* (Medford, Mass: Gordon, 1958).
*These firms include the New York offices of AT&T, Boston Edison, and Stone & Webster.
†This table is compiled from career data submitted by graduates for an anniversary volume published in 1958. Naturally, not all the graduates chose to submit data.

engineer as a corporate leader. Similar arrangements were established with Stone and Webster and AT&T. Jackson discussed the possibility of a course with RCA, and he was instrumental in creating a cooperative program for civil engineering students with the Boston & Maine Railroad. Only in 1927 did Jackson try to expand the GE cooperative course by suggesting that students be permitted to specialize in illumination engineering, but nothing came of his proposal.[64]

In 1932, Jackson found himself again defending the course. Because of the Depression, GE abandoned many of its educational activities and considered withdrawing from the cooperative program. With the help of Timbie and MIT President Karl Compton, Jackson struggled to persuade GE not to terminate the course. For the next several years, to ease the burden on GE, few students were given co-op assignments there. As to financing the course, although the extra expenses in 1934 totaled $3,500, Jackson offered to keep it going if GE would contribute only $750.[65] Clearly, GE was indifferent toward the course as it was willing to drop it to save money.

Although GE was unwilling to invest in Course VI-A, Jackson continued to strengthen his electrical engineering department by adding a concentration in radio engineering, an undergraduate honors course, and a joint research program with the physics department. When combined with his previous institution-building efforts, these new innovations helped make Jackson's department by the mid-1930s the largest and most prestigious electrical engineering department in the United States. Moreover, Jackson trained a large number of electrical

[64]In reporting the success of one cooperative graduate, Timbie remarked to Jackson, "I thought you would like to know that there are a few items on the credit side of the VI-A ledger." See W. H. Timbie to DCJ, October 20, 1923, Jackson Papers, Box 4, Fol. 256. See also DCJ to S. W. Stratton, October 20, 1927, MIT President Records, Box 4, Fol. 327, and G. Hannaeur to S. W. Stratton, January 19, 1928, MIT President Records, Box 8, Fol. 107-1; and DCJ to G. Swope, March 22, 1927, MIT President Records, Box 2, Fol. 327.

[65]K. T. Compton to V. Bush, November 22, 1932, Jackson Papers, Box 4, Fol. 261; W. H. Timbie to DCJ, November 20, 1934, and K. L. Wildes to DCJ, November 21, 1934, both in Jackson Papers, Box 4, Fol. 262. Significantly, when GE initially threatened to withdraw from the program, it was Jackson who reminded GE president Gerard Swope of the importance of education as a part of corporate capitalism: "However, the main question is whether a great industrial organization can draw out of such an arrangement merely because of bad times, without hurting its status. The responsibility of the directors to the stockholders calls for provision of a flow of young men of the highest order into staff and to break thread in a depression (unless under impulsion of dire necessity) means a full stop in the most select stream, the flow of which can never be reestablished again in full confidence." See DCJ to G. Swope, November 2, 1932, Jackson Papers, Box 4, Fol. 261.

engineers as both graduate students and junior faculty. Inspired by Jackson's leadership and institution-building skills, these men became chairmen and professors in dozens of electrical engineering departments; as a result, Jackson's ideas have had a subtle but pervasive influence on engineering education in this century.[66]

Conclusion

The story of the MIT-GE cooperative course reveals much about how engineering education was transformed in response to broad changes in American business and industry from 1880 to 1930. First and foremost, we can see that the process by which engineering education was coordinated with the needs and expectations of industry was hardly smooth and orderly. Leaders at powerful corporations such as General Electric did not succeed in directly rationalizing and standardizing engineering education. They could not simply order entry-level engineers from engineering schools in the same way they ordered other inputs for the industrial machine. As consumers of engineers, corporate America could not assume that the suppliers of engineers, the engineering schools, would deliver the product in the quantity or quality that they wanted; supply did not automatically match demand. In particular, GE found that the MIT cooperative course failed to provide anywhere near the number of entry-level engineers that they needed annually and those that it did provide were overqualified for the essential jobs of manufacturing, testing, and sales. Instead, corporate leaders found that they had to negotiate with engineering educators, who had their own ideas and notions about how engineers should be trained. As a result, the process by which engineering education was matched with the needs of industry was marked by a clash of values and expectations. This is in accord with the general pattern by which historians have interpreted the integration of science and technology with business and industry.

But within this clash of needs and values, it is important to note who played the key role. Previously, Noble has argued that the corporate elite played the predominant role in bringing "both the form and content . . . of engineering education into line with what they

---

[66]On the steady growth of the electrical engineering department, see "Report of the Visiting Committee," June 25, 1925, Thomson Papers, GE Trans. 1915–23, MIT, and Wildes and Lindgren (n. 16 above), pp. 69–72. At an AIEE meeting in 1929, there were over fifty former students and junior colleagues of Jackson in attendance, and all were either college administrators or chief engineers; see Brittain (n. 11 above), p. 355.

perceived to be the immediate manpower needs of industry and the long-term requirements of continued corporate development."[67] In contrast, the MIT-GE course indicates that it was not the leaders of GE who spearheaded the course but rather an engineering educator, Jackson. Throughout the history of the program, we see that GE was reluctant to invest in the course and to hire its graduates. In general, GE officials do not seem to have had any overall plan for the course, and their position can be characterized as passive. Instead, the MIT-GE program was actively shaped by Jackson, who made certain that the course contained a high level of scientific instruction and research and that it was aimed at creating engineers who might become leaders and managers. Because it permitted him to pursue his vision of the engineer, Jackson made the cooperative course a vital part of his institution-building strategy.[68]

Although GE was the passive party in the development of the co-operative course, the firm was hardly powerless. As has been shown, if GE refused to contribute to the course's costs, changed the number of students permitted to enroll, or hired only a few graduates, Jackson had no choice but to accept. In fact, one could conclude that, by assuming a passively aggressive stance, GE maximized its benefits from the course; although the company invested little in the course, it came away with a handful of highly trained engineers who supplemented those trained in the company's in-house programs. Such an approach must have made good sense to corporate managers familiar with vertical integration—who knew that while it was desirable to integrate some functions within the firm in order to realize economies of speed and coordination, there were other risky and costly activities that were best externalized or left outside the firm. I suggest that, although GE provided some training through its test course, it preferred to min-

[67]See Noble (n. 2 above), p. 169.

[68]Significantly, three other scholars have come to similar conclusions with regard to David Noble's claims that engineers and scientists readily adapted themselves to the needs of corporate capitalism. In examining the development of chemical engineering at MIT, John W. Servos (n. 50 above) has argued that, while MIT professors secured industrial support for their research in chemical engineering, by the late 1920s they were also insisting that industry allow them greater freedom in the choice of research problems and in publishing the results. Similarly, Bruce E. Seely (n. 17 above) has found that engineers promoted the development of standards not simply in response to business needs but instead to further their own abstract concept of efficiency. Finally, Peter Meiksins (n. 9 above) has shown through his work on the American Association of Engineers that the engineering profession was hardly unanimous in its support of big business; instead, he has argued that both reformers and rank and file engineers periodically challenge the pro-business stance of the major engineering societies.

imize its involvement in this area and let engineering schools such as MIT assume the burden of risk and cost.[69]

Even though GE benefited from the cooperative course, this does not mean that Jackson was beholden to GE and merely a "stooge" for corporate capitalism. Rather, Jackson had his own unique vision of the professional engineer that complemented but did not match the needs or expectations of industry. American corporations needed scientifically trained engineers to serve within the industrial bureaucracy, but they did not share Jackson's vision of the engineer as leader. In the MIT-GE program, as in his other efforts to promote electrical engineering, Jackson sought to implement this vision rather than simply serve the immediate needs of industry.

In the final analysis, we must return to Jackson's role as academic entrepreneur. In the uncertain and rapidly changing worlds of industry and academia in the early 20th century, he assumed the risks and took decisive action. Industry needed more engineers, and universities were eager to provide new professionals, yet there existed no mechanisms by which this demand and supply could be matched. Through innovations such as the MIT-GE cooperative engineering course, Jackson coordinated supply with demand and thus ensured the sustained growth of American industry, universities, and the electrical engineering profession. He shaped this particular innovation in response to both the marketplace and his own vision of the engineer. Hence it is inaccurate to view the integration of science, technology, and business in America as a deterministic process in which the needs of business simply dictated the outcome. Rather, if we are to understand engineering in 20th-century America, it is necessary to grapple with the complex beliefs and actions of academic entrepreneurs such as Jackson.

[69]For a discussion of vertical integration, see Chandler, *The Visible Hand* (n. 4 above), pp. 287–314, and Chandler, "The Emergence of Managerial Capitalism," *Business History Review* 58 (Winter 1984): 473–503, esp. 479–92. In reviewing the involvement of Charles Steinmetz and GE in the electrical engineering program at Union College, Ronald R. Kline argues that it was indeed GE's policy to leave the teaching of engineering fundamentals to the universities. See "The General Electric Professorship at Union College, 1903–1941," *IEEE Transactions on Education*, in press.

THE "REVOLT OF THE ENGINEERS" RECONSIDERED

PETER MEIKSINS

Recent scholarship on the history of American engineers has focused on the years surrounding World War I as crucially important
to the formation of the modern engineering profession. This is hardly
surprising, since these were the years during which the so-called revolt
of the engineers (to use Edwin Layton's phrase) took place. Mechanical
and civil engineers in particular seemed to be searching for a redefinition of their place in society, to be seeking a more active role in
solving the problems (both technical and social) of the day. Many of
them seemed to have concluded that engineers, by reason of their
training, experience, and social position, could develop a different,
and superior, kind of leadership than that exercised by business. Their
search spawned a number of important "causes" (ranging from the
scientific management movement to efforts to create a unified engineering profession) and often turbulent debates about the meaning
of engineering professionalism. But this activity rather rapidly came
to an end in the probusiness prosperity of the 1920s.

Looking back, historians have concluded that this was a kind of
formative crisis within the engineering fraternity. It was during this
period in their history that the engineers came to grips with fundamental questions about their relationship to business, their rightful
degree of social responsibility, and the appropriate mode of organization for the profession as a whole. Consequently, a study of this
period should yield important insights into the nature and social role
of the engineering profession in 20th-century America.

Unfortunately, there has been no unanimity as to what precisely
those insights might be. In the most important analyses, two subtly

PETER MEIKSINS is associate professor of sociology at State University of New York
College at Geneseo. The research for this article was supported by summer stipends
from the SUNY Research Foundation and the National Endowment for the Humanities, whose assistance is gratefully acknowledged. The author also thanks Lynn
Zimmer, Sarah Harrington, and the *Technology and Culture* referees for their support
and constructive criticism. An earlier version of the paper was presented at the 1984
meeting of the Society for the History of Technology.

This essay originally appeared in *Technology and Culture,* vol. 29, no. 2, April 1988.

different views of the engineers' social role have emerged. Both agree that American engineers eventually came to align themselves with the businessmen who employed them, but they disagree about *why* this occurred and whether it was inevitable.

One school of thought is well represented by David Noble's *America by Design*. Noble takes the position that the engineers' "revolt" in the early 20th century was, in many respects, an anomaly. Although he notes that there was more than one theoretically possible form of professional identity for engineers, he concludes that the possibility of a serious challenge to the "marriage" of engineering and corporate business was small indeed.[1] Engineering reformers such as Morris L. Cooke and Frederick Haynes Newell are characterized as "corporate liberals" who sought to combine reform with the requirements of corporate business. The growth immediately after World War I of the American Association of Engineers, with its reform rhetoric and its concern for the engineers' material welfare, resulted largely from the very temporary, untypical experience of young engineers in the post-war slump. Once prosperity returned, the association quickly declined.[2] In sum, Noble sees no real challenge to business hegemony over engineering in these types of protest. He concludes that the engineers' dependence on industrial organizations made the "impotence" of engineering reformers and their accommodation with American business virtually inevitable.[3] The lesson of early-20th-century engineering history, for Noble, is the forging of this accommodation, the willing subordination of engineering to corporate ends, and the creation of a "domesticated" brand of engineer willing to serve, wittingly or unwittingly, as an arm of "capitalist reason."[4]

A somewhat different view of the engineers emerges from the work of the historians of professional societies, particularly that of Edwin Layton and Bruce Sinclair. While conceding that an accommodation between business and engineering had been struck, they are less willing than Noble to regard the "domestication" of the engineer as natural and inevitable. Sinclair, for example, emphasizes the recurring conflict within the American Society of Mechanical Engineers (ASME) between what he calls its "technical" and its "social" purposes and the instability of business dominance over the society. Debates over the nature of the ASME, and over the meaning of both its technical and its social purpose, continued in the 1930s, well after the decline of

---

[1] David F. Noble, *America by Design* (New York, 1977), pp. 39–41.
[2] Ibid., pp. 62–63.
[3] Ibid., p. 63.
[4] Ibid., pp. 321–24.

the original revolt of the engineers, and arose even in the 1950s and 1970s with a new concern over engineering ethics.[5] Engineers, in Sinclair's view, are an ambiguous, complex group: accommodated to the needs of American business but unsure and uneasy about the consistency of this position with their own professional standards.

Edwin Layton, in his now-classic *The Revolt of the Engineers*, draws a similar picture. In his account of the decline of the early-20th-century revolt, Layton places much of the blame for the failures on the reformers themselves—their inability to "deliver the goods" and their own retreat into conservatism in the complacency of the prosperous 1920s.[6] This capitulation was not necessary and inevitable—engineering professionalism of the kind represented by Cooke, Newell, and others might have been more successful. And, as Layton makes clear in his 1986 preface to the new edition of *The Revolt of the Engineers*, debates over the social responsibility of the engineer did not end in the 1920s. He clearly hopes that the engineers will become a "loyal opposition" in American business, neither capitulating uncritically to nor rejecting out of hand the needs of large corporations. He also clearly believes that it is possible for engineers to move in this direction.[7]

Each of these analyses captures important if partial truths about the role of engineers in American society. However, none of them succeeds in telling the whole story. Noble overstates the inevitability and completeness of the engineers' accommodation to business and is, therefore, unable to explain the periodic flare-ups of engineering dissent to which Layton and Sinclair point. On the other hand, Layton and Sinclair do not adequately emphasize the complexity of engineering dissent. Layton does note that there has been more than one strain of engineering dissent in the United States. But, as this article will argue, the complex relations among these diverse forms of protest and the probusiness engineering orthodoxy must be the *focus* of an analysis of engineering professionalism and the engineers' role in American society.

This article proposes a reinterpretation of the revolt of the engineers and, in the process, a reanalysis of the engineers' position in American society. It argues that the truly significant result of the defeat of this revolt was the emergence of a new kind of engineering protest, centered on the material concerns of the average engineer

[5]Bruce Sinclair, *A Centennial History of the American Society of Mechanical Engineers, 1880–1980* (Toronto, 1980), pp. 21–28; chap. 8.

[6]Edwin T. Layton, Jr., *The Revolt of the Engineers* (Baltimore, 1986), p. 212.

[7]Ibid., p. xi.

and rooted in the changing conditions of American engineers. Ironically, it was this new form of dissent that gave the appearance of strength to the very different reform movement led by prominent engineers such as Cooke and Newell. The engineering "progressives" did have a substantial amount of support. But it was not so much support for their ideas on business as it was support derived of a momentary alliance with a new generation of engineer-employees concerned about their material interests. The alliance of these revolts was extremely unstable—it proved very easy indeed to detach them from one another, leaving both in a weakened condition. In the end, the strand of protest exemplified by Cooke proved deficient in strength. In contrast, the newer strand of protest survived, becoming, at least at times, an important factor within the engineering profession.

At the risk of oversimplification, we can identify three broad strains of engineering ideology in the period 1910–30. First, there was the ideology of the engineering elite. These men, many of whom were consultants and businessmen themselves, dominated the professional associations. On the whole, they resisted the idea of a unified, activist engineering profession and tended to deny that there was any conflict between the interests of business and those of engineers. Second, there was a group of dissident elite engineers who may be termed "the patrician reformers" that included figures such as Newell and Cooke. They envisaged a unified, independent, activist engineering profession that would play a real leadership role in American society. It was some of these men, as part of their effort to forge such a profession, who elaborated the engineering critique of business domination and who argued for its partial supersession by engineering leadership. Finally, there was a large group of rank-and-file engineers, many of whom were young employees of large burgeoning organizations. They did not challenge business's right to rule; but they did demand a more humane, comfortable form of employment and a better chance *as individuals* to rise through the ranks. Let us take a closer look at these three engineering ideologies and at how their complex interaction determined the broad shifts in engineering thought in the early 20th century.

The Engineering Establishment

Although it was engineering dissent that attracted the most contemporary and historical attention, there was a strong, persistent business-oriented viewpoint within the early 20th-century engineering profession. This was particularly true of the national engineering societies, where such opinion was arguably dominant throughout the period in question. Owing to their strict membership requirements, these societies were populated largely by the engineering elite—that is, by prominent educators, successful consulting engineers, and, most notably, by engineer-executives of large corporations (a fact which chagrined Morris L. Cooke no end). This elite did not share the reformers' vision of the engineer's role in society.

At times, it did appear that the elite (at least in the civil and mechanical branches) had become somewhat less orthodox. One can point, for example, to Herbert Hoover's involvement in the report on *Waste in Industry* and to a certain amount of "progressive" rhetoric, such as the following statement by President Carman of the ASME: "Thus we had, growing side by side, two distinct organizations, or classes: one, Industry or Capital, the other Labor Unions or Labor. Not only have both of these organizations increased in size, but each has combined and federated with others of its kind, until both are unwieldly—overgrown, unable to control or direct their followers— each seeking the advantage over the other, and both actually taking advantage of the great consuming class, the Public."[8] One can also point to the activities of moderate figures within the "founder" societies. In the American Society of Civil Engineers (ASCE), men such as Gardner Williams and Arthur P. Davis, although they did not go so far as Newell or Cooke in their calls for engineering independence, did press for a unified profession and for greater engineering involvement in public affairs.[9]

It is undoubtedly true that there were a Left and a Right within the engineering establishment. Nevertheless, the weight of elite opinion seems to have been relatively friendly to business. Most leading engineers rejected the idea that there was a contradiction between the interests of business and those of engineers. And, although they did call for greater engineering involvement in public affairs, they backed away from any form of involvement that organized the profession around an independent, adversarial program.

---

[8]*Journal of the ASME* 44 (1922): 8. Donald R. Stabile, "Herbert Hoover, the FAES, and the AFofL," *Technology and Culture* 27 (October 1986): 819–27, interprets Hoover's notions about voluntary cooperation among business, labor, and engineers as "progressive." However, he concedes that these ideas, as distinct from Hoover himself, were *not* popular in the engineering profession.

[9]For example, at the Third Conference on Engineering Co-operation in 1917, even the "right wing" led by Gardner Williams favored unity. A transcript of this conference is included in the Newell Papers, Manuscripts Division, Library of Congress, Washington, D.C., Box 5. Within the ASCE, Arthur P. Davis fought for affiliation to the FAES and other "progressive" causes. See *Engineering News-Record* 85 (September 16, 1920): 570.

The engineering elite did seem to acknowledge the *distinctiveness* of the engineer and his attitudes and skills, but this did not involve him in conflict with business elites. Prominent engineers such as Dexter Kimball criticized businessmen and their methods, but they felt that the road to improvement lay in engineers' becoming managers themselves.[10] Kimball remarked that there was a need for engineers who spoke the "businessman's language," who could persuade industrial leaders to follow a wiser course.[11] Engineers of this type clearly felt that it was simply a matter of time before the technical man took up his place in industrial management. And they were very concerned that he should be well prepared to handle his new responsibilities— hence, the innumerable conferences, articles, papers, and speeches dealing with the need to provide engineers with training in management and economics. Here we have a vision of the engineer *as* businessman (even if a reforming one) working alongside other industrial elites. As we shall see, this is a far cry from Morris L. Cooke's vision of a militant, organized, independent profession.

Leading engineers also frequently joined in the cry for greater engineering involvement in public affairs. For example, the *Journal of the ASME* approvingly reported in May 1919 a conference of prominent New York City engineers on "The Engineer as Citizen" that included presentations on "The Civic Responsibility of the Engineer," "The Relation of the Engineer to Legislation," and "The Relation of the Engineer to Administration."[12] Yet participants in such discussions were notable for their caution. More often than not, they seemed to have had in mind placing *individual* engineers in important civic positions. Engineering journals were filled with discussions of the engineer as city manager, with praise for Hoover's appointment as secretary of commerce, and with complaints that engineers were seldom considered for appointment to advisory boards and other public bodies. *Engineering News-Record* initiated a series of biographical sketches of engineers in public life with a view to providing positive role models for the younger members of the profession.[13] What was missing in all of this was any serious commitment to the idea of the engineering profession as a whole as an independent, organized force. Indeed,

[10]See, e.g., Dexter Kimball, "Opportunity of the Engineer," *Journal of the ASME* 43 (1921): 280.

[11]"The Engineer's Industrial Opportunity," *Engineering News-Record* 89 (December 14, 1922): 1007. Interestingly, J. P. J. Williams responded a few weeks later, arguing that it was time "to shake off the yoke of the financiers, owners and business executives." *Engineering News-Record* 90 (January 25, 1923): 176.

[12]*Journal of the ASME* 41 (May 1919): 446–51, 496.

[13]"The Engineer in Public Life," *Engineering News-Record* 88 (May 11, 1922): 763.

the engineering elite shied away from organized intervention in public affairs whenever the opportunity to become so involved presented itself. This is why C. E. Drayer, an important engineering "progressive," criticized the founder societies for *talking* about public affairs but staying out of politics.[14]

Calls for greater organized involvement in civic affairs, such as Cooke's "The Public Interest as the Bedrock of Professional Practice,"[15] met with considerable hostility, and the membership of the ASCE in 1920 rejected amendments that would have mandated enlarging the society's scope to include cooperation in economic and civic affairs.[16] Nor did attempts to unify the profession to deal with nontechnical questions meet with much success, as Layton has shown. The Federated American Engineering Societies (FAES) encountered considerable resistance within the engineering fraternity and little ability to take action. By 1922, a mere two years after the organization's birth, the *Engineering News-Record* complained that "In the activities of the Federated American Engineering Societies there has been little to stir the blood. . . ."[17] All in all, the idea of an *organized* engineering presence in public affairs did not appeal to the conservative elite of the profession.

The engineering elite, thus, did call for a greater role for engineers—but as individuals, not as an organized group. The elite sought to gain for engineers a place in corporate administration and political life so that they might work *together* with, not against, business and political elites. What they resented most, as became clear in the affair of Arthur P. Davis and the Reclamation Service, was the suggestion that engineers were not competent to function as businessmen and administrators.

In June 1923, Arthur P. Davis, former president of the ASCE and a leading engineering reformer, was dismissed as head of the U.S. Reclamation Service by Interior Secretary Hubert Work. The secretary declared that there was a need to put a "businessman" in charge of the service in order to put an end to engineering "excesses."[18] The reasons for the dismissal were clearly more complex. Nevertheless, Secretary Work's public denigration of the engineers' administrative

[14]"The Engineer in Politics," *Engineering News-Record* 79 (November 1, 1917): 850.

[15]Morris L. Cooke, "The Public Interest as the Bedrock of Professional Practice," *Journal of the ASME* 40 (May 1918): 382–83.

[16]"Results of Amendment Vote," *Engineering News-Record* 85 (October 7, 1920): 719.

[17]"Society Action and Inaction," *Engineering News-Record* 88 (January 5, 1922): 4–5.

[18]"Davis Removal Part of General Upheaval," *Engineering News-Record* 90 (June 28, 1923): 1139; for a summary account of this incident, see Donald C. Swain, *Federal Conservation Policy 1921–1933* (Berkeley, Calif., 1963), chap. 5.

abilities produced a storm of protest from the engineering establishment. Responses such as this from the leadership of the ASCE revealed the elite's conception of the engineer's relation to business: "The implication that engineers are not competent business administrators is refuted by numerous engineers who today are conducting as executives many of the great railroad systems, public utilities, and industrial enterprises of this and other countries, and the U.S. Reclamation Service is peculiarly an engineering enterprise."[19] What is missing from such responses is any sense of conflict between engineers and businessmen. True, *Engineering News-Record* did, in its editorial comment on the affair, adopt a disparaging tone in its reference to the businessman, but this was directed more against the types of individuals who posed as businessmen in government—that is, those who represented special political interests[20]—than against businessmen in general. Indeed, the same volume in which the editorial appeared included a second editorial arguing that business and engineering were virtually synonymous.[21] The overwhelming tone in both protests is one of injured pride. Engineers were not seeking to oust the businessman or to reject his values and techniques. They were simply saying that they, too, could be good administrators in both the private and public sectors and that they, in fact, resented the suggestion that there was a significant difference in ability between the businessman and the engineer.

In sum, the engineering elite saw no fundamental conflict between the businessman and the engineer and certainly did not countenance the idea of an independent, well-organized profession posing as an alternative to business dominance in society. This is not to say, however, that elite opinion was wholly static in the period under review. There can be little doubt that the profession as a whole, under pressure from the rank and file and from engineering progressives like Cooke and Newell, reformed in the years immediately after World War I, at least temporarily. All of the national societies except the ASCE joined the FAES, ostensibly to give the profession a more united voice in public affairs. The ASCE, although it did not join, became embroiled in an acrimonious debate over the public role of the profession and considered a number of amendments to its constitution (some of which passed) encouraging greater democratization and social and political activity.

[19]"Am. Soc. C. E. Protests Removal of A. P. Davis," *Engineering News-Record* 91 (July 19, 1923): 116.

[20]In this case, Western farmers were at issue. See "Business Man or Administrator," *Engineering News-Record* 91 (July 5, 1923): 3.

[21]"Business Engineering," *Engineering News-Record* 91 (November 8, 1923): 748.

Nevertheless, at no time did engineering leaders consider a unified profession organized in opposition to corporate leaders. The engineering societies consistently refused to evict the National Electric Light Association, a trade association, from the Engineering Societies Building, thus symbolically affirming their affinity for business.[22] And, the concern of engineering leaders with preparing young engineers for careers in business and *as* businessmen never wavered. In 1914, Alexander Humphreys, speaking on "Business Training for the Engineer," addressed the need for the young engineer to accommodate himself to business and its needs, to learn to consider return on investment, to eschew doing things the best possible way if it was too expensive—the engineer had to "learn that what is good enough is best."[23] It is hard to imagine that the engineering leaders of the more conservative 1920s would have disagreed.

The Patrician Reformers

The second major current in engineering ideology, Layton's revolt of the engineers, also had its roots in the elite of the profession. It is rather difficult to draw a precise line between this more reformist current in engineering thinking and the ideas of some of the moderates discussed above. Nevertheless, there were those within the profession who, either implicitly or explicitly, were moving toward a different attitude to business from that of the engineering elite.

In reality, there were a number of loosely associated currents of "protest." What they all had in common was a vision of the professional engineer as *independent*. The engineer was seen as distinct from the traditional businessman, with his own set of values and techniques. He should not be content to be subordinate to business and its values, for he had something to offer of his own, something that might even be superior to the ideas and talents of businessmen. For many of these "dissidents," the implication was that engineers should organize themselves more explicitly *as engineers*, as a unified, cohesive profession which could give voice to engineering opinion on a wide range of issues.[24]

---

[22]Cooke's campaign against the NELA was a long and fruitless one, continuing well into the 1920s. Some of Cooke's correspondence on this issue is collected in the Morris L. Cooke Papers, FDR Presidential Library (Hyde Park, N.Y.), Box 15, File 144, and Box 174.

[23]Alexander C. Humphreys, "Business Training for the Engineer," paper for the Philadelphia Section AIEE, p. 14. In Frederick W. Taylor Collection, Stevens Institute of Technology (Hoboken, N.J.), Box 11D.

[24]Recent theoretical developments in the sociology of the professions have emphasized that professionalism is a form of occupational organization that seeks to win autonomy and high status for practitioners. See Magali Sarfatti Larson, *The Rise of Professionalism: A Sociological Analysis* (Berkeley, Calif., 1979).

Perhaps the first example of this kind of thinking within the engineering community was the scientific management movement of the early 20th century.[25] Monte Calvert has argued that scientific management was the product of a very specific group of engineers—the old "shop culture" mechanical engineers with roots in the small-scale machine shops of the 19th century.[26] These engineers were accustomed to a substantial degree of autonomy, since their careers often led them to become shop proprietors. Yet, as small shops gave way to large-scale industry, more and more engineers from this tradition were finding themselves in middle-level positions in large organizations. Their response was a theoretical redefinition of engineering autonomy, an attempt to carve out an autonomous role for the engineer within the large organization.

Not all of those who espoused scientific management went this far; but the ideas of men like Frederick W. Taylor and H. L. Gantt implied a considerable degree of autonomy for the engineer in the corporation. As envisaged by Taylor, scientific management would place control over the shop—although not the enterprise as a whole—in the hands of virtually independent engineers: "The shop, and indeed the whole works, should be managed not by the manager, superintendent or foreman, but by the planning department. The daily routine of running the entire works should be carried on by the various functional elements of this department, so that, in theory at least, the works could run smoothly even if the manager, superintendent and their assistants outside the planning room were all to be away for a month at a time."[27]

Although, at first glance, the scientific managers appeared to be echoing conservative engineers' demands to be accepted as *part* of management, there were subtle but important differences between the two programs. First, the scientific managers, unlike conservative engineers, were willing to see themselves as a distinct movement and to organize themselves to take concerted action (in addition to the Taylor Society, scientific managers played a significant role in engineering reform). Perhaps more important, scientific managers saw themselves as *different* from managers and sometimes even in conflict with them. For example, combined with the idea of engineering sovereignty was a degree of impatience with certain kinds of employers.

[25]For a more detailed statement of the argument that follows, see Peter Meiksins, "Scientific Management and Class Relations: A Dissenting View," *Theory and Society* 13 (1984): 177–209.

[26]Monte Calvert, *The Mechanical Engineer in America, 1830–1910* (Baltimore, 1967).

[27]Frederick W. Taylor, "Shop Management," in Taylor, *Scientific Management* (Westport, Conn., 1972), p. 110.

Taylor disparaged the corporate elite's preoccupation with "making money quickly" rather than being concerned about making one's company "the finest of its kind."[28] He often seemed to envisage a kind of division of labor between management and the engineers, governed by what he called "the exclusion principle," whereby engineers would control "the shop," while managers concerned themselves with other (e.g., financial) matters.[29]

Significantly, the scientific managers linked their ideas on the position of the engineer in industry to the question of the engineer's role in society as a whole. H. L. Gantt made this connection explicit when he remarked that

> as a class, the engineers are far more capable of promoting efficiency than either the financiers or Labor Unions, and until they hold up their hands and say the end is reached, my feeling is, they should not allow to let go unchallenged the pessimistic statements of others. I think if engineers, as a class, and especially the American Society of Mechanical Engineers, should take this stand, the engineering profession would be brought more prominently before the public within the next year or two, than would be possible in the ordinary course of events for ten years.[30]

Many of the scientific managers echoed this concern to raise the status of the engineer and "bring him before the public" through their involvement in various reform activities within the engineering societies (especially the ASME). Taylor himself, while president of the ASME, undertook a number of administrative reforms designed in part to open the society to more "progressive" influences.[31] He also engaged in periodic skirmishes with the ASME because it was reluctant to drop its purely "technical" orientation and accept more papers on

---

[28]Quoted in Frank B. Copley, *Frederick W. Taylor* (New York, 1923) 1:388.

[29]Frederick W. Taylor, "Shop Management" (n. 27 above), pp. 126–27; Taylor, "Principles of Scientific Management," in *Scientific Management* (n. 27 above), p. 129. In this connection, it is interesting to note that Taylor normally meant *shop* management when he referred to management. Other elite engineers used the term in a larger sense. This can be seen clearly in the exchange between Taylor and Alexander C. Humphreys. See Taylor to Humphreys, December 10, 1910, Taylor Collection, File 54B.

[30]Gantt to Taylor, July 16, 1911, Taylor Collection, Box 121B.

[31]As president, Taylor sought to restructure the secretary's position. In effect, he forced out the old secretary, F. R. Hulton, then proposed a series of reforms that would have limited the secretary's power. However, the appointment as secretary of Calvin W. Rice, ostensibly a "progressive," obviated some of his reforms. See Taylor Collection, Box 7 and Taylor to Fred Miller, June 7, 1906, Box 8A. For a summary of the incident, see Sinclair (n. 5 above), pp. 84–93.

"social" questions—specifically, shop management.[32] Morris L. Cooke, probably the most vocal progressive engineer, was a close associate of Taylor's, being responsible for the administrative reforms he initiated within the ASME. Cooke pushed the idea of reform even further, and Taylor seemed to approve of the direction in which he was moving.[33]

Finally, it is worth noting that the scientific managers played an important role in the movement to unify the engineering profession. Cooke and Fred Miller were prominent in the first years of the FAES and played a leading role in creating the reports on waste in industry and the twelve-hour day.[34] Not all scientific managers would have gone so far as Cooke; but the connection between engineering reform and scientific management was strong.

A second major reform tendency is associated with the name of Frederick Haynes Newell.[35] Newell, a conservationist and former director of the U.S. Reclamation Service, focused his reform efforts directly on an attempt to unify the profession and increase the status and public responsibility of the engineer. Following his departure from government service in 1914, he turned his attention to the idea of engineering cooperation. Despite initial setbacks, by 1917 Newell had succeeded in attracting representatives of over forty engineering organizations to the Third Conference on Engineering Co-operation.[36] At the outset of this conference, he made clear that cooperation was needed to compensate for the specialized engineering societies' neglect of "matters of general or civic importance."[37] The desire to increase the engineer's role in civic affairs was to be one of the hallmarks of Newell's reform activities.

The Conference on Engineering Co-operation did not lead in the direction Newell had hoped. Moderate engineers, especially Gardner Williams of the ASCE, were able to channel the delegates away from a more independent course toward the nascent Engineering Council, dominated by the founder societies.[38] As a result, Newell began to give up on his idea of a new cooperative organization and gravitated

[32]Taylor experienced considerable difficulty in 1910 trying to get his paper "Principles of Scientific Management" accepted by the ASME papers committee. Eventually he withdrew it and had it printed privately. See Taylor Collection, Box 115.

[33]See Taylor to Philitus W. Gates, October 27, 1908, Taylor Collection, Box 9C, where Taylor comments favorably on Cooke's "The Engineer and the People."

[34]Layton (n. 6 above), chap. 8, has stressed this connection.

[35]Newell's career is summarized in Edwin Layton, "Frederick Haynes Newell and the Revolt of the Engineers," *Midcontinent American Studies Journal* 3 (Fall 1962): 17–26.

[36]See the transcript of this conference in the Newell Papers, Box 5.

[37]Ibid., p. 1.

[38]Ibid., p. 86. Gardner Williams successfully pressed a resolution urging that the Engineering Council be made the vehicle for cooperation.

toward an existing engineering society—the American Association of Engineers (AAE). The AAE had been founded in 1915 by a group of Chicago-area engineers, most of whom were in the public employ.[39] Its major objective was to improve the material situation of poorly paid younger members of the profession, although it was strongly opposed to anything that resembled engineering unionism. It was, in effect, a response to the conditions of mass employment in public and corporate bureaucracies that was now becoming a commonplace situation for engineers. After associating himself with this organization, Newell quickly rose to prominence within it, becoming president in 1919. In the process, he grafted its preoccupation with the plight of rank-and-file engineers onto his own concern about fostering engineering cooperation and a greater civic role for the engineer.

Newell did not adopt a critical stance toward business, nor did anyone in the leadership of the AAE. Indeed, they sounded much more like the engineering elite than other reformers in this respect, filling the pages of their journal, *The Monad* (later *Professional Engineer*), with articles and speeches on careers in business, engineering "salesmanship," and such. Where they differed from the engineering elite, however, was on the question of how engineers should be organized. They made it quite clear that they favored a *unified* profession—and one that was not afraid to speak aggressively on behalf of the interests of engineers. One of their objectives in pressing for cooperation was to improve the engineer's material situation; Newell summarized this aspect of "liberal" engineering ideology as follows: "On the part of these liberals it is urged that every man who is making a living by the practice of engineering should become a member and, as such, be educated and impressed with high standards by contact with men within the society, and not be forced to the alternative of joining a labor union if he desires to do his part toward improving the condition of his fellow engineer."[40] Their strategy, then, was not unlike that of the American Medical Association.[41] Eschewing unionism, they sought to claim a *moral right* to high status and compensation through enforcing the universal excellence of practitioners and fostering high standards of professional practice.

At the same time, Newell and his associates also saw cooperation as a vehicle for strengthening the engineer's public role. Throughout his career, he stressed the engineer's responsibility to "serve" the pub-

[39]For an account of the origins of the AAE, see Arthur Kneisel, "A Chapter of Early History," in *Directory of the American Association of Engineers* (Chicago, 1918).

[40]Frederick Haynes Newell, "Ethics of the Engineering Profession," *American Academy of Political and Social Science: The Annals* 101 (1922): 81.

[41]See Jeffrey Berlant, *Profession and Monopoly* (Berkeley, Calif., 1975), especially chap. 5.

lic. And, in his speech as outgoing president of the AAE, he attributed the organization's success to this, urging engineers to respond to what he saw as the growing public need for their "service."[42]

Implicitly, then, Newell and the AAE were calling for an independent professional organization, one that was willing to form a single voice for engineers and to push for public recognition of both their point of view and their material worth. The national societies were recommending, instead, purely individual solutions to these problems. It should be emphasized, however, that the AAE's leaders remained uncomfortable with the implications of the position toward which they were moving. While an independent profession might become embroiled in conflict with business over working conditions or questions of public policy, Newell and his associates wished to avoid this.[43] Theirs was, then, a rather contradictory ideology, far less consistent and clearly thought out than that proposed by Cooke. Nevertheless, they were at least groping toward an independent professional form of organization, one that would be considerably less docile than the founder societies had been.

Finally, and perhaps most important, there is Morris Llewellyn Cooke, probably the most militant of all the patrician reformers. It was Cooke, above all, who explicitly linked the idea of engineering as an autonomous, well-organized profession to the idea that engineers should not allow themselves to be puppets of business interests. And it was Cooke who worked most openly and most vigorously to purge the engineering societies of business influence.

Cooke's approach to the role of the engineer in American society rested on his assumption that engineering was a public trust. As he argued in an early paper for the ASME: "For after all it is public opinion and not the dictum of the engineering fraternity which finally decides the large question of engineering practice. How much better it would be then to join forces with the people, to work out with the people the people's problems and to build up in the lay mind such a confidence in our devotion to the people's cause that they will be willing to let us lead in matters where our training especially qualifies us to do so."[44] The engineer was thus an expert, working for the public as a whole. This might imply a certain elitism, as Jean Christie

[42]*Professional Engineer* 5 (June 1920): 10.

[43]It is highly significant that the speech of the president who replaced Newell was a spirited defense of public utilities corporations. Leroy K. Sherman, "The Year Ahead," *Professional Engineer* 5 (June 1920): 11–12.

[44]Morris L. Cooke, "The Engineer and the People," *ASME Transactions* 30 (1908): 627.

suggests, but it did require that he be independent.[45] Engineers occupied what Cooke referred to as a "strategic position as the defenders of our social order,"[46] being the mediators between labor and capital; under no circumstances should they take sides. Moreover, Cooke went to great lengths to emphasize that there was a distinction between the engineer and the businessman that needed to be maintained. He summarized his view on this question in a letter to Calvin W. Rice, secretary of the ASME, after the heyday of engineering radicalism had passed:

> I really think the basic difference between us is that you see the best future for Engineering as touched more closely by business than I do. To me engineering can be kept quite free from business—influencing business but relatively uninfluenced by business. Except for evanescent advantages—more seeming than real—engineering has little to gain from such a close association with business as forces it to recognize the average practices of business groups as contrasted with the ethical standards of society as a whole.[47]

Maintaining this independence, however, would not be easy. Cooke recognized that an increasing number of engineers had been "reduced" to employee status, many of them in large industrial, transportation, and utility concerns. As a result, he feared, ". . . we may anticipate an even closer control than exists today of engineering thought and action through financial and other non-engineering influences."[48] It followed logically that engineers needed to *organize* themselves to maintain their independence, to use their professional associations to steer engineering toward the "middle way" representative of the public interest. Yet those associations were anything but neutral.

Throughout the years 1910–30, Cooke was a persistent critic of what he saw as business domination of the professional engineering associations, especially the ASME.[49] For example, in 1915 Cooke issued a pamphlet entitled "Snapping Cords," in which he criticized the practices of the utility companies and accused three prominent engineers—Mortimer Cooley, Alexander Humphreys, and George F.

[45]See Jean Christie, *Morris Llewellyn Cooke: Progressive Engineer* (New York, 1983).
[46]Cooke circular letter to ASME section secretaries, November 3, 1919, Cooke Papers, Box 22, File 232.
[47]Cooke to Calvin W. Rice, February 13, 1925, Cooke Papers, Box 15, File 144.
[48]Morris L. Cooke, "The Engineer as Citizen," convocation address, Syracuse University, March 20, 1928, p. 15. In Cooke Papers, Box 298.
[49]See Sinclair (n. 5 above), pp. 100–110; Christie (n. 45 above), especially chap. 2.

Swain—of being their apologists.[50] Cooley, Humphreys, and Swain sought to have the ASME take some kind of action against Cooke. Cooke fired back in 1917 with a second pamphlet, "How About It?" in which he made public the previously secret proceedings against him in the ASME.[51] He also criticized what he termed the "absentee management" of the society, arguing that its administrative structure was such as to allow the utility interests to dominate and concluding that "if we are to provide for the mechanical engineers of this country an organization which shall be at once able and strong in action, democratic in spirit and progressive in its ideals some radical changes will have to be made in our Constitution."[52]

Most important among the reforms Cooke attempted to press within the ASME was a strong, well-enforced code of ethics. He clearly felt that such a code would bind the engineer to the public interest and prevent influence peddling. In his view, the ASME code of ethics was woefully inadequate, being "dictated by private interests rather than by considerations of public and professional welfare."[53] With this in mind, he presented to the spring 1918 meeting of the ASME a paper entitled "The Public Interest as the Bedrock of Professional Practice," in which he argued that the public interest should take precedence in the engineering code of ethics over even private or professional interests.[54] When the ASME Committee on Aims and Organization took up the question of organizational reform in 1919, Cooke drafted a program of "14 Points," centering on a strong code of ethics, the unification of the profession, and increased society democracy, especially for younger engineers.[55] He followed this up in 1921 with a paper, "On the Organization of an Engineering Society," emphasizing the potential leadership role the engineering profession as a whole could play and urging it to unite and dedicate itself to public service with this in mind.[56]

Cooke thus stood for a truly independent engineering professionalism. He joined Newell and the AAE in calling for engineering unity

[50]Morris L. Cooke, *Snapping Cords: Comments on the Changing Attitude of American Cities toward the Utility Problem* (Philadelphia, 1915); pamphlet in New York Public Library.

[51]Morris L. Cooke, *How about It?—Comment on the "Absentee Management" of the American Society of Mechanical Engineers and the Virtual Control Exerted over the Society by Big Business—Notably by the Private Utility Interests* (Philadelphia, 1917); pamphlet in New York Public Library.

[52]Ibid., p. 3.

[53]Morris L. Cooke, "Public Engineering and Human Progress," paper presented before the Cleveland Engineering Society, November 14, 1916, p. 8.

[54]*Journal of the ASME* 40 (May 1918): 382–83.

[55]Morris L. Cooke, letter to the editor, *Journal of the ASME* 41 (February 1919): 165.

[56]*Journal of the ASME* 43 (May 1921): 232–35, 356.

and defending the younger members of the profession. But he pushed the implications of engineering unity further than Newell wished to go. Where Newell stopped short of implying that a unified engineering profession might come into conflict with business, Cooke was emphatic. Engineering and business were *not* the same thing, and they most definitely differed on a number of important social issues. It was imperative that the profession be sufficiently independent to be able to make clear what those differences of opinion were and to represent the public interest.

The Taylorites, Newell, and Cooke all stood, in varying degrees, for the idea of engineering independence. Noble is right in questioning how "radical" their critique of American business really was. Yet, one cannot reduce their reform program to a simple variant of orthodox engineering ideology. While they did not always agree on the relationship between business and engineering, they clearly shared an intolerance for the idea of the engineer as subordinate. Their goal was an autonomous form of professional organization, perhaps not unlike the AMA, with social recognition for the engineer and a leading role in society for technical men. Such a goal was not compatible with the ideas and interests of most engineering and business leaders.

The Second Revolt of the Engineers

Ironically, a good deal of the support for men like Newell and Cooke came not from engineers for whom autonomy was a real possibility but from the rank-and-file members of the profession. Their support was based less on their desire for independence than on their hope that the programs articulated by the patrician insurgents would yield more tolerable conditions of organizational employment.

Before the dramatic expansion of the profession in the early 20th century, engineers had been able to conceive of themselves as something other than employees—many looked forward to careers as consultants, engineer-executives, or proprietors of engineering businesses. By the 1920s, however, most engineers regarded the prospect of "hanging out a shingle" as rather dim. It is difficult to obtain precise data on the number of engineers with employee status. Nevertheless, the Wickenden Report estimated in the late 1920s that only about 25 percent of engineers who graduated prior to 1909 were consultants and proprietors; among recent graduates, at least 75 percent were in subordinate technical, sales, and clerical positions.[57] Theoretically, of course, these young engineers could look forward to rising through

[57]Society for the Promotion of Engineering Education, *Report of the Investigation of Engineering Education, 1923–1929* (Pittsburgh, 1930), 1:232.

the ranks to positions of authority. But, as even the elite were obliged to acknowledge, their chances were decreasing as the numbers of engineers were increasing: "The man of exceptional ability, indeed, may find it worth his while to work for low compensation because of the future awaiting him. But to hold up to the rank and file of technical workers the idea that they can afford to work for insufficient salaries for the sake of some future high position, which they have not one chance in twenty or fifty of obtaining, is a gross deception."[58] Moreover, report after report published by the founder societies and by the AAE showed that large numbers of subordinate engineers earned poor salaries, often taking home less than skilled blue-collar workers in the same enterprise.[59]

By the second decade of the 20th century, rank-and-file engineers had begun to develop a distinctive set of goals and a new view of the engineer's place in society. First and foremost, they had begun to demand better pay and working conditions. As early as 1911, engineers were publicly denouncing their low wages,[60] and this became a constant refrain in engineering periodicals of the World War I era. Nor were they opposed to organizing themselves to press for higher wages; on the contrary, they were strongly in favor of *collective*, rather than purely individual, efforts to deal with the "compensation question." As one anonymous engineer wrote to *Engineering News-Record*:

> It seems to me that the engineering societies ought to go to Washington and arrange a scale for the payment for professional services. They may object to this on the ground that they would thus be getting into the same class as the labor unions. Perhaps it is more desirable to say that you belong to a profession and remain on a high professional basis and starve to death, than it is to belong to a labor union and have your personal affairs conducted in a business-like manner.[61]

Complaints such as this one did not imply that engineers should become independent of their employers. Instead, they seemed to be

---

[58]"Progress by Engineering Council's Compensation Committee," *Engineering News-Record* 82 (June 19, 1919): 1199.

[59]For example, Charles Whiting Baker compiled data on engineers in railway service in 1919 and found that machinists and boilermakers earned more than assistant engineers. He also found that the gap between engineering and nonengineering personnel had narrowed considerably between 1916 and 1919. "Pay and Position of Engineers in Railway Service," *Engineering News-Record* 82 (January 30, 1919): 228–31.

[60]"Concerning Inadequate Salaries to Instrument Men on Surveys," *Engineering News* 66 (October 5, 1911): 412.

[61]"Compensation of Engineers," *Engineering News-Record* 81 (October 31, 1918): 819.

taking an employee status for granted, demanding simply that they be paid a decent wage.

Some engineers, especially in the public sector, elected to pursue better wages through unionization, and some of their fledgling unions affiliated to the AFofL.[62] These engineers apparently had concluded that unionism was the only answer: "An engineering society is an organization in charge of an aristocracy. The subordinate engineer, like the vassal of old, pays his dues and proceeds to do as he is told by the aristocrats of the profession, who dictate the policies of the organization, lay down systems of professional ethics, which work very nicely for the control of the subordinate membership, and otherwise conduct the affairs of these societies in the interest of the aristocracy of the profession."[63] It is probably safe to say, however, that this was a minority opinion; most engineers rejected the association with the blue-collar worker that unionism implied. This was hardly surprising, since one of the major complaints of the rank-and-file engineers was that they had been reduced *below* the level of the lesser, blue-collar worker.[64]

Rank-and-file engineers distinguished themselves from blue-collar workers even further through their emphasis on promotions. They did not, as a class-conscious worker might, spurn promotions as a form of co-optation into the dominant class. Rather, one of their major complaints was that they were mired in dead-end jobs that offered them little authority or chance for promotion. This was clearly a pervasive sentiment, mirrored in the constant lament by engineer-employers that young engineers entered the workplace with unrealistic expectations.[65] One can detect this, too, in the irritation of some

[62]Rumors of union formation had already begun to circulate by 1912. See, e.g., *Engineering News* 67 (January 11, 1912): 75–76. In 1913, an organization called the Chicago Technical League was formed "for the purpose of finding ways and means for the advancement of engineers' salaries and standardizing of positions." *Engineering News* 70 (December 4, 1913): 1143. A variety of other short-lived and/or small engineering unions soon followed.

[63]J. L. Harrison, letter to the editor, *Engineering News-Record* 83 (December 25, 1919): 1071.

[64]For a good example of this type of complaint, see Kenneth P. Armstrong, letter to the editor, *Engineering News-Record* 82 (April 10, 1919): 736.

[65]The Wickenden Report emphasized this problem and even advocated a stratified educational system to deal with it: "The period of maladjustment which marks the entrance of graduates to active life is possibly the result of an educational scheme which aims to fit everybody to get to the top and to fit nobody to be efficient in minor responsibilities, and which ends in an undignified scramble after the stars." *Report of the Investigation* (n. 57 above), 1:68.

engineers when their elders attempted to "instruct" them on the art of getting ahead:

> If Mr. Railroad Official will suggest a method whereby an assistant engineer assigned to work in the drafting room almost exclusively, and who rarely comes in contact with men occupying higher positions than the chief clerk of his own department, or the third or fourth assistant to the junior aide to the general manager's assistant secretary, can obtain the necessary information regarding the "requirements of official positions," I will be very glad to pursue his course of study in an effort to prepare myself for a better position.[66]

In short, rank-and-file engineers were pursuing a "middle way" between the patrician reformers and trade unionism. They rejected the latter in the name of their own high status; but they did not share the former's preoccupation with engineering independence or Cooke's emphasis on freeing the profession of business influence.

Two episodes illustrate the kind of activity these engineers had in mind. First, there is the movement to institute engineering licensing, which began to gather steam around 1910. The rank and file hoped that licensing might place them "on a better footing in their relations to the public and particularly their relations to their employers,"[67] thereby gaining them improved wages. Acting on this hope, they lobbied for the passage of licensing bills in a number of states, urged the engineering societies to endorse the principle of licensing, and later won AAE support for licensing bills. They even had a degree of success, although portions of the engineering elite, who had nothing to gain from licensing, continued to oppose them.[68] One can readily see the appeal of licensing as a strategy. It avoided the loss of status implicit in unionism, legitimated the engineers' claim to privileged status, and offered the possibility of higher wages and better treatment since it provided another bargaining chip and a kind of professional monopoly.

---

[66]Letter to the editor, *Professional Engineer* 5 (June 1920): 22.

[67]"The Proposed License Laws for Engineers in N.Y. State," *Engineering News* 65 (March 9, 1911): 298.

[68]For example, opposition continued even after the passage of licensing laws in many states on the grounds that such laws interfered with the ability to practice across state lines. See "License Laws Begin to Restrict," *Engineering News-Record* 85 (July 1, 1920): 2. Early on, there was some public speculation among younger engineers that the older men resisted because they were themselves employers of engineers and feared higher salaries would result. See "Licensing Engineers," *Engineering News* 69 (March 20, 1913): 585.

Rank-and-file engineers sought more than licensing, however. During World War I and its aftermath, they became involved in moves to reshape professional organizations along more democratic lines that would be more responsive to their needs. Between 1910 and the early 1920s, all of the major engineering societies were under considerable pressure to decentralize, to establish local sections, and to give these at least a certain amount of power. These movements were often led by the patrician reformers (e.g., Cooke in the ASME); but they clearly were supported by younger, rank-and-file engineers who would have been able to influence society policy within a decentralized structure.[69] Probably more important, such engineers were the backbone of the American Association of Engineers. The AAE consistently directed its message to the rank-and-file engineer, and its membership was clearly drawn from this group.[70]

What did the rank-and-file engineers hope to gain through such engineering organizations? On the face of it, they appeared to be responding to the ideal of engineering independence emphasized by their patrician leaders. This argument does not stand up to closer scrutiny, however. In the case of the AAE, for example, it is quite clear that neither Newell's rhetoric of engineering unity nor the society's preoccupation with ethics was what attracted the rank and file.

The timing of membership shifts alone should indicate this. The association remained relatively small until 1918, when it succeeded in gaining a substantial wage increase for railroad engineers.[71] Almost overnight, its membership began to grow, rising to 7,000 in 1919 and reaching a high point of almost 27,000 in 1921.[72] There can be little doubt that it was the compensation question that had attracted these new members. Similarly, it was when the AAE began to back away from this issue, when it began to argue that the compensation question had been resolved and that it was time to move on to other issues,

[69]The connection between the section movement and the desire to shift power to the less elite members of the profession and to improve their material situation is well discussed in Layton (n. 6 above), chap. 5.

[70]1918 membership data indicate that 60 percent of the members were under thirty-two years old. *The Monad* 3 (June 1918): 6. Data on the age of members for later years are not available, but President A. N. Johnson indicated in 1922 that he felt compensation was the key to the membership rise. "An Opportunity for Service," *Professional Engineer* 7 (June 1922): 4.

[71]"Increases in Compensation Granted Railway Men," *The Monad* 3 (May 1918): 22. For more information on the history of the AAE, see William Rothstein, "The American Association of Engineers," *Industrial and Labor Relations Review* 22 (1968): 48–72, and Peter Meiksins, "Professionalism and Conflict: The Case of the American Association of Engineers," *Journal of Social History* 19 (1986): 403–21.

[72]*Professional Engineer* 6 (July 1921): 17.

that its membership began to decline.[73] One railway engineer protested this shift in no uncertain terms: "I object at this time to instilling the issue as the paramount issue of raising the ethics of engineering when we have the crying need before this Association of increasing the compensation of engineers. That, after all, is the main thing. That has been the backbone of this Association up to this time."[74] Somewhat later, a Missouri highway engineer protested his dues notice, declaring that "I wouldn't have the heart to collect a past due bill for $8.75 if I had done no more to earn it than the A.A.E. has for me or the rest of the engineers here in Missouri for that matter."[75]

In sum, the rank and file did not follow the patrician insurgents because they shared their priorities. There was, if the following letter is at all typical, even an element of impatience in their attitude toward issues such as autonomy or ethics: "If you are going to raise the standards of ethics very much higher, you will have to provide it with an oxygen tank so that it does not start frothing at the mouth. What you need is an organization provided with 42-centimeter anti-aircraft guns to puncture the bag of ethics and bring it a little nearer to the earth."[76] The rank and file clearly hoped that Newell and Cooke would help them gain material improvements, but they did not share their vision of the engineer's social position. For them, the engineer was an employee in a corporate or public bureaucracy whose primary problems were wages, conditions, and promotions. In this context, visions of engineering independence and leadership did not have a great deal of meaning.

The Decline of Engineering Reform

We are now led to the question of why: Why was it that for a few years in the World War I era patrician reformers such as Cooke and Newell were able to dominate the stage, convincing many observers that the engineering profession as a whole had become a progressive force that posed an alternative to the twin selfish interests of capital and labor? Clearly, it was not because the patrician reformers represented a majority opinion within the engineering societies. On the contrary, they were notably unsuccessful in their attempts to get these societies to change. Engineering unity remained a dream, as did the

---

[73]See e.g., the speech by George W. Hand, in *Professional Engineer* 5 (April 1920): 9. The AAE declined almost as rapidly as it had risen; its membership had dropped to 8,000 by 1924 and continued to decline more slowly thereafter.

[74]*Professional Engineer* 5 (May 1920): 41.

[75]*Professional Engineer* 9 (January 1924): 24.

[76]*The Monad* 1 (May 1916): 20.

ideals of a strong code of ethics, the removal of business influence within the profession, and the goal of a large role for organized engineers in public life.

What made the patrician reformers seem strong (stronger than they actually were) was their ability to gain support from the rank and file of the profession. As we have seen, this support was more in the nature of an alliance. Cooke and Newell stressed, as part of their programs, the idea of engineering unity, hence their view of the engineering profession included the rank and file. The latter, in turn, although finding little appeal in discussions of ethics, autonomy, and service, did hope that calls for unity implied better conditions for them. The alliance of these two movements strengthened both—the rank and file gained leaders with access to the corridors of professional power, while the patricians gained mass support. Pooling their efforts, they were able to place a substantial amount of pressure on the conservative leaders of the profession, even succeeding in gaining a measure of reform. At the same time, this alliance was a source of weakness. It is considerably easier to defeat a large but divided movement than a united one of comparable size.

The defeat of engineering reform involved two related responses on the part of establishment forces. First, the engineering elite responded by retreating tactically. By yielding on some issues, they succeeded in defeating more radical proposals and even in co-opting portions of the reform program. Second, the elite counterattacked, successfully driving a wedge between the patrician reformers and rank-and-file engineers. They were not solely responsible for the collapse of this alliance, but the fact that it did fall apart was crucial to the demise of engineering reform.

The founder societies did adopt some of the reforms suggested by the engineering progressives but in a rather watered-down form. When substantial portions of the profession called for engineering unity in order to strengthen the engineers' public role, the elite responded with the Engineering Council and the FAES. As we have seen, the extent of their commitment to unified action was limited, but these organizations, under founder-society control, were preferable to potentially more militant and uncontrollable societies such as the AAE.[77] As it turned out, the FAES siphoned off some of the support for engineering unity yet was too weak to be very active. It was engineering unity with which the engineering establishment could live.

---

[77]Layton (n. 6 above), p. 124, has suggested that the Engineering Council was *created* to forestall the AAE. This may be an exaggeration, but it is certainly true that both the Engineering Council and the FAES were preoccupied with undercutting the AAE.

Demands for democratization were met with the formation of local sections in the founder socities.[78] But since these were given little power, and since efforts to give them more were resisted, this was little more than a gesture.[79] Similarly, Cooke's call for stronger codes of ethics were met by the establishment of committees to consider the question in the founder societies. The result was a great deal of discussion and a variety of proposals for revised codes. All of this was insipid and failed to satisfy men like Cooke,[80] but it did succeed in taking some of the steam out of their attacks on ethical questions.

Perhaps the most interesting example of partial reform and co-optation was the case of the scientific mangement movement. As we have already seen, scientific management implied a somewhat altered relationship between the engineer and the businessman. We have also seen that it was not embraced wholeheartedly by the founder societies. Yet, by the mid-1920s scientific management had become a kind of orthodoxy within both engineering and managerial circles. The key to this development was the watering-down of the program itself, removing its more extreme proposals. In industry, as Dexter Kimball pointed out, managers learned to apply scientific management *selectively*, adopting only those methods which suited their needs.[81] Similarly, the ASME did create a committee in 1912 to report on "The Present State of the Art of Industrial Management," under the direction of Leon Alford.[82] The report, however, was relatively mild, causing Taylor to complain that "it is so colorless and insipid that it seems hardly worth discussing."[83] Subsequent efforts by the ASME to deal with the questions raised by the scientific management movement were equally weak; those who wished a stronger version of Taylorism had to look outside the ASME to the Taylor Society. As in the case of other engineering "reforms," the development of a moderate, watered-down version of scientific management may have weakened the movement calling for stronger measures.

[78]"Local Sectional Meeting of National Engineering Societies," *Engineering News* 68 (October 24, 1912): 781. The ASCE proved most resistant to the idea of local sections.

[79]"Further Amendments to Civil Engineers' Constitution," *Engineering News-Record* 87 (November 24, 1921): 873; "Am. Soc. C. E. Amendments Rejected by Meeting," *Engineering News-Record* 88 (January 19, 1922): 125–26.

[80]*Journal of the ASME* 41 (July 1919): 601–3.

[81]Dexter Kimball, "Has Taylorism Survived?" *Journal of the ASME* 49 (June 1927): 594.

[82]See Taylor Collection, Box 12, for a copy of this report.

[83]Taylor to General William Crozier, November 14, 1912, Taylor Collection, Box 11A.

In addition to this strategy of tactical retreat, the elite vigorously resisted the engineering dissidents, successfully separating the two reform movements. There are many reasons why the alliance between the patrician reformers and the rank-and-file engineers collapsed. The Red Scare of 1919–20 engendered negative perceptions of all progressive movements; the return of prosperity in the mid-1920s must have weakened discontent, although, as we shall see momentarily, not all engineers were satisfied with their material situation. What must not be forgotten, however, is that the engineering elite also played an important role in defeating the reform movement.

The most important part of the elite "counterattack" was directed at the AAE, by far the largest and most dangerous organized expression of the reformist alliance. A concerted effort was made to bully the AAE into submission by attacking it as an example of trade unionism. Morris L. Cooke described the pressure on the AAE as "relentless."[84] Speaker after speaker appeared before the AAE urging its members to move away from the compensation issue, with its dangerous overtones of unionism, toward less controversial issues: "The question of salary increases has been very well disposed of, I think, for the time being. . . . I think the most important thing for us is to devote our attention to the opening of new channels, places where engineers can utilize their talents that heretofore have not been open to them."[85]

All in all, the AAE came under a considerable amount of the sort of pressure that is difficult to resist at the best of times. But what made it particularly effective in the case of movements such as the AAE was the fact that they were divided. The patrician reformers were extremely sensitive to the criticism that the AAE was a union—they were not directly concerned themselves with the compensation issue and their high status made the idea of unionism particularly abhorrent. Under pressure from the engineering establishment, they pulled back from the compensation issue, strongly condemning unionism in the process.[86] The rank and file rejected unionism too, but they also protested when their leaders backed away from the compensation issue, and there were periodic attempts within the AAE

[84]Cooke to F. E. Schmitt, November 15, 1920, Cooke Papers, Box 81, File 294.

[85]*Professional Engineer* 5 (April 1920): 9.

[86]Thus, in *The Monad* 4 (December 1919): 4, the AAE leadership published a "Statement of the A.A.E. on Trade Unionism in the Profession." It argued that it was impossible to subscribe simultaneously to the tenets of professionalism and trade unionism.

to put material gains back at the center of the agenda.[87] Since the leadership of the movement did not share such goals, however, this did not prove possible within the AAE. It was this internal division that proved fatal to engineering reform. Its momentary strength proved illusory, as the differences between the two movements were exposed.

Conclusion

What can we conclude from this argument regarding the nature of the shifts in engineering opinion? First and foremost, there clearly was a shift. Between 1910 and 1920, men like Cooke and Newell did develop a vision of the independent engineer, playing a leadership role in society and (for Cooke at least) declaring his independence of business influence. And this type of thinking did lose strength after 1920. But it is equally clear that this trend did not signify a massive change of heart on the part of engineers. We are not witnessing a situation where large numbers of engineers, who once cherished independence and the distinction between themselves and businessmen, suddenly decided that they had been wrong. Instead, what lay behind this ostensible change of opinion were shifting alliances and the changing demographics of the engineering profession, specifically the increasing numbers of employee-engineers concerned about their material position. For a time, the rank and file followed Cooke and Newell, thereby strengthening the latters' militant professionalism. When they discovered that they did not have the same goals, they withdrew, leaving behind only the relatively small number of engineers whose professional experience attracted them to the idea of independence.

This analysis challenges the notion found in Layton and Sinclair that the defeat of the revolt of the engineers was temporary. The experience of the 1920s clearly shows that the engineering profession had become infertile soil for the type of professionalism advocated by the patrician reformers. The days of the independent consulting engineer and the small machine-shop proprietor were past. Now as the majority of engineers became middle- and lower-level employees

---

[87]In 1922, the railway engineers sought to establish a separate railroad department within the AAE because they were dissatisfied with the manner in which their interests were being handled. See "Railroad Department Sought," *Professional Engineer* 7 (April 1922): 4. The proposal was defeated at the June meeting of the AAE. Subsequently, the Chicago chapter in 1923 proposed the establishment of a quasi-autonomous Railway Men's Technical Council to represent engineers. This, too, was rejected as "trade unionism," and the Chicago chapter was eventually expelled. See *Engineering News-Record* 90 (May 17, 1923): 891 and *Professional Engineer* 9 (June 1924): 20–21.

in corporate and other bureaucracies, the idea of independence, especially independence from business, became increasingly anachron-
istic. There are ways in which employees of business can oppose their
employers—through various forms of collective bargaining and
through advocacy of public ownership of corporations. But the type
of professionalism advocated by the patrician insurgents, with its emphasis on an association of *independent* practitioners, no longer made
much sense.[88]

Yet, the weakness of this form of engineering protest should not
lead us to conclude that engineers are, inevitably, the willing servants
of business that Noble's view suggests. It is undoubtedly true that most
engineers have responded to the prospects of six-figure salaries and
successful corporate careers by accepting the legitimacy of the large
corporation and the managers who control it. But one must be careful
to remember the difference between someone who already has these
rewards (the position in which elite engineers find themselves) and
someone who is merely a prospective recipient (the position of the
average engineer). While the former clearly will embrace the corporation and its goals with enthusiasm, the latter's identification with his
employer is more complex. He is, after all, not yet a manager—he
must be "motivated" by the offer of high rewards, an offer that does
not always buy unconditional loyalty (as contemporary studies of engineers clearly show)[89] and that cannot always be guaranteed. It should
not surprise us, therefore, that the rank-and-file engineers' acceptance
of their lot has been conditional.

Thus, it is worth noting that even after the defeat of the early 1920s,
rank-and-file engineers did not completely abandon the idea of collective conflict with their employers about the terms and conditions
of labor. Railroad engineers, as we have seen, continued to grumble

---

[88]The National Society of Professional Engineers (NSPE), with its traditional emphasis on AMA-style licensing, embodies some of the elements of this older professionalism. But even the NSPE, because of the ineffectiveness of licensing in the context
of organizational employment, has been obliged to search, so far not very successfully,
for alternative means of pressing the engineers' case. See Robert Zussman, *Mechanics
of the Middle Class* (Berkeley, Calif., 1985), pp. 164–67, for a useful discussion. It is also
highly significant that the "new professionalism" of the late 1960s and early 1970s,
with its apparent focus on ethical issues and "whistle-blowing," was in fact a movement
deeply preoccupied with material issues. See A. Michal McMahon, *The Making of a
Profession: A Century of Electrical Engineering in America* (New York, 1984), chap. 8, for
a discussion.

[89]The most important recent research for the United States is Zussman (n. 88 above).
For a British comparison, see Peter Whalley, *The Social Production of Technical Work*
(Albany, N.Y. 1986), which develops the concept of "trusted workers" for the analysis
of corporate engineers.

over their poor material position. Municipal engineers of various kinds organized themselves once again in the mid-1920s in Chicago, Boston, New York, and several other cities, even conducting a walkout in Chicago in 1925.[90] From time to time, the rank and file also attempted to revive the idea they had seen in the AAE.[91] The movement had clearly been reduced since the heyday of the AAE. Yet one must remember that this was a movement that had been effectively decapitated just a few years earlier. One must also remember that these attempts to organize took place in a period marked by violent opposition to unionism of any kind and to engineering unionism in particular. Under the circumstances, the fact that such organizing activities took place at all is testimony to the lasting character of rank-and-file concerns with the material question.

It is also significant that the next time the engineers "revolted" in large numbers, in the unionism movement of the 1940s and 1950s, it was the rank-and-file concerns that came to the fore.[92] This movement, for all its complexity, bore a strong resemblance to the rank-and-file insurgency of the early 1920s. Once again engineers sought to make material gains while emphasizing their difference from the working class below.

The lesson of the revolt of the engineers, then, is not that engineers have been wholly domesticated once and for all. Rather, it is that the nature of conflict within the profession and between engineers and employers was changing. The reformism of Cooke, Newell, and others, including their critique of business, was dying as engineers became accustomed to bureaucratic employment. In its place, there developed a pattern of often muted but occasionally overt conflict between engineers and their employers over material issues.

[90]"Chicago Technical Staff Suspends Work for Half a Day," *Engineering, News-Record* 95 (July 9, 1925): 78; "More City Engineers Join in Pleas for Increased Pay," *Engineering News-Record* 95 (August 6, 1925): 234. This controversy filled the pages of the *News-Record* for several months afterward.

[91]In late 1925, municipal engineers from five cities met in New York to establish a *new* national organization promoting the welfare of engineers. *Engineering News-Record* 95 (December 25, 1925): 1045. Similarly, an organization known as the American Society of Engineers dedicated to improving the economic welfare of engineers appeared in Chicago in the mid-1920s. A flyer from this organization is among the Cooke Papers, Box 5. The same box contains correspondence dated mid-1925 between Cooke and C. E. Drayer regarding this organization.

[92]For engineering unionism, see Herbert Northrup, *Unionization of Professional Engineers and Chemists* (New York, 1946).

INDEX

Abbot, Henry, 17, 92–94; Mississippi River study, 97–99, 121
Abert, John J., 94–95
"Academic entrepreneur." *See* Jackson, Dugald C.
Accelerometer, 326–27
Accreditation Board for Engineering and Technology, 185–86
AC/DC. *See* Rotary converter
Aeronautical and aerospace engineering, 180
AF of L, 417
Ailnolth, 31
Air-cooled engine, 285. *See also* Copper-cooled engine
Alabama A&M, 136, 145
Alden, George I., 134–35
Alexander, Magnus, 379–80, 382–84
Alford, Leon, 422
Alternating current, 170, 209–10, 373; Pupin's work with, 71
American Association of Engineers (AAE), 187, 371, 400, 411–12, 414, 418, 426; membership, 419–20; reform "counterattack," 423–24
American Bell Telephone Company. *See* American Telephone and Telegraph
American by Design (Noble), 400
American Chemical Society (ACS), 183–84, 346–47, 358; opposition to chemical engineering, 345–46, 355
American Electrochemical Society, 354
American engineers, 9–11, 13, 35–41; birthplaces of, 35–36; family backgrounds of, 36–38; educational backgrounds of, 38–41. *See also* Engineers
American Institute of Chemical Engineers (AIChE), 183–84, 353, 362; early history of, 355–61; membership requirements of, 355, 363. *See also* Chemical engineering
American Institute of Electrical Engineers (AIEE), 24, 263, 272, 362, 382–83; *Transactions*, 108, 263, 276
American Institute of Mining Engineers, 24

American Mathematical Society, 276
American Medical Association, 411
American Revolution, 10
American Society for Testing Materials, 353
American Society of Civil Engineers (ASCE), 24, 101, 361–62, 403, 405–6; debates on jetties, 108; call for metals testing, 239
American Society of Mechanical Engineers (ASME), 24, 101, 135, 163, 400; administrative reform, 409–10, 414, 422; fiftieth anniversary of, 252
American standard screw thread, 25, 151–65; design of, 153; competition with British standard, 153–56; acceptance of, 158–65; role of Franklin Institute, 152, 159; role of U.S. Navy, 155, 159–61; role of industry, 159–62; professional engineering societies' role in, 163; international consequences of, 162–64. *See also* Sellers, William
"American style" of engineering, 25
American Telephone and Telegraph (AT&T), 175–76, 261, 263–64, 273, 370, 378; cooperative engineering, 395; dispute with Pupin and Stone, 272; loading coil dispute, 279–80. *See also* Campbell, George A.
American West, 16, 101
Anderson, Joseph Reid, 48
"Angle of repose," 54–55
Anthony, William A., 373
Appian Way, 29
Applications of the Science of Mechanics to Practical Purposes (Renwick), 232
Apprenticeships, 2, 9, 13–16, 22, 32, 40, 71n, 180
Arlington Experiment Station. *See* U.S. Bureau of Public Roads impact studies
Army Corps of Engineers. *See* U.S. Army Corps of Engineers
Arnold, Chester A., 270
Ashby's Gap Turnpike, 45, 47
AT&T. *See* American Telephone and Telegraph

427